Key Papers on Fish Populations

Edited by D.H. Cushing

◇ IRL PRESS

Oxford · Washington DC

IRL Press Ltd.,
P.O. Box 1,
Eynsham,
Oxford OX8 1JJ,
England

British Library Cataloguing in Publication Data

Key papers on fish populations.
   1.   Fish populations
   I.   Cushing, D.H.
   597'.05'24   QL618.3

ISBN 0 904147 58 4

Dr.D.H.Cushing, 198 Yarmouth Road, Lowestoft, Suffolk NR32 4AB, England.

Cover drawing taken from 'Sea Fisheries: their Investigation in the United Kingdom', M.Graham (editor), Chapter II, Fisheries of the United Kingdom by H.Wood.

Printed in England by Information Printing Ltd., Eynsham, Oxford.

Contents

Key Papers on Fish Populations

ACKNOWLEDGEMENTS   i

INTRODUCTION   1

Some theoretical considerations on the 'overfishing' problem   3
E.S.Russell

Modern theory of exploiting a fishery, and application to North Sea trawling   23
M.Graham

The theory of fishing   37
R.J.H.Beverton and S.J.Holt

Some aspects of the dynamics of populations important to the management of the 111
commercial marine fisheries
M.B.Schaefer

A study of the dynamics of fishery for yellowfin tuna in the Eastern Tropical   125
Pacific Ocean
M.B.Schaefer

Stock and recruitment   152
W.E.Ricker

Handbook of computations for biological statistics of fish populations.   217
Appendix I. Development of model reproduction curves on the basis of a theory of
predation at random encounters
W.E.Ricker

INTRODUCTION TO THE LATER PAPERS IN THE DEVELOPMENT OF   227
FISHERIES SCIENCE

Estimation of mortality rates   231
J.A.Gulland

Population dynamics of the Arcto-Norwegian cod   243
D.J.Garrod

An investigation of accuracy of virtual population analysis using cohort analysis 291
J.G.Pope

A generalized stock production model   303
J.J.Pella and P.K.Tomlinson

The effect of fishing on the marine finfish biomass in the Northwest Atlantic   343
from the Gulf of Maine to Cape Hatteras
B.E.Brown, J.A.Brennan, M.D.Grosslein, E.G.Heyerdahl and R.C.Hennemuth

Contents continued overleaf

The concept of the marginal yield from exploited fish stocks 365
J.A.Gulland

Scientific advice on catch levels 372
J.A.Gulland and L.K.Boerema

Stock and recruitment 375
D.H.Cushing

Acknowledgements

I am grateful to the following for permission to reproduce papers:-

Dr H. Tambs-Lyche, Secretary General of the International Council for the Exploration of the Sea, Copenhagen (Russell, Graham, Gulland);

Dr V. M. Hodder, Assistant Executive Secretary of the North West Atlantic Fisheries Organization, Dartmouth, Nova Scotia (Pope, Brown *et al.*);

Dr C. C. Peterson, Assistant Director of the Inter American Tropical Tuna Commission, La Jolla, California (Schaefer, Pella and Tomlinson);

Mrs. Edith Graham, Lowestoft (Beverton and Holt);

Michel Gosselin, Manager, Commercial Operations Group, Supply and Services, Canada, Ottawa (Ricker, Garrod);

Margaret A. Walsh, University of Wisconsin Press, Madison, Wisconsin (Cushing);

Dr Carl J. Sindermann, Sandy Hook (Gulland and Boerema);

and to Dr W. E. Ricker, Nanaimo, R. J. H. Beverton, CBE, FRS, Swindon, Dr S. J. Holt, Rye, Dr J. A. Gulland, Rome, Dr J. J. Pella, Auke Bay, Dr B. E. Brown, Woods' Hole, Mass. and J. G. Pope, Lowestoft, for personal permission to reproduce their papers.

Introduction

D.H.Cushing

The essential papers on the development of fisheries science lie concealed in somewhat inaccessible literature. One is an annex to a working group report of the International Council for the Exploration of the Sea. Neither report nor annex have been published in the regular publications of the Council and copies lie in the archives of Government laboratories in Europe and North America. Other papers have been published quite properly in journals of the Fisheries Commissions which do not always reach the libraries of universities where ecology is taught. Hence there are good reasons for a more extensive dissemination.

Because many fishes can be aged quite easily and because some collections of vital statistics extend back for two to seven decades, stocks of fish are probably the most well studied of all wild populations. The fraction of each stock landed on the quays is usually a fairly large one. Consequently, each paper includes argument in mathematics. None of it is difficult, although the train of thought sometimes may be long.

There are fourteen papers of which five provide the basis of the science and the rest represent the subsequent developments since the fifties of the present century. The basic five are those of Russell, Graham, Schaefer, Ricker and Beverton and Holt. There are four developments: (a) Gulland's exploitation of the catch equation to make virtual population analysis, or cohort analysis, Garrod's application of it to the Arcto-Norwegian cod stock and Pope's estimate of errors in this method; (b) Pella and Tomlinson's generalization of the Schaefer method and the extension of Brown *et al.*, to all species caught by the trawl, the simplest multispecies model; (c) Gulland's marginal yield which defines an economically feasible yield and his definition of the fishing mortality ($F_{0.1}$) which generates that yield and (d) Cushing's (1981) account of the present state of the problem of the dependence of recruitment on parent stock, unfinished business.

In the more recent set of papers there are reflections of current controversies. The concept of the Maximum Sustainable Yield (MSY) was stated in different words by Russell, Graham and Schaefer; Ricker (1946) called it the Maximum Sustained Yield and it was written into certain treaties. During the late sixties and early seventies, when quotas were being introduced in the North Atlantic, it was realized that MSY might be exceeded merely because of errors (Pope and Garrod, 1975). More recently, the problems of management under different conditions of environmental variability have been investigated by Beddington and May (1977) and by Horwood and Shepherd (1981). They all concluded that stocks should be managed at values of fishing mortality somewhat less than that which generates MSY. Further, economists had shown that the best economic yield was necessarily less than MSY; Gulland's definition of $F_{0.1}$ (that at which the slope of the yield

1

curve is one tenth of that at the origin) represents the best estimate of economic yield for divergent economies across the exchanges. Ideally, fishermen would then take the greatest catch at highest profit for as long as possible.

A more recent controversy concerns the use of multispecies models. Any trawl haul includes on average more than one species, two or three in the Barents Sea, six or seven in the North Sea, twelve in the Irish Sea and fifty in the Gulf of Thailand. They eat each other and in the so-called single species models, estimates of recruitment and of natural mortality should include the effects of such predation. They do not because the right information is lacking although attempts are now being made to obtain it. The solution proposed by Brown *et al.* (1976) represents the best with the single species approach.

Because recruitment is highly variable, its dependence upon parent stock remains a central unsolved problem. Nearly thirty years ago Ricker (1954) pointed out that the essential processes probably occur during the larval or early juvenile stages. Now that the fish larvae of some species can be aged by daily rings on their otoliths (Brothers *et al.*, 1976), such material should be collected by decades with estimates of growth rates, death rates, food and predators. Unfortunately, such work is expensive in research vessel time.

From 1926 to 1976 fisheries were managed internationally. Since January 1 1977 most stocks have been managed by the coastal state, although the source of scientific advice frequently remains international. Management had been established on an international frame because fish live in the open sea and scientific evidence could provide a base for agreement between nations. During the period of international management, two factors became important: (a) distant water vessels were seen too frequently off shores foreign to them where people in the third world often lived and (b) international commissions reacted too slowly to scientific advice because too many nations were represented and because the science was sometimes weak. Since the coastal state has become responsible for management there have been both improvements and failures.

I am grateful to Professor N.A.Mitchison FRS who suggested to me that such a compilation was needed.

References

Beddington,J.R. and May,R.M.: 1977, 'Harvesting natural populations in a randomly fluctuating environment,' *Science,* **197**, 463-465.

Brothers,E.G., Mathews,C.P. and Lasker,R.: 1976, 'Daily growth increments in otoliths from larval and adult fishes,' *Fish. Bull.,* **74**, 1-8.

Pope,J.G. and Garrod,D.J.: 1975, 'Sources of error in catch and effort quota regulations with particular reference to variations in the catchability coefficient,' *Res. Bull. Int. Comm. Northwest Atl. Fish.,* **11**, 17-30.

Ricker,W.E.: 1946, 'Production and utilization of fish populations,' *Ecol. Monogr.,* **16**, 373-391.

Horwood,J.W. and Shepherd,J.E.: 1981, 'The sensitivity of age-structured populations of environmental variability,' *Math. Biosci.,* **57**, 59-82.

Some theoretical considerations on the 'overfishing' problem

E.S. Russell

Journal du Conseil International pour l'Exploration de la Mer, **6**, 3-20, 1931.

Editor's introductory notes

Russell does not refer to Petersen (1894) who examined the plaice stock in the Skagerak and distinguished growth overfishing from recruitment overfishing; the first is the loss of yield at high fishing rate because too many big fish have been caught and the second is the loss of yield because there is not enough stock to generate the incoming or recruiting yearclasses. Nor did he refer to Garstang (1903-04) who showed that stock in the North Sea had declined by half under the pressure of fishing after the advent of steam trawling in the last two decades of the nineteenth century.

Russell had visited Raymond Pearl and their discussions of the logistic curve might well have defined the scientific problems more precisely. In this paper the axioms of fishery science are established, for example, the self-contained stock in weight is separated into the catchable stock of larger animals and the non-catchable one of smaller fish. The catch rate is defined as the number caught in unit fishing time by a standard vessel. The conditions of equilibrium are stated in simple terms.

The objective of rational exploitation as the maximum sustainable yield is stated. The state of recruitment overfishing is recognized but considered unlikely. The conditions under which the stock declines under the pressure of fishing are stated and consequently if the size at first capture (or mesh size) is increased 'with equal intensity of fishing, the yield would be greater'. It is the problem of growth overfishing, but Russell recognized that he could not solve it.

The argument with Kyle is no longer of interest, but today we are sometimes chary of the catch rate because the catchability coefficient, q, may vary inversely with abundance (Ulltang, 1976; MacCall, 1976; Garrod, 1977). An important axiom states that $F = qf$, where F is the instantaneous coefficient of fishing mortality and f is fishing effort or time spent fishing. As part of his development of his simple and famous equation, Russell raises the possibility of density dependent growth and density dependent natural mortality at high stock. Fifty years later little progress in these directions has been made. Russell's great achievement was to state the axioms of fish population dynamics and they reverberate throughout the subsequent literature.

In the Western world, Baranov's (1918) work remained unknown until it was translated by Russell in 1938. It had no influence then because of the advent of the second world war. It had none after that war because a better formulation of the yield equation had been reached by Hulme, Beverton and Holt (1947). There is no evidence that Baranov's work had any influence at all in his own country.

D.H.Cushing

References

Baranov,F.T.: 1918, 'On the question of the biological basis of fisheries,' *Nauchnyi issledovatelskii ikhtiologischeskii Institut Isvestia,* **1,** 81-128.

Garrod,D.J.: 1977, 'The North Atlantic cod,' in Gulland, J.A. (ed.) *Fish Population Dynamics,* John Wiley & Sons Ltd., London, New York, Sydney, Toronto, pp.216-242.

Garstang,W.: 1900-03, 'The impoverishment of the sea,' *J. Mar. Biol. Assoc. UK. N.S.,* **6,** 1-70.

Hulme,H.R., Beverton,R.J.H. and Holt,S.J.: 1947, 'Population studies in Fisheries Biology,' *Nature,* **159,** 714-715.

MacCall,A.D.: 1976, 'Density dependence of catchability coefficient in the California Pacific sardine, *Sardinops sagax caerulea,* purse seine fishery,' *Rep. Calif. Coop. Oceanic Fish. Invest.,* **18,** 136-188.

Petersen,C.G.J.: 1894, 'On the biology of our flatfishes and on the decrease of our flatfisheries,' *Rep. Dansk. Biol. Sta.,* **4,** 146 p.

Ulltang,O.: 1976, 'Catch per unit of effort in the Norwegian purse seine fishery for Atlanto-Scandian (Norwegian spring spawning) herring,' *FAO Fish. Tech. Paper,* **155,** 91-101.

Some theoretical Considerations on the "Overfishing" Problem.

By

E. S. Russell,
Ministry of Agriculture and Fisheries, London.

I. Elementary Formulation of the Problem.

1. To decide in any given instance whether fishing operations are or are not being carried out in a manner ultimately wasteful to the stock is admittedly a difficult task, for the conditions to be taken into account are extremely complex and extremely variable, and the data available are as a rule incomplete and not always easy to interpret in an unequivocal way. Yet notwithstanding this difficulty of application to the concrete case there are certain general principles about which there can be no reasonable doubt and about which everyone should be agreed. It is my aim here to formulate in a simplified and general way, and without mathematical treatment, the broad facts of the case, to state in simple language those elementary principles that are at the back of everyone's mind who deals with the problem of the rational exploitation of the fisheries. Most of the truths to emerge will appear as obvious truisms, but there is, one feels, a distinct advantage to be gained by formulating in a simple way the essential conditions of the problem, as a help towards its more detailed study.

2. Let us simplify the problem down to its bare essentials by considering a completely self-contained stock of fish of one particular kind living in a large area which is systematically fished. Let us further assume that the fishing gear used is standardised and is such that all the fish reaching a length l are liable to capture and that none of length less than l are caught. The total stock in the area may then be divided into those of l and upwards in length and those less than l, and we shall call these respectively the catchable stock and the non-catchable stock.

Let us consider what will happen to a given catchable stock of initial total weight S_1 over a period of time which we may take for convenience as one year. We will assume that the individuals comprising the total stock grow, i. e. increase in weight, during the period, and that the total stock is recruited in the normal way by annual broods.

Considering first the catchable stock with which we start (S_1), we may say that each individual either survives to the end of the year, having grown in the interval, or is caught, with a growth-increment depending on the length of time it has survived, or is otherwise eliminated after a varying period, also with the equivalent growth-increment. The catchable stock will, however, during the year receive additions due to the non-catchable stock growing up to the limit l. The individuals comprising this added stock will, as soon as they enter the catchable stock be subject to the same chances as those making up S_1, i. e. will either survive to the end of the year, or be caught, or die from other causes before the end of the year, in each case with the appropriate growth-increment. We can accordingly deduce the weight of the catchable stock at the end of the year (S_2) from the weight at the beginning of the year (S_1) as follows.

On the credit side we must start with S_1, and to this add the sum of the initial weights of all individuals (new stock) transferring from the non-catchable to the catchable category, as they reach the length l. Let us call this sum Σa; to this we must add the sum of the growth-increments of all individuals surviving at the end of the year (Σg) whether belonging to S_1 or to the new stock; from this total we must subtract the sum of the weights of all fish caught during the year (Σc) and the sum of the weights of all fish which have died from natural causes during the year (Σm). Σc and Σm of course include the growth-increments of the fish which have been caught or have died.

We get then $S_2 = S_1 + \Sigma a + \Sigma g - (\Sigma c + \Sigma m)$ or, using capital letters for the sums,

$$S_2 = S_1 + (A + G) - (C + M).$$

S_2 will therefore be $> =$ or $< S_1$ according as $(A + G)$ is $> =$ or $<$ $(C + M)$. In other words, if more is taken out of the catchable stock in a year $(C + M)$ than is replaced by natural processes $(A + G)$ the total weight of the available or catchable stock will diminish; if the loss balances the gain the available stock will be the same at the end of the year; if the natural replenishment is greater than the loss due to fishing and other mortality the available or catchable stock at the end of the year will have increased. This is self-evident, and the sole value of the exact formulation given above is that it distinguishes the separate factors making up gain and loss respectively, and is therefore an aid to clear thinking.

(A) conveniently represents in terms of weight the influx of smaller individuals (belonging for the most part to younger year-groups) which have just reached catchable size. (A) accordingly separates out mainly the effect of "fluctuations", due to good or bad brood-years, upon the catchable stock. (G) represents the total growth-increment of the surviving individuals — of the remnant of S_1 and A. It will presumably vary according to the conditions in the area as regards e. g. food and temperature. (C) is a datum to which an approximately accurate numerical value can in certain cases be assigned. It does not necessarily coincide with the total landings from the area, for to the landings must be added the weight of all individuals caught but discarded.

3. In this formulation of the problem we have essentially to do with a number of processes, the rate of which may vary. This applies to all those factors which were treated above as summations, namely A, G, C and M. The value of A depends upon the rate of introduction of new individuals to the catchable stock, which is in turn dependent (1) on the number growing up, and (2) on their rate of growth, while belonging to the non-catchable stock; G clearly varies according to the rate of growth; C varies according to the rate of catching, M according to the rate of natural mortality. It is important then to consider what is the effect upon the stock of variation in the rate of these processes, having regard also to their necessary inter-connections.

Here it is important to make quite clear in which terms, whether of number or of weight, we state the rate of change of the processes whose summated values we have so far treated only in terms of weight. It is clear that rate of growth should by definition be stated in terms of increment of weight. Rate of capture, rate of natural mortality, and rate of introduction of new stock, might be stated in terms either of numbers or of weights, but there is an obvious advantage, for clarity of thought, if we state them in terms of numbers, for these rates will then be independent of variations in rate of growth. We shall accordingly in the following discussion use rate of capture, rate of natural mortality, and rate of introduction of new stock as referring to numbers.

4. We shall consider first the effect of variation in the rate of catching. This we define as the number caught per unit of time by the standard fishing gear. There seems no objection to defining the rate in terms of number and not of weight, because the chances of capture (above length l) are, so far as we know, independent of the size of the fish. We may con-

veniently start from the theoretical case where in a specified year $A + G$ just balances $C + M$: a condition of stabilisation, in which $S_2 = S_1$. The rate of catching (number) is such as to give the value C (weight). We may call this Case (1). Let us now assume that in Case (2), the rate of growth, the rate of natural mortality and the rate of introduction of new stock remain constant, but that the rate of catching is lower than in Case (1). It follows immediately that in Case (2) $C + M$ must be less than in Case (1), for total mortality is by hypothesis less. Similarly G will be increased in value, since the number of survivors is greater in Case (2) than in Case (1). The total weight and number of the stock will therefore be greater in Case (2) than in Case (1); the stock will therefore have increased, owing to the decreased rate of catching. In the above argument we deal with the individuals making up both S_1 and A. If we now assume that in Case (3) the rate of catching is increased we arrive by the same simple reasoning at the conclusion that the stock in this case will diminish in total weight and total number.

But it is important to note the assumptions on which these conclusions depend, namely the constancy of rate of growth, rate of natural mortality, and rate of introduction of new stock. Are these assumptions justifiable on biological and mathematical grounds? Have we the right to assume that the rate of capture may vary independently of each and all of the other factors? Taking A first of all, it is difficult to see how variations in the intensity of fishing could in the course of a year have any direct effect upon the numbers of the new stock growing up i. e. upon the numbers of the non-catchable stock — unless of course the species were cannibalistic. Even if fishing were so intensive as to wipe out all potential spawners, the direct effect of this would under present conditions of fishing not be seen till at the earliest 18 months after the spawning season (taking the case of the haddock as a typically fast growing fish). An indirect effect through the rate of growth of the non-catchable stock is, however, conceivable, for a vast clearing out of catchable stock might leave more food available for the non-catchable stock and increase their growth-rate so that they reached size l more rapidly. A in such a case might be increased, if more survived to length l owing to their passing through the vicissitudes of early life more rapidly.

This brings us to a consideration of the possible effect of variations in the rate of capture upon the growth-rate of survivors. It is quite conceivable that increase in capture-rate should result in increase of growth-rate of survivors, since the amount of food made available would, *ceteris paribus*, increase. One condition must, however, be fulfilled before this conclusion follows — the supply of food must not be in excess of requirements. If the food supply were superabundant, the growth-rate in all three cases considered would be at the maximum possible under the other conditions prevailing; variation in rate of capture would then be of no effect on rate of growth. If, however, — and this seems the more likely case — the food supply is relatively limited and there is consequent competition for it by many species, it seems reasonable to suppose that increase in rate of capture might result in increase of growth-rate, both of catchable and of non-catchable stock. Whether it is possible in such circumstances for the increase in G, due to increased growth-rate to equal or exceed the decrease in G due to increased capture-rate is a complicated problem which cannot be considered here. It may, however, be remarked in passing that any food "released" by increase of capture-rate would be eaten by other species besides the one considered.

As to the effect of variations in capture-rate upon the rate of natural mortality it is impossible to say much from the biological point of view. If any part of the natural mortality were due to starvation, or to cannibalism, then increase in capture-rate would tend to lower natural death-rate. There is of course on purely arithmetical grounds a certain inverse relation between C and M; $C + M$ represents total mortality (in terms of weight); if C increases it will diminish M, since some individuals will be caught which would otherwise have died a natural death; similarly if M is increased it must diminish C as some individuals will die which would otherwise have been caught.

5. We have established the fact that A, G, C and M are or may be inter-connected biologically. It is therefore a little dangerous to attempt to treat the variations in rate of each of these factors singly. We have attempted to do so with the rate of capture, but it is necessary to bear in mind the qualifications and corrections to be applied. We may now consider the broad effects of variation in the other rates, namely rate of introduction of new stock (expressed in numbers), rate of growth (expressed in weight) and rate of natural mortality (expressed in numbers). It is clear first of all that A is a highly important source of variation. We know that in practice the numbers of the on-coming new brood may vary enormously from one year to another. Starting out from our standard Case (1) and assuming the other rates specified to be constant, a big increase or decrease of A would clearly make a considerable difference to resultant stock (S_2). In practice a big increase might depress rate of growth, but, leaving this out of consideration, it would certainly increase C, M and S_2, as compared with their values in Case (1).

An interesting result emerges if we assume that, for any reason unconnected with variation in capture or mortality rate, the growth-rate is increased. As compared with our standard Case (1), C and M would be increased, and G also, and the increase in G would over-compensate the increase in C and M, so that S_2 would be greater than S_1, though there would be no increase in numbers. The truth of this conclusion is easily checked by considering that in both cases the number of survivors is the same, but their weight is greater in the second. So too the number caught and the number naturally eliminated are the same in the two cases, but in weight they are greater in the second case than in the standard case. Increase in growth-rate in such conditions means increase in weight of catch and at the same time increase in weight of surviving stock. The importance of high growth-rate in increasing the yield of a fishery is clearly brought out by these considerations.

The effect of variations in rate of natural mortality has already been dealt with in part in the preceding section, where it was shown that M would increase partly at the expense of C. The effect of a large increase in natural mortality rate in decreasing resultant stock is obvious and needs no detailed comment.

The main object of the above discussion has been to show that in considering changes in catchable stock from one year to another there are several other factors to consider besides variation in the rate of capture, and an attempt has been made to indicate the direction in which these other factors would work, whether as increasing or as decreasing the resultant stock.

6. From the theoretical point of view it is of interest to discuss the two extreme cases of variation in rate of capture — (1) where the rate is nil, and (2) where it increases to an indefinitely large degree. Let us take (1) in its most extreme form and let us consider the case of an area which has never been fished at all. Knowing that the total amount of available food cannot be unlimited, we can deduce that there is a maximum value for S which cannot be exceeded. Let us assume that on a virgin ground the stock tends to approximate to this maximum and that the amount of food remains constant from year to year. S will under these conditions remain stationary, and for any year we may write $S_2 = S_1$. It follows also that $A + G = M$. This means that the mortality rate and the rate of replenishment and growth of the catchable stock must vary together; if M is small the annual increase $(A + G)$ must likewise be small.

Let us see how this would work out in practice. Assuming the supply of food to be limited in relation to existent stock and the stock eating up the food as fast as it is produced, there must inevitably be competition between the fish for the available food. There is a broad difference between the old fish and the young fish in respect of their utilisation of the food — the young fish put relatively more to growth and less to maintenance. If therefore on the whole the older fish win, the stock will gradually become composed of large old fish which utilise all the food for mere maintenance and do not grow, and M will consist mainly of small fish which have been starved out of existence. If on the contrary the younger fish manage to annex more than their share they will grow at the expense of the older fish, which will die of starvation; M will then consist mainly of the larger fish. Without some knowledge of the actual facts it is impossible to say what theoretical formulation comes nearest to the actual conditions found on a virgin ground.

Even the few theoretical conclusions arrived at above as to the state of things on a virgin ground, may not correspond accurately to what is actually found; in particular the assumption that a maximum is reached and steadily maintained may be a wrong one. This much is, however, certain, that there is a theoretical maximum to S and that mortality and replenishment must in some way balance one another — possibly in a cyclical manner.

Coming now to the second limiting case of variation in rate of capture, it is easy to see that an indefinite increase in the rate would lead to a virtual extermination of the stock, first by reducing to indefinitely low numbers the existent stock and, through destruction of the spawners, reducing A in course of time to nearly zero. This is of course a limiting case, never actually met with so far as fish are concerned, but it brings out the possible danger to stock through undue destruction of spawners brought about by very intensive fishing.

7. If now we consider the size-distribution of the catch, under different rates of capture and different rates of growth, certain obvious deductions can be drawn. Other factors being constant, an increase in rate of capture will mean a diminution in the average size of the catch, as compared with that in the standard case, since the individual fish will on the average be caught earlier, i. e. at a smaller size, than if fishing is less intensive. An increase in rate of growth will obviously act the other way — if capture-rate is constant, but rate of growth increases, the average size of those caught will be greater. Decrease in growth-rate will give the same kind of result as increase in capture-rate — a decrease in the average size of those caught.

8. Coming back now to our starting point, we may say that the formula stated in paragraph 2 above represents a balance-sheet. We start with a working capital S_1; to this is added in the course of a year

$(A + G)$, and from it is taken away $(C + M)$. At the end of the year our working capital is S_2, which will be greater than, equal to, or less than S_1, according as income $(A + G)$ has exceeded, equalled or fallen below expenditure $(C + M)$. S_2 is what is carried over as capital from one year to the next. It does not represent merely the difference between income and expenditure but that difference added to or subtracted from S_1. S is therefore a continuing value which alters according to the difference between income and expenditure.

It is clear that if expenditure is consistently higher than income, S will be a diminishing quantity, and that if the amount of fishing remains constant during this process C must also fall. The practical problem then appears to be to keep S at such a level, or to bring S to such a level, that the maximum value of commercially utilisable fish can be drawn from it annually without causing a progressive diminution of S. If the annual increment represented by A were fairly constant from year to year, the problem would be one of obtaining a constant maximum yield. It might be theoretically possible to evaluate this optimum yield and to estimate the amount of fishing required to obtain it. A stabilised fishery would be the result. It is, however, common knowledge that A is a fluctuating quantity and that in certain important fish, e. g., haddock, the fluctuations are very great.

It appears therefore that the ideal of a stabilised fishery yielding a constant maximum value is impracticable. It might, however, be practical politics to attempt to adjust the amount of fishing each year to the variations in the stocks of particular fish in particular regions, as even now fishing does shift to some extent according to the abundance or scarcity of fish in particular regions. If such variations in abundance could be foretold a year or so in advance this adjustment could be made more rapidly and with more certainty of success.

9. The problem of rational exploitation is, however, an exceedingly complicated one. Let us arbitrarily simplify it by assuming that A is constant from year to year, i. e. that fluctuations do not exist. (Probably if we take a sufficiently long series of years fluctuations do average out). It is clear that a condition of stabilisation exists when $C + M = A + G$; S_2 in this case equals S_1, and the stock remains constant from year to year. But it is also clear that this stabilisation may take place at various levels, depending on the magnitude of C. If $C + M$ is small, $A + G$ will also be small, and the annual product of the fishery will be well below the maximum possible. But the aim of rational exploitation is to get the maximum yield annually, compatible with maintaining stocks at a steady level.

Regarding M as a constant factor, and in any case as outside control, our problem is to increase C as much as possible while keeping S constant from one year to another.

An increase on the credit side $(A + G)$ would be achieved if it were possible to increase the rate of growth (as can be done for example to a limited extent by transplantation in the case of plaice). This would result, if other conditions were constant, in an increase in C, as was shown in paragraph 5 above.

With regard to the debt side $(C + M)$, clearly much depends upon l, the theoretical limit between catchable and non-catchable stock. Hitherto we have assigned no particular value to l, but in practice, as we all know, l tends to approximate to, but to be less than, the size at which the fish becomes of marketable value. In what follows we shall assume that the value of l corresponds roughly to something less than the commercial minimum.

It would seem that C might be increased by lowering l, but clearly this would be of no value commercially, and it would of course decrease A, and lead in the long run to a diminution of S. Let us consider what would be the effect of increasing l by a moderate amount, say the amount represented by one year's growth in length (n).

If the limit were increased to $l + n$, the catch would at first be less, but $A + G$ would increase. Now consider the state of affairs after one year, when the fish of length l had grown to $l + n$. If M were low there would not be many fewer fish of length $l + n$ at the beginning of the year than there were of length l a year before, and, since weight increases as the cube of length, their weight would be very much greater. The catchable stock at the beginning of the second year would be increased in weight, and C would therefore be greater during the second year. A and possibly G would also be increased. The net result should be that with equal intensity of fishing the yield would be greater. But the process of increasing l could not be carried very far without entailing drawbacks. There would be an increasing wastage from natural mortality among the non-catchable stock, and it is at least conceivable that rate of growth would be diminished; it is probable also that as heavier stocks of non-catchable fish were accumulated there would be less room and food for the incoming new brood, and renewal of stocks would be slowed down.

It is clear that this question of the best size at which to commence capture raises problems of great complexity which require a fullness of treatment which cannot be attempted here. In particular the effects, both beneficial and harmful, of the "thinning" of the stock above and below the commercial minimum size demand most careful analysis.

11. We have attempted in the preceding paragraphs to give a simple and general formulation of the problem of "rational exploitation"; it is obvious, however, that an abstract formulation is of little value if it cannot be applied in practice. This raises at once the question, can we measure or estimate the changes in the stocks of fish? I propose to consider this question in the second part of this paper, leaving to a later occasion the concrete study of the "overfishing" problem in the light of the available statistical evidence.

II. Can we Measure Changes in the Stock?

The measure commonly adopted of the abundance of fish on a ground is the catch — more properly the landing — per day's absence from port. In Great Britain of recent years a more accurate measure, the landing per 100 hours' actual fishing, has been employed. The landings are of course given in terms of weight.

It is clear that these values do not give any definite indication of the absolute quantity of catchable fish on the grounds visited, and do not allow of any actual value being allotted to S. They represent the weight of commercially valuable fish taken over a certain period of time with certain fishing implements, but what proportion this weight bears to the total catchable stock (S) on the grounds fished remains unknown[1]).

They have, however, relative or comparative value. We may assume that a trawl of a certain spread, a certain height of head-line, a certain cod-end mesh, fished at a certain speed, takes on the average a constant proportion of the total number of fish of catchable size present at the time, whether the fish be numerous or scarce. If this fundamental assumption is sound, then the amount of the catch per unit of time is a valuable index of the total weight of catchable fish on the ground, and may be used for studying the variations of S in space and in time. Without information as to the size of the fish caught it cannot of course be used as an index of the number of fish of catchable size on the grounds.

As has already been indicated, we have to deal in the commercial statistics not with catches, but with landings. This makes our index less accurate, for a varying amount of fish is caught which is not brought to market, owing to its negligible commercial value. On occasions the weight of the discarded fish may amount to a considerable fraction of

---

[1]) Total catchable stock might be estimated in a rough way from the results of extensive and successful marking experiments.

the total catch, if fishing has taken place in areas where undersized fish abound. Nevertheless, with due precautions, the landings per unit of fishing time may be used for comparative purposes as roughly equivalent to the catches.

It will, however, be apparent that certain conditions must be fulfilled before the landing, or even the catch, per unit of fishing time, can be safely used as an index of the comparative abundance of the catchable stock.

The single haul, or group of hauls, on the same ground on the same day, gives an index of abundance which is valid only for that particular ground at that particular time. If we wish to have a reliable index of the abundance of the stock over a large area we must have numerous hauls well distributed over the whole extent of the area in question, especially if, as is almost invariably the case, the abundance of the fish varies with locality. Furthermore, since the stock (weight) of the fish in a given area is a varying quantity, constantly affected by migration into or out of the area, by the effect of fishing operations, by increase of weight due to growth, by the introduction of new stock, and by other causes, the value of the catch per unit of fishing time, deduced for the area as a whole, during a given short period of time, is valid only for that period.

It follows then that for the catch per unit of time to have general validity for any area the sampling must be adequately distributed over that area; even then, the values can be accurate only over limited stretches of time, since the stock is a changing one. The catch (or landing) per unit of fishing time should therefore, strictly speaking, be used as an index only over short periods of time, since the quantity of which it is an index is a constantly changing one. Practical considerations make it difficult, if one is dealing with large masses of data, to use a shorter period than a month, and our British fishery statistics are accordingly for the most part tabulated by months. In practice, however, it is found that the landing per unit of fishing time can be usefully employed even over such periods as a year, but of course only when very extensive and adequate data are available. Used over such long periods it gives an average index for the weight of catchable stock during the period and in the area considered, i. e. a value which is or may be accurate only for a short time during the period. Nevertheless this average index has a perfectly definite meaning mathematically. Its meaning may also be expressed in practical terms as follows — it gives the relation between yield and expenditure of time over the period of a year's fishing in the area considered. This is of course a matter of primary importance from the commercial point of view.

In dealing with the commercial statistics of landings we do as a matter of fact commonly apply the landing per unit of fishing time over periods of a year, which incidentally are natural cycles in the life-history of the fish. It is calculated as a rule on total landings by vessels of a particular class divided by the total number of hours' fishing. It would be possible also, and in some respects preferable, to calculate it by months, and take an average of the monthly values. Actually, when fishing is large in amount and well distributed in time and space, the values obtained by either method are practically the same. The following example may be given:— In 1927 the average landing per 100 hours' fishing by British steam trawlers of all bottom fish from the North Sea was 141 cwts., calculated on total landings divided by total number of hours' fishing. The monthly values (in cwts.) of the landings per 100 hours' fishing were 127, 134, 151, 152, 153, 145, 136, 125, 130, 153, 144 and 135, giving an average value of 140.4 cwts. Taking a single rectangle, namely G6, which is well fished throughout the year, the landing per 100 hours' fishing was 89 cwts. calculated in the first way, and 88.5 cwts. calculated in the second way. It will be seen that the differences are insignificant.

Where the data are adequate, the landing per unit of fishing time even over a period of a year may accordingly safely be used as an index of the average abundance of the catchable stock, as affected by the amount of fishing and by all other modifying factors, such as the influx of new stock, and the growth of the fish. It is, mathematically speaking, an average of the landings per unit fishing time which have been made continuously throughout the year.

It is clear from this discussion that the primary condition for the validity of the catch (or landing) per unit fishing time is that the sampling shall have been adequate, having regard to the particular area and period considered. To get any broad and reasonably accurate picture of the comparative abundance of stocks over wide areas it is necessary to have at one's command accurate statistics of a large and well distributed fishery. Data which relate merely to a small and restricted fishery will be apt to give an inadequate and probably a misleading picture. Where an area is inadequately sampled, as for instance the Barents Sea in the early years of its exploitation, no great reliance can be placed on the value of the landings per unit fishing time.

There is a second condition of great importance, especially when comparisons are made over a long range of time, namely that the fishing power employed, as gauged by size of vessel, and type and catching power of the gear employed, must have remained approximately constant during the period considered. This is a point of rather special interest, in view of the developments which have taken place since the war in the size and effectiveness of trawling gear. The general shift over from the ordinary otter trawl to the VIGNERON-DAHL trawl in one or other of its numerous modifications has undoubtedly made the interpretation of any changes in the landings per unit fishing time more difficult and uncertain. The general direction of the change due to this development of trawling gear is of course known, but its exact effect is difficult to estimate accurately, owing to the great diversity of types of trawling gear now in use.

A third point which must be taken into consideration in appraising the value of the landing per unit fishing time as an index of stock is the possible variation in the requirements of the market. The statistics relate to the quantities landed for sale; if the demand for fish grows the small fish previously rejected at sea as unsaleable may acquire a marketable value and be brought to port to swell the landings. So too, when we are considering the landings of all demersal fish taken together, and comparing the landings per unit fishing time over a series of years, our deductions may be affected by the fact that species previously discarded at sea are now brought to market, having acquired a marketable value. It is common knowledge that in the course of the last twenty years and more, several kinds of fish formerly rejected as of no commercial value have gradually come on the market, and so have helped to increase the landings per unit fishing time of total demersal fish.

The three conditions of validity mentioned are the most important, though other minor difficulties may arise, as everyone knows who has attempted to deal with fishery statistics. But if these three conditions hold good, or if the effect of any alterations in them can be allowed for, the landing per unit fishing time may be expected to afford a generally reliable index of the comparative abundance of the catchable stock, so far at least as weight is concerned.

It will be convenient and helpful at this stage to consider certain detailed objections raised by KYLE[1]) to the use of this measure of the relative abundance of the stock. Some of these objections are perfectly valid, and he has performed a useful service in emphasising the dangers of an uncritical use of this measure of abundance; we shall see, however, that his criticisms do not invalidate the use of the criterion, provided it be applied with due caution.

[1]) Die Statistik der Seefischerei Nordeuropas, Handbuch der Seefischerei Nordeuropas, Bd. X, Heft 4, Stuttgart, 1928.

He sets out to prove that the methods for determining the quantitative distribution of fish recommended by ARCHER (average catch per voyage, per hour, etc.) are inadequate, and can lead to no scientific conclusions (p. 23).

After referring to the landings per day's absence by German and English vessels in 1925 from the various Regions — which as a matter of fact show considerable agreement — he goes on:—

"From the English statistics for the same year the quantities per hour's fishing can also calculated, which give to a certain degree a still more accurate picture.

	kg per hour's fishing
Barents Sea................	453
Iceland	521
North Sea	62
Channel...................	51
South of Ireland	125
Biscay....................	109

It appears therefore that the North Sea, in respect of its richness in fish, is not comparable with Iceland or the Barents Sea, since both of these yield 7—8 times as much fish per hour. It is, however, necessary to regard such figures with reserve.

The regions are for one thing not equally intensively fished: although for example the German and English data for the Barents Sea in respect of total catch agree with one another, the number of hauls is too small and limited to too few months for them to be regarded as representative.

In the second place, there may be great differences between the catches, when the areas are very large; thus the German catch, i. e. per day's absence, for the North Sea is considerably greater than the English, simply because the Germans fish relatively little in the North Sea, and then mostly in the best spots and in the best months, while about 600 English steam vessels are almost constantly at work in the North Sea.

Thirdly, if 600 vessels fish an area, their individual catches cannot be expected to be as high as when only 150 are at work. This is about the relation between the English fishery in the North Sea and in Icelandic waters. The North Sea is accordingly fished 4 times as intensively as the latter, and on this ground alone one might expect that the Icelandic catches would be at least twice as large as the North Sea catches.

Again, the catching power of the vessels is very different, and it is easily seen that the size of the catches increases with distance from the North Sea ports; that is to say the vessels are larger and have a greater catching power. The Iceland steamers for instance catch at least 50 per cent. more than the ordinary North Sea steamers.

Fifthly, the choice which the fishermen find on the distant grounds affects the statistical data. As will be further shown below, the Germans bring very few plaice but many coalfish to their home markets, because they get in England a better price for their plaice but a worse price for their coal-fish than in Germany; this probably explains the lower German average catch shown for Icelandic waters.

These various difficulties and others to be mentioned later are here specially brought into prominence, because great importance has been attached to these factors. It has been thought that from such data conclusions can be drawn as to the exploitation (overfishing or otherwise) of an area, and so they can — but whether these conclusions are always valid is a matter for grave doubt" (pp. 24—25).

Of these five objections, all but the third are perfectly valid, and they have already been recognised in principle in our discussion above. The first, second, and fifth are merely particular instances of inadequacy of sampling, which would be looked for and allowed for by anyone having experience in the difficult art of treating fishery statistics. The fourth caveat is also one which we have specifically mentioned.

More interest, both theoretical and historical, attaches to the argument outlined in the third objection. It is here pointed out that the difference in the average landing per day's absence between Iceland and the North Sea is in part accounted for the fact that the North Sea is fished much more intensively — a perfectly correct conclusion. The amount of fishing is obviously one of the principal factors affecting the magnitude of S. During the War, the amount of fishing in the North Sea was greatly reduced and the l. p. d. a. from this area was in 1919 nearly double the pre-war average. Similarly, if the intensity of fishing in Icelandic waters were to be suddenly increased one would expect to find a diminution in the l. p. d. a. The l. p. d. a. is in fact a running index of the weight of commercially valuable stock left on the grounds, and one therefore expects it to be influenced by the actual amount of fishing, since fishing is one of the main factors in reducing stock. Accordingly KYLE's objection has no weight whatever against the validity of the l. p. d. a. as a comparative index of commercial stock remaining on the grounds. It gives an indication of the state of S as affected by the amount of fishing and by the other factors, G, M and A, which modify it. It does not purport to do more.

I said above that the question was one of historical, as well as theoretical interest. I had in mind the important evidence given by GAR-

STANG so long ago as 1904 before a Hourse of Lords Committee[1]) in which he expressed the opinion that "it is impossible merely from the evidence of the decline in the average catch of the fishermen with an increasing number of boats to conclude that there has been an improverishment of the grounds". He has recently discussed the question again in a most interesting way in his third Buckland Lecture[2]) and has come to the conclusion that the average catch is not invalidated as a measure of abundance. "It registers the abundance truly enough, and when it falls with an increase in the intensity of fishing, it does so because, pending replenishment, there is greater depletion, the greater the rate of capture; and there is an inevitable lag in the process of replenishment" (p. 29).

The factor of replenishment is of course in practice extremely important — as indicated in our general formula by G and A — but the catch per unit time would be valid even if we supposed this factor to be absent.

At the risk of being tedious, let us consider this point again. Let us imagine an area of limited size, populated by a stock of catchable fish of a certain definite number and evenly distributed over the area. We shall assume, in order to simplify the case as much as possible, that this stock is not added to by the introduction of new stock growing up. Let us consider what will happen if this stock is fished (1) by 10 vessels, or (2) by 100 vessels, of equal size and fishing power, over the same period of time, say one month. If our original assumption is correct, that a vessel with standard gear, fishing under standard conditions, takes on the average a fixed proportion of the catchable fish present on the ground covered, then it is apparent that in the first few hours of fishing the catch per hour will be practically the same for the 100 vessels as for the 10 vessels. But the 100 vessels will take more away from the stock than the 10 vessels; hence in case (2) the catch per hour will fall at a greater rate than in case (1), and by the end of the month the catch per hour by the 100 vessels will be definitely less. The average catch per hour over the whole of the month will also be less for the 100 vessels than for the 10. This difference indicates that the stock at the end of the period is less when 100 vessels have been fishing than when only 10 vessels have been at work — which is self-evident. The point is in fact so elementary as hardly to require proof. But it does bring out the cardinal fact that the average catch per hour at any particular

---

[1]) Report on Sea Fisheries Bill (H. L.), London, H. M. Stationery Office, 1904, p. 115.

[2]) The Fishing News, Aberdeen, Vol. XVIII, No. 888, June 7th, 1930.

time is an index of the stock remaining on the ground at that time. Taken over a period of time, the catch per hour indicates the average state of the stock during the period, as affected by the amount of fishing going on.

The catch per unit of fishing time is clearly affected by the amount of fishing which is being carried on contemporaneously. But this is simply because the amount of the stock is being affected by the amount of fishing, which again is self-evident.

KYLE in the paper referred to above appears to think that this characteristic of the catch per unit time invalidates its use as an index of stock. He apparently argues as follows. If we compare an area during a period when fishing has been slight with the same area during a period of intensive fishing, and find that in the second period the catch per unit time is less, we cannot conclude that the stock has diminished in the second period as compared with the first — it may simply be that a stock of the same size as the original stock has been shared out among a larger number of vessels. The argument is perfectly sound up to a point, but it does not take all the facts into account. It is true only as regards the stock to start with, before fishing operations have begun. In illustration, let us take our case of the 10 and the 100 vessels, and let us assume that the stock on which the 100 vessels set to work is not necessarily the same in magnitude as the stock fished by the 10. If the catch per unit time by the 100 vessels is less than that by the 10 we cannot infer that the stock to start with was less — it may have been equal to, or even greater than, the original stock which the 10 vessels started to exploit. Hence a difference in the catch per unit time does not *per se* indicate a corresponding difference in original stock.

But the decreased catch per unit time by the 100 vessels does indicate that the final and the average magnitude of the stock, under the influence of fishing, was less than when the 10 were fishing. That is all it can show, but it is quite sufficient for all practical purposes. We rarely encounter a case where we can speak of an original stock prior to fishing. The nearest approach to such a case is when large shoals of fish approaching the shore in a body become the object of a coastal fishery, as in the Norwegian cod fishery. If in such a fishery there are large fluctuations in the amount of fishing from year to year it is difficult or impossible to draw conclusions merely from changes in the catch per boat per unit time as to the relative abundance of the successive yearly stocks, as they come in, before they have been affected by fishing.

In actual practice, however, in dealing with the modern trawl-fishery, which is in most areas both intensive and continuous, the question of "original" stock simply does not arise; we have to deal with stocks which are constantly and considerably being reduced by fishing operations, and the catch per unit time shows us sufficiently well what is actually taking place. Under present-day conditions, fishing is one of the most potent factors affecting the magnitude of stocks, and our index, the catch per unit time, takes account, and rightly, of this as well as of the other factors concerned. The amount of the stock is as we have seen, a constantly fluctuating one, depending upon the amount of fishing, the rate of natural mortality, influx of new stock, and rate of growth — as indicated in our elementary formula. As we have to deal with actualities, with stocks subjected constantly to intensive fishing, it is of little interest to speculate as to what the state of stocks would be if this factor were eliminated, or reduced to what is was twenty, thirty or fifty years ago. What we have to study are the changes in S under the present conditions of intensive fishing, and our best guide to these changes is the catch or landing per unit fishing time.

KYLE prefers as an index of productivity the total yield of fish, but makes no attempt to relate this with catching power, — although it is obviously in the main a function of the amount of fishing, — for the reason that total catching power cannot be estimated accurately. Overfishing will be indicated, he considers, if there is a progressive diminution in quantity or quality of the total yield over a period of years — "If in a long series of years the yearly quantities decrease, or if the quality becomes constantly poorer, we can then say that the productivity of the area in fish is too small for the intensity of the fishery" (pp. 29—30).

Total yield is undoubtedly a most important datum, corresponding approximately to C in our elementary formula, but it gives merely what is taken out, and affords no indication of what is left in the sea. For an index of remaining stock it is absolutely necessary to have recourse to some index relating catching power to weight of catchable fish on the ground. The catch or landing per unit fishing time can, if certain conditions be fulfilled, be used as such a measure or index.

Modern theory of exploiting a fishery, and application to North Sea trawling

M. Graham

Journal du Conseil International pour l'Exploration de la Mer, **10**, 264-274, 1935.

Editor's introductory notes

This paper is important because for the first time data from the fisheries are used to show that fisherman might catch as much if fewer of them practised their dangerous trade at sea. The work in this paper was never applied in detail but its most important consequence was the Overfishing Convention of 1946 from which stemmed the International Commissions in the North Atlantic.

Russell had defined C as the annual decrement of stock in weight due to fishing i.e., $C = FB$, where B is average annual biomass, the product of numbers and weight. In equilibrium Graham assumes $C = V$, where V is the growth rate of biomass. Graham tried to obtain values of M, the instantaneous rate of natural mortality, but we have to admit that the attempt failed.

Table 1 shows that with fishing mortality reduced to zero there is an immediate increment of numbers and a doubling of biomass, which is what happened with the relaxation of fishing effort during the first world war (g is the instantaneous growth rate, w weight in grams and n numbers of cod per unit of effort and m is the instantaneous coefficient of natural mortality).

Graham compared the catches in 1913 and in 1919 and concluded that the stock nearly doubled, but his comparison of catches per days' absence shows that they had more than doubled by a little. He calculated an effective respite during the first world war of $2\frac{2}{3}$ years. With these parameters he was able to model the observations in stock before and after the first world war and from the yield curve to establish a maximum at half the greatest biomass.

Graham used the logistic equation implicity in which the natural rate of increase 'is directly proportional to the difference between the weight of stock at that moment and the maximum weight the area will support'. Ricker (1975) put Graham's ideas formally: –

$$Y = VB - (VB_\infty)B^2$$

where Y, of yield, and B are equilibrium estimates;
B_∞ is the maximum biomass the environment will carry.

Graham believed that W. F. Thompson had made considerable progress in his analysis of the stock of Pacific halibut. Thompson showed the inverse dependence of catch per unit of effort on fishing effort (as Garstang had done) and he developed a form of the catch equation (Thompson, 1951). However, he wished to return a stock to a 'normal' stock density and did not try to establish a maximum.

References

Ricker,W.E.: 1975, 'Computation and interpretation of biological statistics of fish populations,' *Bull. Fish. Mar. Ser 5,* **191**, 382p.
Thompson,W.F.: 1950, 'The effect of fishing on stocks of halibut in the Pacific,' *Publ. Fish. Res. Inst.*, Univ. Washington Seattle.

Modern Theory of Exploiting a Fishery, and Application to North Sea Trawling.

By

Michael Graham,

Fisheries Laboratory, Lowestoft.

———

THERE have recently been three important papers on the theory of most effective fishing. R u s s e l l's of 1931[1]) laid the theoretical foundation. H j o r t, J a h n and O t t e s t a d[2]), in 1933, detected the significance of the point of inflexion in a sigmoid population curve. To T h o m p s o n and B e l l, 1934[3]), belongs the credit for emphasizing the importance of the rate of fishing, from which follows the peculiar attraction of the modern theory, that the benefit of efficient exploitation lies more in *economy of effort* than in increase of yield, or preservation of future stocks, though both of these purposes may also be served. Thus T h o m p s o n and B e l l's argument has been put into practice, and has in fact led to a progressive economy of effort during three years, in which the yield of the Pacific halibut has been artificially stabilized at a somewhat low level, which it had reached during the depression of 1931[4]). Their theory, however, is incomplete. The present work has had a double aim, that of correcting and continuing the theory, and of applying it to the North Sea fisheries.

There are three ways of approaching the problem. They all use data, different data in each case, and they all lead to the same conclusion, that an economy of effort is desirable in the North Sea, that is, *that a certain proportion of the time and money of the fishermen is at present devoted to reducing their catch, or is at least wasted.* It also follows that in unrestricted fishing, proceeding as it usually does at ever increasing intensity, as grounds and habits become better and better known and gear more and more improved, there must come a time when new inventions are harmful. Nevertheless, once the new

———

[1]) Journ. Cons. Int. Expl. de la Mer. VI. Copenhagen.
[2]) Hvalrådets Skr. 7. Oslo.
[3]) Rep. Int. Fish. Comm. 8. Seattle.
[4]) *Loc. cit.* p. 12.

invention has done its harm in reducing the productivity of the stock, its use must be continued, because the old gear will not pay expenses on the less productive stock. So the fishermen are left with the expense of the invention, with no compensating increase in yield.

It is tempting to suggest practical ways of making the improvement in exploitation, because the peculiar virtue of the modern theory is that the measures needed involve Economy and Leisure, both of which are intrinsically good, whereas measures we have thought of before were devoid of any attraction in themselves. However, it would be altogether premature to   discuss practical measures, whilst the theory and evidence of its applicability have not stood the test of criticism.

If the theory is sound there are many practical implications, but they are not the concern of this paper.

The essentials of the theory are two. **A.** If the rate of mortality be reduced, as by reducing the rate of fishing (hence increasing economy and leisure), the average age of the population will be raised, and *vice versa.* **B.** There is a most profitable age to harvest any growing crop. For example, the hay crop, of species maturing in different weeks, must yet be cut at a certain age which takes advantage of as much growth as possible and avoids as much "mortality" by seeding as possible. The pasture should carry only a certain number of sheep, neither too many nor too few for the best yield; that is, there is an optimum rate of cropping. A growing steer must theoretically be sold at a certain most profitable age, depending on rate of growth and rate of food consumption.[1]) It should be noted that none of these cases takes account of the necessity of leaving sufficient adults for reproduction, with which indeed this paper does not deal, except inclusively under (3) below. But we note that anything we do for better exploitation in the North Sea should be helpful also as regards reproduction.

It follows from these two considerations, A and B, that all we have to do, to be satisfied that economy and leisure are available, is to demonstrate that the present age of our stock is below the most profitable age. Actually it seems that, near its maximum, the yield of the North Sea is comparatively stable for quite considerable changes in rate; it is expressed by a rather flattish topped curve, and it is therefore most practical to say that it will pay to reduce fishing so long as the yield is not thereby reduced. Also, the maximum yield is not exactly the most profitable. Some further economy can still be made by reducing fishing, depending on the ratio of overhead costs (requiring a certain turnover) to running costs per ton of fish.

We shall now apply the theory enunciated in A and B, above, to the North Sea Fisheries, in three independent ways.

Our three solutions use varying amounts of the theory, and each has its share of approximations and assumptions. No. 2 seems the best founded.

---

[1]) J a m e s W i l s o n. The Principles of Stockfeeding. London 1927.

1. Evidence of General Statistics.

In this section we take it as an accepted fact that the stock in 1928—1932 consisted of smaller fish than in 1909—1913. There are statistics which support this, but they are open to some objection and I prefer to take this fact to be generally admitted. Now, we find that the total yield in 1909—1913 was fully as great as in 1928—1932. But, if the fish were larger, then it is legitimate to think that they were older; so an older stock gave, for five years within which we know of no great change in fishing rate, as great a yield as did a younger stock for another five years, within which we also know of no great change in fishing intensity. We use the landings of all North Sea species, except herring, by all fishermen, in metric tons (000 omitted)[1]:

> 1909—1913: 419, 416, 440, 463, 433.   Mean: 434
> 1928—1932: 413, 431, 458, 428, 411.   Mean: 428

Actually the *catches* in the earlier period were certainly higher than in the later, because fish were then rejected at sea which are now landed. The evidence of these figures simply is that the catch *was* no less before the introduction of the Vigneron-Dahl trawl and other changes of which the nett result has been to lower the age of the stocks.

So far as it goes, the evidence of general statistics is, therefore, that the yield would not be less if the stock were stabilized at a lower fishing rate.

2. From Growth and Age-Census Data.

a. A r g u m e n t.

According to R u s s e l l's equation a stock will be in equilibrium with fishing when the "logarithmic" rates (r in compound interest equations, such as $w_2 = w_1.e^{rt}$) tending to increase it are equal to those tending to decrease it, or

$$C = A + G - M \quad \text{where C is capture,}$$
A is recruitment,
G is growth,
M is natural mortality;

say

$C = V$, calling V the "rate of natural increase." .... (1)

Consider a stock N_1W_1, where N stands for number and W for average weight, becoming N_2W_2 at a year older.

Then

$$N_2W_2 = e^V N_1W_1 \quad .... (2)$$

Now, in equilibrium, the yield is Y in
$$Y = NWC$$
$$= NWV$$

---

[1] Bulls. Stats. Cons. Int. Copenhagen.

So

$$\frac{Y_2}{Y_1} = \frac{N_2 W_2 C_2}{N_1 W_1 C_1} = \frac{e^V N_1 W_1 V_2}{N_1 W_1 V_1} = \frac{e^V V_2}{V_1} \quad \dots\dots\dots \quad (3)$$

So

$Y_2 - Y_1$ will be positive

so long as

$e^V V_2 - V_1$ is positive

or, *a fortiori,* since V_1, V and V_2 must be in descending order of magnitude, as

$e^{V_2} V_2 - V_1$ is positive,

which, expanding e^{V_2}, discarding minor necessarily positive terms, substituting the constituents of V, and putting in numerical values for A and G from the data (which will be described later) reduces, for the North Sea cod, to the statement that

$Y_2 - Y_1$ will be positive, so long as

0.456 at least equals $(M_2 - M_1) + 1.516 M_2 - M_2^2$ $\dots\dots\dots\dots$ (4)

in which $(M_2 - M_1)$ and M_2 are considered to be unknown.

If we put in trial values for these unknowns we find graphically that either the change in natural mortality rate or the new natural mortality rate has to be quite ridiculous, from the information we have, for the right half of equation (4) to be equal to 0.456. We therefore conclude that there is no doubt that the new yield will at least equal the old, and that there is plenty of margin for error in the approximations we have used as to A and G. The values obtained are as follows.

If $M_2 - M_1$ were (%)	M_2 would have to exceed (%)
40	3
30	11
20	19
10	29
0	41

Now, from our census data, we know that we are only proposing to raise the average age from $2^1/_2$ to $3^1/_2$, in a fish which is first mature at about 5 years of age, and it is really very unlikely that in these young fish the natural mortality rate would thereby be raised. In fact we would expect it to fall. Yet if we make the change only zero, we find that the new mortality rate has to be as high as 41 per cent., which is quite impossible from what information we have, and even from consideration of the natural span of life, which is about 8 years even under fishing, and likely to be about 20 without fishing. But if we make M_2 something reasonable, as 11 per cent. or less, we would have to believe that the change was positive by the absurd amount of 30 per cent. or more. We see from inspection that continuation of the series of the trial values in either direction can only result in more ridiculous values, either of one unknown or the other.

Similar results are obtained from the data on North Sea haddock and plaice. The three species, cod, haddock, and plaice, together make up more than 60 per cent. of the North Sea trawl catch, whether by weight or value[1]). There is no reason to believe that the other species would show compensating changes, sufficient to prevent the conclusion applying to the total. It is therefore concluded that the yield would be no less, were the fishing effort reduced so as to allow the fish to become one year older.

b. Data and Approximations.

The data have all been published elsewhere. For the cod I have used those of this laboratory[1a]), the haddock data are from the work of H a r o l d T h o m p s o n of the Scottish investigations[2]). For the plaice we use B ü c k m a n n's data[3]). The general method was to calculate "logarithmic" rates of growth at various ages by applying the compound interest law to the data given. Where interpolation was necessary this was done graphically. It is hoped that the footnotes will enable the computations to be repeated. Table 1 is the calculation sheet for the cod, using the census of September[4]). Another, using the annual census, gives similar results. Little letters are used for rates, etc. at a given age.

Table 1.

	$1^{1}/_{2}$	$2^{1}/_{2}$	$3^{1}/_{2}$	Age $4^{1}/_{2}$	$5^{1}/_{2}$	$6^{1}/_{2}$	$7^{1}/_{2}$	Total
g	1.89	1.01	0.60	0.44	0.39	0.29		
w[5])	250	950	2000	3350	5010	7180		
wg	472	960	1200	1474	1954	2082		
n_1[6])	47	32	8	9	3	1	1	$101 = N_1$
$n_1 wg/100$	222	307	96	133	59	21		$838 = N_1 W_1 G_1$
$n_1 w/100$	118	304	160	302	150	72		$1106 = N_1 W_1$
								$\therefore G_1 = 0.758$
								$A_1 = 118/1106$
								$= 0.107$
m	0.20	0.13	0.08	0.05	0.05	0.05		
n_2[7])	47[8])	40	29	8	9	3		$136 = N_2$
$n_2 wg/100$	222	384	348	118	176	62		$1310 = N_2 W_2 G_2$
$n_2 w/100$	118	380	580	268	451	215		$2012 = N_2 W_2$
								$\therefore G_2 = 0.651$

---

[1]) Bull. Stat. 1932 (1934) p. 18 Mean 1912—1932.
[1a]) Graham, Sea Fish. Invest. XIII. 4. London 1934.
[2]) Rapp. Cons. Int. LIV. Copenhagen 1929.
[3]) Rapp. Cons. Int. LXXX. Copenhagen 1932.
[4]) Graham 1934, p. 139.
[5]) By interpolation. Weights at whole years from G r a h a m, 1934, p. 137, March, and R u s s e l l, Sea Fish. Invest. V. 1. 1922. p. 75. Also, for age 1, length from G r a h a m 1934, pp. 42 and 67, and weight from R u s s e l l, 1922, by integrating length cubed with respect to length from average length at 0 to average length at 1 year old and so on, and multiplying by a condition factor. A simpler method would have served for the present purpose.

Lines g, w and n_1 are from the data. The calculation as far as G_1 is simple multiplication, addition and division. The line m requires explanation. Strictly speaking we only know the order of magnitude of the natural mortality rate, not any values as accurate as those given. We do, however, only use these values for weights in the calculation of G_2, and we rely on the experience that errors in weights are of minor importance in calculating an average. As to our knowledge of the order of magnitude, we have the total mortality rate for haddock from T h o m p s o n's census of age[1]), and we have the fishing mortality rate for plaice from large-scale marking experiments. Thus B o r l e y reported that something like 30 per cent. per annum of marketable plaice were returned in the years before the war[2]). If we subtract fishing mortality rate from total mortality rate we are left with natural mortality rate. With these data in mind, I have simply written in some probable values for cod. The remainder of the calculation can be followed from the notes, but there are two more approximations that should be pointed out. Firstly it has been assumed that the same number of recruits would be found in the line n_2. Secondly, it has been taken as a first approximation that g and w would be unchanged in the older stock. It will be remembered that the defence of these approximations is that there is ample room in the final conclusion for error that they may have introduced[3]).

The conclusion of this digest of the results of fishery research is also that a lower fishing rate would give as great a yield, when the stock became stabilized at that rate.

3. From Consideration of the Effect of the War on Landings.

(i) The principal assumption of this section is that V, the "logarithmic" rate of natural increase of the stock at a given moment, including rate of reproduction, is directly proportional to the difference between the weight of the stock at that moment and the maximum weight the area will support. We can be pretty sure that V is some positive function of that difference in weight, but the assumption of direct proportion may, of course, be too simple. It is, however, a reasonable first approximation.

---

[6]) Graham 1934, p. 139.
[7]) $n_2 = n_1 . e^{-r}$, where r is the assumed rate of natural mortality at a given age in whole years.
[8]) Assuming new entry unchanged.
[1]) *loc. cit.* p. 157.
[2]) Sea Fish. Invest. III. 3. London 1916. p. 66.
[3]) Some further notes will be necessary for repeating the calculations for cod (annual census), for haddock and for plaice. In the cod calculation it was necessary arbitrarily to repeat n_2 at $2\frac{1}{2}$, owing to disturbance by net selection. It can be shown for all three species, that disregard of net selection has probably caused an underestimate. For haddock I used T h o m p s o n, 1929, p. 146, and R u s s e l l, Sea Fish. Invest. I, 1914, Pt. I, p. 130 and integrated. For plaice I used Bückmann's figures, p. 13. Stock A.

(ii) From the *Bulletins Statistiques* we can express the landings of all bottom species (that is excluding herring) as percentages of the average of 1909—1913. We obtain

1919	1920	1921	1922	1923	1924	1925
105	128	115	117	88	89	100

That is, the drain of 1920—1922, namely 28, 15 and 17 per cent. above the pre-war average could not be sustained, although the stock was augmented by the war respite.

(iii) We have unpublished figures for the landing in each month by 1st class British steam trawlers landing their own catch in England. We also have the corresponding number of days absence. The earliest period in 1919 in which the number of days absence approached the pre-war figure was at the end of the year, from October to December. The data are

	Days absence (00 omitted)		Landing (cwts; 000 omitted)	
	1913	1919	1913	1919
Oct.	137	106	177	348
Nov.	119	100	162	266
Dec.	131	113	172	262
Total	387	319	511	876

These figures are not accurate indices of the relative weights of the stock, but they can be used as a rough guide. Thus 876/511 is 1.71. Alternatively, (876/319) : (511/387) is 2.08. We conclude that the stock at the end of 1919 was rather less than double the weight of the pre-war stock.

(iv) From (iii) we make the assumption that the upper limit of stock is not less than twice the pre-war weight.

(v) We have all in our minds the idea that the stocks had considerable respite during the war. I try to estimate this by saying that it was equivalent to so much time of complete cessation of fishing. The working has to be crude. Foreign fishing was not so much curtailed as English fishing. Also English fishing was more curtailed than the number of days absence indicates. The statistics are:

	1913	1914	1915	1916	1917	1918	1919
Eng. days absence (000 omitted) ..	147	124	76	41	37	56	92

Because of information collected soon after the end of the war as to war-time interference with even the reduced fishing, I think a more comparable series would be as follows.

	1913	1914	1915	1916	1917	1918	1919
Corrected Eng. days absence	147	120	60	30	30	50	85

(This correction is not necessary to the conclusion).

If we multiply each of these figures by the corresponding ratio of the total landing to the English landing, we obtain another series, which seems the best we can achieve, as an index of intensity of fishing.

	1913	1914	1915	1916	1917	1918	1919
Relative intensity	346	335	191	157	109	141	230

Finally, I call the respite of 1914 $\dfrac{346-335}{346}$ of a year, and so on. Adding up the series of such fractions we obtain about two and two-thirds years. This gives some numerical idea of how much respite the fish had, but obviously one wishes it were better.

(vi) Knowing that many marks are shed or not sent in, we use the information from B o r l e y[1]) to assume that the pre-war rate of natural increase (equals fishing rate in a stable stock) was more than 30 but distinctly less than 50 per cent. per annum in 1913 and is 50 per cent. now.

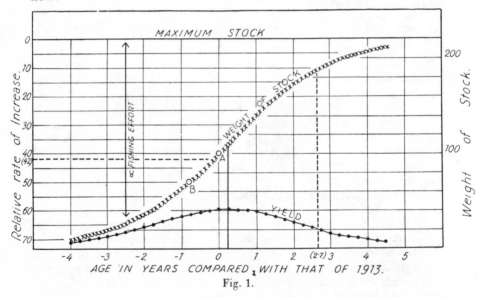

Fig. 1.

(vii) Under (1) we have seen that the landing of recent years is hardly less than in 1909—1913, the difference in catches being only what we can allow for rejected fish in pre-war days. So we assume that the pre-war catch was not more than 20 per cent. greater than the present catch.

Thus Fig. 1 was drawn. Starting at A with stock of weight 100 we chose a rate of 40 per cent. per annum and a maximum stock of 220. One tenth of the chosen rate was applied at simple interest to give the stock one tenth of a year later. The rate was then changed in proportion to the new vertical distance from the maximum and the

---

[1]) *loc. cit.*

next point was found. So proceeding upwards and downwards the whole curve of weight of stock was drawn. It clearly resembles the sigmoid curves of H j o r t, J a h n and O t t e s t a d and of earlier writers. Judging by the difficulty encountered in drawing the trial attempts at Fig. 1, very little latitude has been introduced by the graphical method. This is the only curve encountered that would fulfil the numbered requirements already given.

Fig. 1 gives a complete representation of the theory, although, as we have been at some pains to show in our arrangement of this paper, the basis for action does not rest alone on the fundamental assumption as to rate of increase. The main curve AB in the figure traces the natural growth of a stock, starting from a low value and not subject to fishing. At any point the stock may be stabilized by a fishing effort which is proportional to the vertical distance between the point on the curve and the maximum. The yield under the condition of stability is given by the lower curve, of which the ordinates are differential coefficients of the stock function, that is, they measure the slope of the main curve.

Taking the stock at A as the pre-war stock and that at B as the present stock and numbering the conclusions of the curve according to the requirements, we have:

(ii) that the maximum stable catch is only a little above that of the stock at A, so that the drain of 1920—1933 could not be maintained.

(iii) and (v) that the stock increased to 1.9 times its pre-war weight in 2.7 years.

(iv) that the maximum stock is 2.2 times the pre-war, or about 3 times the present stock.

(vi) that the pre-war rate was 40 per cent. per annum.

(vii) that the pre-war stock could maintain a catch 14 per cent. greater than could the present stock.

These fulfil the requirements.

If we accept these calculations as a rough approximation, we find that the maximum stable yield is 15 per cent. greater than the present yield and that this yield would be obtained by 75 per cent. of the present effort.

Such as it is then, the estimate is that about one quarter of the fishermen's time and expense is at present devoted to reducing their catch.

By way of illustration we finally work out what would be the yield in successive years as a result of reducing fishing time by one-sixth. The present rate of natural increase is assumed to be 50 per cent., balanced by a fishing rate of 50 per cent. In the new circumstances the stock would start to increase at 50—42 or 8 per cent. per annum. As the stock increased it would have a new rate of natural increase, less than 50, until finally the weight increased to the amount shown in Fig. 1 for the rate of 42.

The calculation is carried out in steps of one-tenth of a year:

Table 2. Abridged.

Time	Stock	Rate of Increase	Catch 0.042 of Stock
0	70		2.94
		0.50 — 0.42 = 0.08	
0.1	70.5		2.96
		0.08	
0.2	71.0		2.98
		0.08	
0.3	71.5		3.00
		0.49[1])— 0.42 = 0.07	
0.4	72.0	.	3.02
.	.	.	.
.	.	0.07	.
.	.		.
0.9	74.5		3.13
	Yield in 1st year	=	30.35
1.0	75	0.48[1])— 0.42 = 0.06	3.15
.	.	.	.
.	.	0.05	.
.	.		.
1.9	79		3.32
	Yield in 2nd year	=	32.43
	Yield in 3rd year	=	34.13
	Yield in 4th year	=	35.48
	Final yield[1])	=	39.50
	Present yield	=	35.00

Against our saving of about 16 per cent. in coal and some other expenses, we have therefore to debit a loss in turnover of 13½ per cent. in the first year, 7½ per cent. in the 2nd, 2½ per cent. in the 3rd, and thereafter no loss, but finally a gain of 13 per cent.

It is to be noted that growth is assumed to take place equally throughout the year. This is not so. To ensure as little loss in turnover as possible such an operation should begin in March or April.

Statistics of the war-time respite therefore agree with the other two sources of data, in showing that a reduction of fishing-rate would not harm the yield. They also provide an estimate of the benefit.

Conclusion and Acknowledgement.

This paper is an attempt at more precise formulation of the opinion, that has been long held by many fishermen and scientists, that it would pay to give the fish a chance to grow. Most fishermen and some scientists would also add "and a better chance to breed." I am

---

[1]) From Fig. 1.

particularly indebted to Mr. W o l l a s t o n, among the colleagues who have been so helpful in discussing this paper, for his emphasis on the possibility that the stocks of some species are not in equilibrium with fishing, but, owing to shortage of eggs, may be in a state of decline, a subject which he is investigating further. Much might be written as to pros and cons in this matter, but a decision is hampered by lack of crucial evidence. In this paper it is only necessary to point out: (i) that even a stock declining in numbers can be better fished, from the point of view of economy, leisure, and yield, if the fish are allowed to grow; (ii) that any reduction of fishing rate would slow down the decline; (iii) that, for all we know, a reduction of, say, 16 per cent. in the fishing rate might indeed be sufficient to arrest the decline altogether. Only this should be made clear, that if the stock be in such a state of decline, one must not be too sanguine as to increase of yield over the present yield, but should rather think of it in terms of the yield that would be taken if the fishing rate were not reduced.

Finally, the correction proposed to the theory, as developed by T h o m p s o n and B e l l, may be put into words: It will pay to reduce the fishing rate at any rate so long as the stock will thereby grow in weight sufficiently for the product of the new reduced fishing-rate multiplied by the new augmented stock to be no less than the product of the old higher rate and the old smaller stock.

The theory of fishing

R.J.H.Beverton and S.J.Holt

Sea Fisheries: their Investigation in the United Kingdom, M.Graham (editor), Arnold, London, pp. 372-441, 1956.

Editor's introductory notes

This paper is a short version of the longer one published in 1957. There are two important innovations, the estimation of rates of growth and mortality from observed age distributions and the use of the von Bertalannfy equation to describe the growth of fishes. Their main achievement was the formulation of the yield per recruit as function of fishing mortality, which avoided the difficulties raised by the variability of recruitment. Further they were able to predict the short and long term changes following an increase in mesh size. In other words given good estimates of growth and mortality the problem of growth overfishing can be solved quickly. They were also able to model the effects of density dependent growth and density dependent natural mortality upon their yield per recruit curves.

Ricker (1946) anticipated some of the work of Beverton and Holt, but he did not exploit the age distributions of the fishes to the same degree. Like Baranov he formulated the catch equation, $C = (F/Z)R'(1\text{-exp-}Z\lambda)$, where C is catch in numbers, Z, the instantaneous coefficient of total mortality, R' is recruitment in numbers and λ the number of exploited age groups. Like Ricker, Beverton and Holt used the development of the catch equation to show that catch per unit of effort was a proper index of stock. But that conclusion must be modified if, q, the catchability coefficient varies inversely with abundance.

Beverton and Holt applied their methods to marking or tagging experiments, in which individuals of a small sample of the real population are marked or tagged with numbered buttons, or Petersen tags. Such experiments are used for a number of purposes, but Beverton and Holt used the results to estimate fishing mortality as Graham (1938) had done earlier. They also developed an equation relating recruitment to parent stock which depends upon density dependent mortality during a critical period in early life history.

Beverton and Holt laid the foundations of fishery research as we know it today. As will appear below, the same ends are achieved today with somewhat different methods. From their work the North Atlantic Commissions were able to come to practical and agreed conclusions and all nations there agreed optimal mesh sizes by the early sixties. But that did not prevent further increases in fishing effort, nor the development of industrial fisheries in the North Sea which generated new forms of growth overfishing.

References

Graham,M.: 1938 'Rates of fishing and natural mortality from the data of marking experiments' *J. Cons. Inst. Explor. Mer.,* **13,** 76-90.

Ricker,W.E.: 1948 'Methods of estimating vital statistics of fish populations' *Indiana Univ. Publ. Sci. Ser.,* **15,** pp. 101.

CHAPTER IX

The Theory of Fishing

TABLE OF CONTENTS

PAGE

INTRODUCTION 372

Comparison with agriculture—necessity of population studies, where the operative factors are not under human control—empirical and analytical approaches, as in Chapter VII—the theoretical model—references to population dynamics—open systems and steady states—models for a particular fishery—the trawl fishery for plaice and that for haddock.

THE PRIMARY FACTORS AND THEIR COMBINATION IN A THEORETICAL MODEL . . . 375

Recruitment—limiting age—mortality—mesh selection—knife-edge selection—entry to exploited phase—fishing mortality—fishing power, effort, and intensity—yield in numbers and weight—growth—Bertalanffy's equation—expressions for the yield—for biomass—general behaviour of the model in accordance with previous ideas on fisheries.

ESTIMATION OF PARAMETERS 386

Total mortality—from age-composition of plaice—weighting—fishing mortality (F) or natural mortality (M) separately—when fishing has varied significantly—Silliman's determination for California sardine—determination of M for plaice—and of fishing mortality—by marking experiments—theory for losses—for variation in rate of fishing—method of eliminating steady losses—determination of loss on marking—use of two marks—dispersal of marked fish—age at capture—selection ogive—alternate hauls or covers—knife-edge selection —age at recruitment—rejection at sea—growth parameters from data—weight and length—cubic relationship—life-span—relative recruitment—absolute recruitment.

PREDICTIONS IN CONTROLLED FISHING 408

Yield-intensity and yield-mesh curves—maximum at intermediate intensity—at intermediate mesh—effects of changed parameters—Baranov important—Ricker and Silliman—Graham's synthesis—Thompson's halibut work—Burkenroad's criticisms—aims of regulation—eumetric fishing—economic considerations—profit curve—the best dependent on circumstances—former usages of terms—number of species—methods of regulation—reduction essentially to two—expediency—equivalence—immediate effects—the transitional period—small steps.

MORE COMPLEX THEORY 421

Density dependence and spatial variation—growth, data available—linear relation—model for relation to food supply—Dawes' experiments—food for maintenance—food for growth—grazing—effects on the simple model for yield—linear relationship for natural mortality—effects on yield model—growth and mortality—recruitment—expressions for total egg-production and per recruit—pre-recruit mortality—probability of compensation—relation postulated—example in haddock—self-generating model—simultaneous density-dependence of all parameters—spatial variation in factors—effective overall fishing mortality coefficient—transport coefficient—random dispersal—searching efficiency—tactics of fishing.

CONCLUSION 440

Behaviour of fish populations and of man as predator.

INTRODUCTION

The preceding chapters have led to a requirement to state in one comprehensive equation the main factors affecting the yield of the fishery. It may be noted that this requirement is

INTRODUCTION

characteristic of sea-fishing, as compared for example with agriculture, and derives from the fact that fishing is a kind of hunting rather than of cultivation.

Man cannot, as yet, do much to replenish the stocks of sea-fish in order to compensate for the yield he takes, and he has to rely on the natural resilience of the fish populations to make good the depletion caused by fishing. This is so despite spasmodic attempts to build up reserves of young fish by releasing fry reared in hatcheries and to increase the growth rate of older fish by transplanting them from areas where growth is slow to areas of higher productivity. Thus the emphasis in research at the moment must necessarily be on a study of fish populations in the natural state and subject to the many disturbing influences of the environment. Nevertheless, it will become clear that there is promise of advantage in projects which involve more drastic interference than hitherto with the marine community including perhaps, in the future, large-scale transplantation.

The main task is therefore to learn about the behaviour of fish populations as statistical aggregates, and in particular how they react to the various amounts and kinds of fishing activity to which they have been, or might in the future be, subjected. It should be clear now why population studies form a basic feature of fishery research but play little part in agricultural science. In many ways a herd of cattle is not a self-contained population : its numbers can be altered at will ; individuals do not necessarily have to compete for food ; if the average age becomes too great, younger stock can be introduced ; and it is not necessary to rely on the reproductive capacity of a particular herd to maintain its numbers. All these factors and others besides will be recognized as determining the commercial value of a herd and the supply of milk or meat produced by it. They are equally relevant in marine fish communities, but instead of each factor being under control, separately and at will, they are all interrelated and man has, in effect, one means only of influencing all of them, namely, by the process of fishing itself.

We may summarize the differences thus : the way in which man exploits a natural fish population affects the productivity of that population in a complex manner that has no parallel in agriculture. So we arrive at the main problem posed in preceding chapters ; how should a given fish population or community be exploited, and the amount and kind of fishing activity be adjusted, to obtain the best results ? Unfortunately, the individual fisherman is not in a position to make this assessment himself, and it is only relatively recently that organized scientific research, both in this country and elsewhere, has enabled a partial solution to be found.

At this stage the question must be left in a rather vague form because it would be premature to attempt a more precise definition of what are the " best results ". It may, however, be noted that this phrase means roughly, the " greatest sustained yield ", but that the economic structure of the fishing industry must also be taken into account. It will be found that these two requirements are to some extent conflicting and that what is " best " is usually something of a compromise between them.

Empirical and analytical approaches. In a general sense two approaches to our problem can be distinguished ; these can be called the *empirical* and the *analytical* methods. Ideally, if there is available enough information relating to the effort put in and the yield obtained from a fishery in the past—effort being measured in terms of numbers and sizes of ships and the amount of time spent fishing, and yield as the annual catch or market landings —then one could perhaps find by inspection of the statistics the conditions which led to the best yield being obtained. In Chapter VII it has been shown that such an enquiry shows only trends. In practice this method is seldom of much use on its own though an extension of it, namely, controlled experiments in which changes are made in the fishing intensity and the resulting changes in the catch are observed, has been used, notably in the Russian fishery of the Caspian Sea. Even then there are difficulties of interpretation. For example, it is necessary to know how to get the best *sustained* yield, and it may well be that the amount of fishing has not been stabilized long enough for this to be reached and measured. Moreover, as we saw in Chapter VII, fish populations show very wide natural fluctuations in abundance which may greatly exceed the changes caused by man's operations, and it is difficult if not impossible to distinguish them by means of commercial statistics of catch and effort alone. A final objection to the empirical method is that it provides no means of judging whether the conditions which gave the best yield in the past would continue to do so in the future, or whether there might not be some other way in which the population could be exploited to give even better results.

These difficulties are largely overcome by the analytical approach, which is to investigate the causes of change in population size and yield, and to break them down into the separate components such as the mortality caused by fishing and by natural events such as predation, disease, senility, etc. ; the rate of growth in weight of individual fish ; and the reproductive rate. If enough is known about such factors and their interrelationships, a theoretical model of the population can be constructed and its reaction when various factors and values are changed can be observed, attention being directed particularly to changes in the amount of fishing activity. It is clear that the actual process of analysis is the first—and in some ways the most difficult—stage, and it is followed by synthesis to provide an hypothetical model to be compared with the real population. This method is both the most practical and the most fundamental, since by it we have more chance of finding *why* things happen and thus of gaining more control over our problem, which is indeed one of the main functions of science.

Theoretical model. This technique of constructing a theoretical model and investigating its properties is familiar enough in the physical sciences, and it is not an exaggeration to say that it is the main method used in theoretical physics at the present time. It is also a valuable tool in operational research (Goodeve, 1948). The procedure is less common in biology, partly because biological phenomena are more complex than physical ones, and in many cases insufficient is known to enable the analytical stage of the process to be completed, but the field of population dynamics is one branch of biology where mathematical models have been used for a long time. Good examples of these are given by Kostitzin (1939). Rafferty (1950) and von Bertalanffy (1950) have dealt with the principles on which such models of biological systems depend, and the latter author has also pointed out that most biological processes are ' open ' ones. By this term is meant that there is a continuous flow of energy into and out of the system, and it is a characteristic of open systems that they usually tend to attain a condition of dynamic equilibrium, or *steady state*. A living cell is an example of such a system, taking in energy in the form of oxygen and nutrients and giving it out as excretory products and the maintenance of life-processes. An exploited fish population can also be regarded as an open system ; the young fish, or recruits, which enter the population each year and the growth in weight of the adults together represent the intake of energy, while losses take the form of natural deaths and the yield obtained by man. For the population to be in a steady state the inflow and outflow must balance each other, but accord-

41

ing to how much fishing there is so the balance may occur at various levels of population abundance or density, corresponding to various magnitudes of yield. This fact has been simply expressed symbolically and in slightly different ways by Graham (1935) and Russell (1942). There are certain other important characteristics of biological open systems which are found in fish populations, notably a tendency for an externally applied force to be opposed and equilibrium to be regained after a disturbance, the reasons for which will become clear at a later stage.

Our procedure will therefore be to set up a mathematical model of a fished population, treating it as part of a predator-prey system, and then to see whether its behaviour can provide an answer to the problem of how that population *should* be fished to produce the best results, or at least to provide some improvement. Clearly, the value of any such model and the validity of any conclusions drawn from it will depend on how similar its properties are to those of the system of which it is an analogy. Whether this similarity can be tested directly depends on how much information is available concerning the past history of the system in question, and often such a test will be difficult, if not impossible ; what can be done is to make sure that each of the main factors responsible for the observed behaviour of the system is represented as faithfully as possible in the model. We cannot, therefore, set out to construct a generalized model which would be directly applicable to any fishery, but must base our model on a specific case. We shall not attempt to derive any universal ' laws ', though we shall indicate where relevant the extent to which the model derived in this chapter might be applicable to other cases. In this connection it will be found necessary at first to make many simplifying assumptions that in themselves may never be strictly true in any particular case, but nevertheless represent sufficiently well for the present purposes the characteristic in question. The view will therefore be adopted that the simplest hypotheses that take account of all the available observations, both specific and general, without contradiction or inadmissible discrepancy, will be considered to be satisfactory until it is shown by fresh evidence that additional factors and relationships must be allowed for and the model made that much more complex.

The example which will be used primarily is the trawl fishery based on the North Sea plaice of which the biology has been described in the preceding chapters. The haddock has also been dealt with in an earlier chapter, and this species also is susceptible to population study, because of the abundance of data.

THE PRIMARY FACTORS AND THEIR COMBINATION IN A THEORETICAL MODEL

A number of authors have published theoretical models of fish populations, to which reference is made later. Some of these are relevant and in part applicable to our chosen example, but the methods here discussed are based on the recent work of the present authors, and for proofs of many of the statements made and a fuller treatment of some of the problems raised, the reader is referred to that work (Beverton and Holt, in the Press). The four primary factors—recruitment, growth, natural and fishing mortality—have already been mentioned, and, since these will form the basis of our model, it is appropriate to consider them in turn and to formulate each in mathematical terms.

Most marine fish spawn during a relatively short period each year, such as in early spring, and it is useful to trace the history of the batch of eggs laid by the plaice population during one spawning season. In Chapters VI and VIII it has been shown that in the very early stages the eggs, and the larvae developed from the fertilized surviving eggs, are pelagic ; later, the fry adopt a demersal habit—for plaice in the North Sea mostly on the nursery grounds near the shores of the eastern coasts, being carried there by the north-easterly current which prevails in the spawning area. Not until plaice are two or three years old do they enter the heavily fished area of deeper water and become liable to encounters with fishing gear and hence to capture, and this movement defines the process of *recruitment* in this population. The number of fish which is thus recruited, and hence enters the *post-recruit phase* of the life-history at age t_p years (measured from the date of laying down the otolith nucleus and taken to be April 1st) is denoted by R. They are the survivors from the original batch of eggs, E in number, but are far fewer—perhaps less than 1 per cent. Hence the recruitment rate is not the same as the reproductive rate of the population, and this distinction is important.

Mortality. It is convenient at this point to define an age t_λ beyond which no adult fish will survive, that is to say, if a fish reaches that age without having been caught, eaten by predators, or killed in any other way, it is assumed that it will die immediately. The necessity for this assumption will become clear at a later stage, and it is to a certain extent justified by the observation that very few plaice above the age of fifteen years are found during the age-analysis of market samples and almost none over twenty.[1] For certain purposes it is possible to put $t_\lambda = \infty$ in the final expressions without introducing large errors, and this may simplify the mathematics considerably ; but it must be done with care, and the fixing of a finite age-limit avoids the difficulties and dangers of extrapolation to higher ages for which there are little or no data. It should be noted that the difference between t_λ and t_p defines the duration of the post-recruit phase.

We are not concerned now with the mortality and growth of fish in the pre-recruit phase, but after they have been recruited it is necessary to know what happens to them, since they may not at once be liable to capture though they continue to be exposed to the chance of death from other causes. Little is known about the specific causes of natural mortality in the plaice, and the simplest acceptable hypothesis is that a certain constant proportion of all fish present die during a given time. The number that actually die is therefore the product of that proportion and the total number present. Now this latter is decreasing continuously, but if a small enough time interval is considered the actual decrease will be very small. Thus we arrive at the idea of an instantaneous rate of change, a concept which is the basis of the differential calculus and of great importance in developing theoretical population models of the kind we are considering. If the number in the brood at any age t is denoted by N_t, the rate of decrease in numbers of the brood at any moment is

$$\left(\frac{dN}{dt}\right)_M = -MN \qquad . \qquad . \qquad . \qquad . \qquad . \qquad (1)$$

[1] This assumption is still useful even if it were true, as Bidder (1925) has suggested, that the natural life-span may in fact be unlimited.—Ed.

where M is called the coefficient of natural mortality. It should be noted that the reason why there is a minus sign on the right-hand side of this equation is that numbers are decreasing but it is simpler to refer to positive values of the coefficient M. A possible implication of this formulation is that the causes of death act randomly, so that any individual fish is equally likely to die at any time.

The above differential equation is of a kind that will be met on a number of occasions in this chapter, and since limitations of space will usually preclude any detailed mathematical argument being presented, it may help at this stage to show how (1) can be manipulated to the required form. For subsequent purposes we are interested in the number of fish, N_t, that survive to any age t greater than t_ρ, rather than in the instantaneous rate of decrease $\dfrac{dN}{dt}$, and the derivation from (1) of an expression for N_t is achieved by solving the equation or, in effect, transforming it to the integral form of which it is the differential. In mathematical textbooks (1) would be described as an ordinary linear differential equation of the first order and the first degree, and it may be solved by a process known as " separating the variables ", these latter being N and t in the present case. Thus dividing both sides of (1) by N, and multiplying both sides by dt, produces the separation required, since (1) then becomes

$$\frac{1}{N} dN = - M \, dt$$

Both sides can now be integrated directly, obtaining

$$\log_e N_t = - Mt + C \qquad . \qquad . \qquad . \qquad . \qquad (2)$$

where C is the constant of integration. Note that $\log_e$ denotes the natural (Napierian logarithm. It remains to evaluate the constant C. The number present at age t_ρ is R by definition, i.e., the number of recruits ; thus (2) must hold for the special case in which $N_t = R$ and $t = t_\rho$, and making these substitutions gives

$$\log_e R = - Mt_\rho + C$$

from which

$$C = \log_e R + Mt_\rho$$

Substituting for C in (2) gives

$$\log_e N_t = \log_e R - M(t - t_\rho)$$

and taking antilogarithms gives the required expression for N_t, viz.

$$N_t = Re^{-M(t-t_\rho)} \qquad . \qquad . \qquad . \qquad . \qquad (3)$$

in which e is the base of natural logarithms.

Although members of the young brood are at this stage suffering encounters with fishing gear, it is supposed that they are sufficiently small to pass through the meshes of the cod-end (and Chapter VIII has mentioned the evidence that it is through this part of the trawl that most escape takes place). Mesh experiments, such as those in which trawls with very small and with normal meshed cod-ends were used alternately on the same grounds (*vide* Jensen, 1949, also Ch. VIII and below, p. 400), have shown that the proportion of fish in each centimetre size-group that are retained gradually increases with increasing size until eventually, when the *maximum girth* of the fish is just larger than the average perimeter of the mesh lumen, retention is practically complete. The full range of length between almost certain escape and complete retention may be about 10 cm., but most of the change occurs within about 5 cm. or less. However, for the simple form of our theory it can be shown that no appreciable error is introduced if it is supposed that all fish less than a certain size (and hence age) escape, while above it they are always retained. This has been called the assumption of *knife-edge selection*, and we shall denote the threshold age by $t_{\rho'}$. The number of young fish attaining this age is obtained from (3) by substituting $t = t_{\rho'}$, thus giving

$$N_{t_{\rho'}} = Re^{-M(t_{\rho'} - t_\rho)}$$

For convenience we may put $N_{t_{\rho'}} = R'$, and $t_{\rho'} - t_\rho = \rho$, so that

$$R' = Re^{-M\rho} \qquad . \qquad . \qquad . \qquad . \qquad . \qquad (4)$$

From age $t_{\rho'}$ onwards the brood will be subject to a fishing, as well as to a natural, mortality ; that is, they will have entered the *exploited* phase of the population.

It can be assumed as a first approximation that each haul of the net takes a certain proportion (perhaps all, but not necessarily) of the fish that happen to be in the path of the trawl, and during all the hauls that are being made at any given moment by all the actively working ships of the fleet a certain proportion of the population is removed. Fishing mortality can therefore be represented, in a manner analogous to that used above, as a constant instantaneous coefficient, and we have

$$_F\left(\frac{dN}{dt}\right) = -FN \qquad . \qquad . \qquad . \qquad . \qquad (5)$$

this expression holding from age $t_{\rho'}$ to t_λ so long as the following two conditions are satisfied :

(i) That the fishing mortality coefficient (F) does not vary with age, as would happen, for example, in a drift-net fishery where very large fish cannot be retained by the gear, or if there is any considerable escape of very large and active fish from the *mouth* of the trawl.

(ii) That F does not vary with the abundance of fish. This implies that the gear is not " saturated " by a large catch of fish blocking the cod-end, reducing the flow of water through the trawl or reducing the speed at which it is towed, and thus reducing the effectiveness of the gear.

These assumptions—at least their approximate truth—have been implicit in the discussions in previous chapters of the trawl as a sampling implement, and of the use of the catch per unit effort in study of stocks. For some fisheries they may well be in doubt, but both requirements are satisfied, at least approximately, in the North Sea trawl fishery for plaice which we are considering. Hence the rate of change of numbers, between ages $t_{\rho'}$ and t_λ, is given by

$$\frac{dN}{dt} = {}_M\left(\frac{dN}{dt}\right) + {}_F\left(\frac{dN}{dt}\right) = -(F + M)N \qquad . \qquad . \qquad . \qquad (6)$$

This equation may be solved by the same procedure as described for (3) above, and we have

$$N_t = R'e^{-(F+M)(t-t_{\rho'})} \qquad . \qquad . \qquad . \qquad . \qquad (7)$$

It will be noticed that (7) gives the numbers surviving in the exploited phase as an exponential function of age. Let us suppose that the total number in the brood surviving to any age t in the exploited phase, or a proportional index of this quantity, can be measured.

Then taking natural logarithms of both sides of (7) and rearranging gives

$$\log_e N_t = [\log_e R' + (F + M)t_{\rho'}] - (F + M)t \qquad . \qquad . \qquad . \qquad (8)$$

which is an expression linear in t. Thus our theoretical model, if fishing and natural mortality have been formulated correctly, predicts that the logarithms of the numbers at successive ages should fall on a descending straight line. The information required to test this is available ; it consists of age-composition data in which the abundance of a year-class at each age of its life is expressed in units bearing a constant and proportional relationship to the actual abundance of the fish. These are obtained either by sampling the catch on the market or by using research vessels to sample at sea (see Chapters VII and VIII). In Fig. IX.1 are plotted against age the logarithms of indices of the average abundance of a

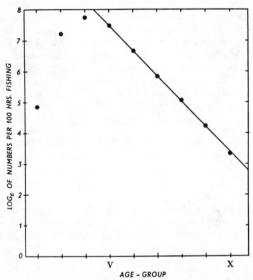

Fig. IX.1.—Mortality Rate of Plaice.

Natural logarithms of average number caught of each age-group of North Sea plaice per 100 hours' fishing, 1929–38. The slope of the line gives the estimate $(F + M) = 0.83$. The first three points are lowered by absence of fish from the exploited area (some individuals still remaining in the nursery areas) and, possibly, by rejection at sea.

number of year-classes of plaice, and it will be noted that the straight-line relationship predicted by (8) fits the data very well indeed. The same is found when data referring to North Sea haddock are treated in this way, and a similar characteristic is present in many other species and fisheries, thus confirming that the simple model postulated above is adequate in a general sense as well as in the special case of the North Sea plaice. It should be pointed out, however, that if F and M were of the same order of magnitude both (1) and (5) would be confirmed with reasonable certainty by the data plotted in Fig. IX.1 since the chance of their sum taking an exponential form if individually they did not is remote ; in fact, during the period to which the data refer, fishing mortality was very much greater than natural mortality. The conclusion that the hypothesis is satisfactory must therefore be taken in the main to refer to death caused by capture.

So far the coefficient F has been interpreted in terms of the mortality caused in the fish population, but it must now be related to the amount of fishing which generates that mortality. From the previous discussion it will be clear that F must be proportional to the number of vessels working and to some quantity which might be termed the efficiency, or *fishing power*, of each. Also, the chance of a fish getting caught is proportional to the time during which it is exposed to capture, so that F is proportional to the total number of hours spent fishing by all vessels, with allowance being made for the differing fishing powers and gears of each, the resulting index being called the *fishing effort*. The fishing power of a trawler has been found, by comparing the catches of a large number of vessels of different sizes fishing near to each other at the same time, to be roughly proportional to its gross tonnage. Hence fishing effort can be measured in the units " ton-hrs. fishing per year ", but if it should be necessary to compare the fishing activities in different water basins or in different parts of the same water basin, the *fishing intensity* must be computed, that is the fishing effort per unit area. This is necessary because in the above derivation of F it has been supposed that each vessel in each haul takes a certain proportion of the *whole* population, but this proportion depends on how much of the total population is concentrated in the path of the trawl during that haul. Thus a given number of vessels exerting a given fishing effort could take a greater proportion of the total population per unit time if they concentrate where fish are most dense than if they spread out over the whole area.

A word about the magnitude of F, which can take any positive value from 0 to $+ \infty$, is not out of place here. If $F = 1$ it does not mean that the whole of the population is removed in a year (or in whatever unit of time is used), but that a number of fish equal to the population at any one moment would be caught during the course of the year *if that population was continually replenished* and thus kept at a constant level of abundance. To catch all the population in the year, without replenishment, would require a value of F approaching infinity.

Using equation (7) as a starting point, it is now possible to deduce an expression for the yield from the year-class over its fishable life-span, the latter being given by $t_\lambda - t_{p'}$ and denoted by λ. Now the rate of yield in numbers is the same as the rate of decrease due to fishing, but with the sign changed ; thus

$$\frac{dY_N}{dt} = FN_t$$

and to obtain the yield in numbers accumulated between any two ages this expression would be integrated with respect to t with those two ages as limits. However, we are more interested in obtaining the yield in weight, Y_W, and if the average weight of fish of age t is denoted by w_t, the rate of yield is given by the product

$$\frac{dY_W}{dt} = FN_t w_t \quad . \qquad . \qquad . \qquad . \qquad . \qquad (9)$$

Growth. In order to obtain from (9) an equation for yield in weight it is necessary to know how weight changes with age, that is, to express w_t in mathematical terms, and growth in weight is in some ways the most difficult to formulate of the primary factors.   In fish there is a characteristic and almost universal pattern of growth.   Thus the curve of growth in overall length typically starts near the origin and approaches an upper limit, being convex upwards throughout.   Fig. IX.2 shows the average length of North Sea plaice plotted against age, the data being obtained from market samples and referring to the pre-war years 1929–38.   The weight of an animal growing isometrically, i.e. which stays the same shape, is roughly proportional to the cube of any linear dimension, and is exactly so if the specific gravity of the organism remains constant.   Hence, if the cubes of the lengths are plotted against age an asymmetrical sigmoid (S-shaped) curve is obtained, the inflection being at a value of the cube of the length less than half the asymptotic value.   This is in fact the type of curve that is found when the weight of plaice is plotted against age (Fig. IX.3), although

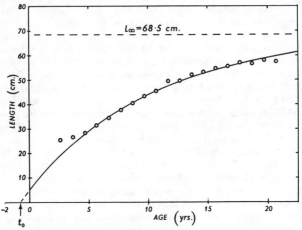

FIG. IX.2.—GROWTH IN LENGTH OF PLAICE.

Average length of each age-group, 1929–38, male and female fish combined.   The curve is that given by the von Bertalanffy equation (12) for growth in length (see p. 383).

the inflection and bend towards the asymptote cannot be demonstrated very well in this population owing to the lack of fish old enough—itself a consequence primarily of heavy fishing.   The same general pattern is found in nearly all fishes—apart from minor differences due to disturbing factors such as variations in the supply of food available to fish of different ages, and so on—and its consistency suggests that there is one basic growth mechanism underlying it.   A number of authors have put forward theories for this and growth equations to represent it.   At least, any acceptable equation must predict a weight-at-age curve similar to that of Fig. IX.3.   Some (e.g. Glaser, 1938) have no inflexion, and others (e.g. Robertson, 1923) give a symmetrical sigmoid.   The equation we shall use throughout this chapter is that due to von Bertalanffy (1938, 1949) ; not only does this give a growth curve closely similar to that shown in Fig. IX.3, but the underlying concepts are, in our opinion, much the most satisfactory of those which have so far been put forward.   It should, however, be pointed out that the validity of the population models to be developed does not depend directly on the truth of von Bertalanffy's theory and, in fact, with the important exception of the treatment of the dependence of growth on food supply and population density, any equation which represented the weight-at-age data equally well could be used throughout. For example, an empirical polynomial with three or more terms, could be used.

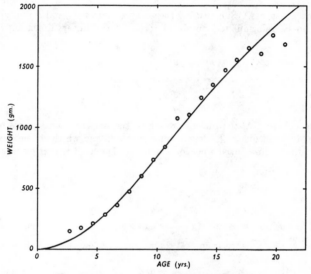

Fig. IX.3.—Growth in Weight of Plaice.

Average weight of fish of each age-group, 1929–38, and the fitted von Bertalanffy equation (14) for growth in weight. See Fig. IX.7 and p. 406 for parameter values and method of fitting.

The theory proposed by von Bertalanffy regards the weight of an organism as being subject at any moment to two opposing forces : the processes of anabolism or synthesis which tend to increase the weight, and those of catabolism or breakdown which tend to decrease it. In the simplest form of his theory, the rate of anabolism is taken as proportional to the magnitude of the " resorbing surfaces " of the animal, while the rate of catabolism is regarded as being proportional to the body weight itself. Since anabolism and catabolism are proceeding continuously throughout the life of the animal, the difference between them at any moment defines the rate at which the weight of the animal is changing at that moment. This idea can be formulated as a differential equation defining the instantaneous rate of change of weight by writing

$$\frac{dw}{dt} = Hs - Dw \qquad . \qquad . \qquad . \qquad . \qquad . \qquad (10)$$

where H denotes the coefficient of anabolism
       D ,,   the coefficient of catabolism.

Now the magnitude of the resorbing surfaces, i.e. the value of s, cannot be measured directly, but it can reasonably be taken to be proportional to the square of the linear dimension (l), if growth is isometric. Similarly, body weight can be regarded as proportional to the cube of l, so that we can put

$$s = pl^2$$
$$w = ql^3$$

where p and q are constants. Substituting these expressions for s and w in (10) enables the latter equation to be reformulated purely in terms of the linear dimension l, thus

$$\frac{dw}{dt} = \frac{d(ql^3)}{dt} = 3ql^2 . \frac{dl}{dt}$$

and hence

$$\frac{dl}{dt} = \frac{Hp}{3q} - \frac{D}{3}l$$

By transforming (10) to give the rate of growth in the linear dimension, we now have an equation that is homogeneous in the dependent variable (l) and which can be solved directly. It is an ordinary linear differential equation similar to (1) except that it contains a constant term $\frac{Hp}{3q}$, and its solution is

$$l_t = \frac{Hp}{Dq} - \left(\frac{Hp}{Dq} - l_0\right)e^{-\frac{D}{3}t} \qquad . \qquad . \qquad . \qquad . \qquad . \qquad (11)$$

where l_0 is the integration constant, taken as the length of the organism at age $t =$ zero. As t increases, $e^{-\frac{D}{3}t} \to 0$; hence from (11) the asymptotic length towards which the fish is growing is

$$L_\infty = \frac{Hp}{Dq}$$

Putting also $\frac{D}{3} = K$, for brevity, enables (11) to be written as

$$l_t = L_\infty - (L_\infty - l_0)e^{-Kt} \qquad . \qquad . \qquad . \qquad . \qquad (12)$$

Now since $l_t = \left(\frac{w_t}{q}\right)^{1/3}$, we can put $L_\infty = \left(\frac{W_\infty}{q}\right)^{1/3}$ where W_∞ is the asymptote of the curve of growth in weight, so that von Bertalanffy's equation for growth in weight is

$$w_t = \{W_\infty^{1/3} - (W_\infty^{1/3} - w_0^{1/3})e^{-Kt}\}^3 \qquad . \qquad . \qquad . \qquad (13)$$

where w_0 is the weight at zero age. For most purposes a more convenient form of (13) is that which contains t_0, the age at which the weight is theoretically zero, instead of w_0. Thus putting $w_t = 0$, and $t = t_0$ in (13) gives

$$w_0 = W_\infty(1 - e^{Kt_0})^3$$

and substituting this in (13) we have

$$w_t = W_\infty(1 - e^{-K(t-t_0)})^3 \qquad . \qquad . \qquad . \qquad . \qquad (14)$$

For subsequent manipulation it is simpler to expand this cubic and write the equation in a summation form as

$$w_t = W_\infty \sum_{n=0}^{3} \Omega_n e^{-nK(t-t_0)} \qquad . \qquad . \qquad . \qquad . \qquad . \qquad (15)$$

where

$$\Omega_0 = +1$$
$$\Omega_1 = -3$$
$$\Omega_2 = +3$$
$$\Omega_3 = -1$$

It is important to note the value of t_0 obtained when this equation is fitted to data will not necessarily be zero (this being partly due to sampling errors in the data and partly because the assumption of isometric growth is not valid at very low ages, but this does not matter in the present case because no yield is being obtained from these very young fish.

The continuous curves of Figs. IX.2 and IX.3 are derived from growth equations (12) and (14), and it will be seen that the fit to data is good,[1] but more evidence than this is necessary to establish the validity of von Bertalanffy's conception of the fundamental growth processes. In particular, the coefficients of anabolism and catabolism, which are contained in (14) in a derived form as the parameters W_∞ and K, must have a direct physiological interpretation. To this end von Bertalanffy (1934) took the rate of excretion of endogenous nitrogen in a starving animal as a measure of the catabolic rate, and showed that this agreed well with the value of K obtained by fitting the equation to growth data referring to the same animal.

Expressions for yield and population size. Returning now to expression (9) and substituting in it (7) for N_t and (15) for w_t, we have

$$\frac{dY_W}{dt} = FR'W_\infty e^{-(F+M)(t-t_{\rho'})} \sum_{n=0}^{3} \Omega_n e^{-nK(t-t_0)}$$

The yield obtained from the year-class over its whole fishable life-span is found by integrating this with respect to t between the limits $t_{\rho'}$ and t_λ; and remembering from (4) that $R' = Re^{-M\rho}$, the answer is

$$Y_W = FRe^{-M\rho}W_\infty \sum_{n=0}^{3} \frac{\Omega_n e^{-nK(t_{\rho'}-t_0)}}{F+M+nK}(1 - e^{-(F+M+nK)\lambda}) \qquad . \qquad . \qquad (16)$$

Now if it is supposed that R recruits enter the population, or rather the post-recruit phase of it, each year, the total yield obtaining during the course of one year from all year-classes will comprise the yield from the youngest year-class during its first year in the exploited phase, that from the next youngest during its second year, and so on. Hence the total annual yield from the population is the same as the yield from one year-class during its entire fishable life-span, and is therefore also given by (16).

One of the characteristics of most, if not all, fish populations is that the number of fish which are recruited each year, i.e. the value of R, varies very greatly. However, if the values

---

[1] The first two points are not on the curves because these ages were not sampled representatively, as will be explained later when discussing the mechanism of recruitment in plaice, whereas the last two or three are based on very few fish and are much less accurate than the rest.

of R fluctuate about a constant mean, having no trend, then it can be shown that putting the *mean* recruitment, $\bar{R}$, in place of R in (16), gives approximately the expected *mean* yield, $\bar{Y}_W$. Thus a phenomenon which might at first sight seem to present a considerable difficulty in formulating a useful theoretical model of a fishery, can for many purposes be safely neglected, though of course it could not if it were necessary to develop a model to predict the yield in any one particular year, as must be done sometimes. It can also be shown that it does not matter whether a given annual number of recruits enter the post-recruit phase all at one time in the year or there is a continuous influx of them.

Population models giving statistics other than the total annual yield must now be considered, perhaps the most important being that giving the *annual mean biomass of the exploited phase*, $\bar{P}'_W$. This is the quantity which would be measured if it were possible to obtain a series of estimates of the total weight of the exploited phase of the population during the course of a year, and the mean value computed. An expression for $\bar{P}'_W$ may be obtained most simply (though not in the most rigorous way) by dividing both sides of (16) by F, since the rate of capture at any time is the product of the fishing mortality coefficient and the total weight of fish present at that time ($N_t w_t$), and the catch over a whole year is proportional to the annual mean abundance of the exploited phase of the population. Thus we find

$$\bar{P}'_W = Re^{-M\rho}W_\infty \sum_{n=0}^{3} \frac{\Omega_n e^{-nK(t_{\rho'}-t_0)}}{F+M+nK}(1 - e^{-(F+M+nK)\lambda}) \qquad . \qquad . \qquad (17)$$

It can be shown similarly that the biomass of the pre-exploited phase is

$$RW_\infty \sum_{n=0}^{3} \frac{\Omega_n e^{-nK(t_\rho-t_0)}}{M+nK}(1 - e^{-(M+nK)\rho})$$

so that the annual mean biomass of the whole population ($\bar{P}_W$), which in ecological studies is sometimes termed the average standing crop, is therefore given by the sum of the above two expressions, i.e.

$$\bar{P}_W = RW_\infty \sum_{n=0}^{3} \Omega_n e^{-nK(t_\rho-t_0)} \left\{ \frac{1 - e^{-(M+nK)\rho}}{M+nK} + \frac{e^{-(M+nK)\rho}(1 - e^{-(F+M+nK)\lambda})}{F+M+nK} \right\} \qquad . \qquad (18)$$

It will be necessary to use (18) when predicting the effects of a change in the mesh-size of gear with a model in which growth varies with density, since such a change will alter the age limits of the pre-exploited and exploited phases (see p. 423).

We mentioned above that the fishing mortality coefficient F is proportional to the fishing intensity, i.e. to the concentration of a given fishing effort in space, the effort being measured in standard units of fishing power. For a fishing area of a given size, and with the effort distributed over it in a roughly constant manner relative to the fish concentrations, the fishing mortality coefficient can be taken to be proportional to the total fishing effort. In these circumstances it is seen from (17) that the mean biomass of the exploited phase is proportional to the *catch per unit effort*, so that the latter is an acceptable index of the former. In Chapters II, VII and VIII the usefulness of the catch per unit effort has been demonstrated, as a rough measure, even under less rigorous conditions than have been stated here. It is a particularly important statistic, since in addition to providing an index of population density it is also a measure of the efficiency with which the fishery is operating, and thus has far-reaching economic implications.

s.f.—c c

One further characteristic which is conveniently deduced at this stage is the annual mean numbers in the population, $\bar{P}_N$. This is obtained in the same way as (18) for $\bar{P}_W$ except that reference to weight is omitted, and we find

$$\bar{P}_N = R\left\{\frac{1 - e^{-M\rho}}{M} + \frac{e^{-M\rho}(1 - e^{-(F+M)\lambda})}{F + M}\right\} \qquad . \qquad . \qquad . \qquad (19)$$

No specific deductions can be made from the above expressions until estimates have been made in a particular case of the values of the various parameters they contain. At the same time, the general behaviour of a system in which numbers are continuously decreasing but the individuals are increasing in weight and also in which there is replenishment at yearly intervals can readily be seen. It would be expected that if fishing is very intense most fish will be caught whilst still relatively young and small. On the other hand, if fishing activity is slight a higher proportion of each year-class will die naturally and the annual catch would comprise a smaller number of fish, but the average age and size of these will be greater. Hence there is the possibility that with a certain intermediate amount of fishing there will be the best compromise—fish being allowed to grow to a reasonable size, but enough being caught before too many die from natural causes and before their growth slows down as it inevitably does in old age. It is evident that the same general kind of effect might be produced by adjusting the age at which capture begins, and hence the duration of the exploited phase. An increase in mesh size, for example, would restrict the fishing mortality to older fish and thus, in effect, decrease mortality in the population as a whole. There are, however, important differences in detail between the results of increasing the size of mesh and decreasing fishing intensity which can be assessed only by using a theoretical model of the population. These general concepts have been appreciated for a long time, though they have not become generally accepted until comparatively recently. As was mentioned in Chapter II, the size of mesh, for example, was a matter of concern in the reign of Edward III, which was vigorously revived by Frank Buckland in the latter half of the last century. For an account of the work and ideas of Buckland on this subject the reader is referred to Graham (1948), and more recent work by Davis and others has been recapitulated in Chapter VIII.

ESTIMATION OF PARAMETERS

If a fishery is studied over a period of years under conditions where it can be assumed that the population is in a steady state, and values of the relevant parameters are determined, expression (16) allows the average annual yield to be computed, which can be compared with the observed yield during the period in question. However, we are more interested in a prediction of the yield that would be obtained if the population were exploited in a different way, and changes in fishing intensity and size of mesh, which can be represented in (16) by appropriate changes in the values of the parameters F and $t_{\rho'}$, have a special importance. If our simple model is used for this purpose and it is assumed that the remaining parameters R, t_ρ, M, K and W_∞ stay unchanged, there will almost certainly be an error in prediction owing to the fact that changes in F and/or $t_{\rho'}$ will inevitably cause changes in population density to take place, which in turn alters the values of some, and perhaps all, of the other parameters. On the other hand, these errors would not be expected to be large provided only small changes in F and $t_{\rho'}$—and hence in density—are considered ; consequently, the best procedure is to start by determining values relating to the plaice population during a period of about ten years immediately before the second world war when it was in a reasonably steady condition, and assume that the parameters other than F and $t_{\rho'}$ stay constant and are, in particular, independent of density. Then at a later stage it will be possible to introduce density-dependence and other complications into the simple population models, and to assess their probable effects.

Certain special techniques are required for handling the available data in order to estimate the various parameters contained in (16), and some attention will have to be devoted to this aspect of our problem.

Mortality Coefficients

Age-composition data. These and (8) provide a basis for estimating the total mortality coefficient $(F + M)$, but only indices of the density (e.g. the catch per 100 hours' fishing) are available and not the absolute abundance. Furthermore, these data frequently refer to means over a considerable time period, usually a year, since attempting to measure directly the instantaneous density can introduce considerable sampling variation. The problem is therefore to derive from (8) an expression from which a value of $(F + M)$ can be computed using indices of the annual mean density instead of the absolute instantaneous abundance.

Denoting by $_\theta N'$ an hypothetical index of the abundance of a year-class on the birthday on which it enters age-group θ (that is, at the beginning of its $(\theta + 1)$th year of life), and by $_{\theta+1} N'$ a similar index of the numbers surviving exactly one year later, then we have

$$_{\theta+1} N' = {_\theta N'} e^{-(F+M)}$$

and the index of the mean abundance of the θth age-group during that year will be

$$_\theta \bar{N}' = \frac{_\theta N'}{F + M}(1 - e^{-(F+M)}) \quad . \qquad . \qquad . \qquad . \qquad (20)$$

[*cf.* (19)]. Now if k is the constant multiplier relating the hypothetical index to the true abundance we have from (7)

$$_\theta N' = kR' e^{-(F+M)(t_\theta - t_{\rho'})}$$

where t_θ is the age of the year-class at the beginning of its $(\theta + 1)$th year of life. Hence we can write (20) in the form

$$_\theta \bar{N}' = \frac{kR' e^{-(F+M)(t_\theta - t_{\rho'})}}{F + M}(1 - e^{-(F+M)})$$

Taking logarithms gives

$$\log_e {_\theta \bar{N}'} = \left[\log_e\left\{\frac{kR'(1 - e^{-(F+M)})}{F + M}\right\} + (F + M)t_{\rho'}\right] - (F + M)t_\theta \quad . \quad (21)$$

Since the only variable on the right-hand side of this equation is t_θ, it is found that the natural logarithms of indices of the annual mean density of each age-group are related linearly to the age of fish when they first enter that age-group, the slope of the straight line having a value equal to minus $(F + M)$[1]. Thus the annual mean abundance of a year-class in two successive years of life can be used to estimate $(F + M)$ in the same way as can the instantaneous abundances at yearly intervals, though it must be remembered that doing so implies that both F and M are constant over both years.

The plaice age-composition data presented graphically in Fig. IX.1 do not refer to the abundance of a specific year-class at different ages but to the average abundance of each age-group over the years 1929–38, and thus include a number of year-classes. This conversion has effectively removed the influence of fluctuations in recruitment, which would otherwise have obscured the general trend. The first few points, however, clearly do not lie on or near the descending straight line that fits well the remaining points, and we may note here that when such departures are beyond the range of sampling variation, as in this case, they must be due to real differences in the values of the factor k for the age-groups in question. With the younger age-groups the most likely cause is that these fish are not fully represented in the catches owing to the effects of gear selectivity or because they are incompletely recruited to the exploited area, and we shall show later how these possibilities can be distinguished in the plaice data. The first three points must therefore be ignored in fitting the straight line, and the slope of the latter gives an average value of the total mortality coefficient of the plaice during the pre-war period of

$$(F + M) = 0.83$$

For some purposes it is necessary to estimate the total mortality coefficient at each age of each year-class separately, and then to combine these to obtain an estimate of the coefficient for the whole population. In this case the data will probably be more variable than those shown in Fig. IX.1 and problems arise in connection with the method of fitting the best straight line, since the older age-groups contribute fewer fish to the samples and indices of their densities are therefore less accurate than those of young fish. Some form of weighting method is therefore required, and if it is found that there is no change of $(F + M)$ with age, that is, if there is no sign of curvature in the regression, the slope can be estimated from the equation

$$(F + M) = \log_e \left\{ \frac{\sum_{\theta=1}^{\lambda-2} {}_\theta \bar{N}'}{\sum_{\theta=2}^{\lambda-1} {}_\theta \bar{N}'} \right\}$$

in which a value of $\theta = \lambda - 1$ refers to the oldest age-group in the exploited phase. This is equivalent to weighting each age-group in proportion to its apparent abundance, and the reader is referred to Deming (1948) for a discussion of the statistical aspects of problems such as this, which arise frequently in the analysis of data from population studies. Because the data of Fig. IX.1 are mean values for a number of years the variation is so slight that any method of estimation and weighting gives virtually the same result.

Estimation of the total mortality coefficient is but the first stage in the analysis of

---

[1] It will be seen that although it is not necessary to know the actual value of the factor k, it must be constant. This is not the case in *relative* age-composition data, in which the number in each age-group is expressed not in terms of an index of true abundance but only as a percentage of the total sample, and for this reason such data cannot be used to give true estimates of mortality. Nevertheless, they have to be used in certain fisheries when nothing else is available.

mortality since it is now necessary to obtain separate estimates of F and M. There are several possible ways of doing this but the principle of them all is the same ; it is to estimate the combined coefficient $(F + M)$ and also the value of either F or M separately, the other component being found by subtraction. The choice of method depends on the particular circumstances and the kind of data available, but it should be noted that if either component is very much smaller than the other a large percentage error may be introduced if the smaller component is determined by subtraction.

The most general method involves measuring total mortality under conditions where the fishing mortality is varying in a way known from commercial statistics of effort, and to establish it one must return to the basic theory of mortalities, and introduce a more complex notation. Denoting an index of mean abundance of age group θ in the year X by ${}_\theta \bar{N}_x$, the abundance of the same year-class in the following year by ${}_{\theta+1}\bar{N}_{x+1}$ (i.e. when it comprises the $(\theta + 1)$th age-group), and the fishing intensity in the year X by f_x with a corresponding mortality $F_x = cf_x$ (where c is a constant), we have at once from (20)

$$ {}_\theta \bar{N}_x = \frac{{}_\theta N_x}{cf_x + M}(1 - e^{-(cf_x + M)}) \qquad . \qquad . \qquad . \qquad . \qquad (22) $$

In this equation ${}_\theta N_x$ denotes the index of the abundance of the year-class in question when it enters the θth age-group at the beginning of the year X. Similarly,

$$ {}_{\theta+1}\bar{N}_{x+1} = \frac{{}_\theta N_x e^{-(cf_x + M)}}{cf_{x+1} + M}(1 - e^{-(cf_{x+1} + M)}) \qquad . \qquad . \qquad . \qquad (23) $$

and for the moment the value of M is assumed to be the same in both years. Dividing (22) by (23), taking logarithms and rearranging, gives the expression

$$ \log_e \left\{ \frac{{}_\theta \bar{N}_x}{{}_{\theta+1}\bar{N}_{x+1}} \right\} + \log_e \left\{ \frac{(cf_x + M)(1 - e^{-(cf_{x+1} + M)})}{(cf_{x+1} + M)(1 - e^{-(cf_x + M)})} \right\} = cf_x + M \ . \qquad . \qquad (24) $$

The parameters we wish to estimate in this expression are c and M, since if the former is known, the fishing mortality coefficient in each year can be computed at once by multiplying by the appropriate fishing intensity. Now estimates of these cannot be obtained directly from (24), but it will be noted that the second logarithmic term will always be much smaller than the first, and in fact it approaches zero as f_x becomes equal to f_{x+1}. To a first approximation the second logarithmic term can therefore be ignored, whereupon the expression becomes much simpler, reducing to

$$ \log_e \left\{ \frac{{}_\theta \bar{N}_x}{{}_{\theta+1}\bar{N}_{x+1}} \right\} \simeq cf_x + M $$

Hence plotting logarithms of the ratios of abundance indices for a year-class in successive pairs of years against the fishing intensity in the first year of each pair should give a straight line with slope c and an intersect on the y-axis of M. These estimates would be approximate only, since the second logarithmic term has so far been neglected, but they can now be used to compute approximate values for the latter. If these are then added to the existing ratios of abundance, the whole of the left-hand side of (24) can be plotted against f_x, and second and more exact estimates of c and M thereby obtained. This process of solution by iteration is continued until successive estimates of c and M show no important differences. If the value of M during the period in question is not truly constant but is fluctuating independently of the changes in F, the intersect will give an estimate of the mean natural mortality coefficient, and the magnitudes of deviations from the fitted straight line may provide some indication of the natural mortality during individual years.

The method outlined above cannot be used to find the natural mortality of plaice in the North Sea before the war, since variations in fishing intensity from year to year were so small that their effect on the total mortality was swamped by the relatively large sampling error in estimates of the latter. A satisfactory application has, however, been made by the authors in other cases, notably the Fraser River salmon using data presented by Rounsefell (1949).

A special instance of the method is provided by the California sardine (*Sardinops caerulea*). Silliman (1943) showed that there had been two periods, each covering several years, during which the fishing intensity applied to the population had been fairly steady but at a different level in each. He was able to estimate the ratio of the two fishing intensities (y) and the total mortality coefficient during each period. Denoting these by $(F + M)_A$ and $(F + M)_B$, and the fishing intensity in the earlier period of f_A, we have

$$(F + M)_A = cf_A + M$$
$$(F + M)_B = ycf_A + M$$

assuming the natural mortality to be the same in both periods. From these equations we have at once

$$M = \frac{y(F + M)_A - (F + M)_B}{y - 1}$$

The values given by Silliman are $(F + M)_A = 0.52$, $(F + M)_B = 1.61$, and $y = 4$. Substitution of these in the above expression gives the estimate $M = 0.16$, and subtracting this from the two total mortality coefficients gives the estimates $F_A = 0.36$ and $F_B = 1.45$ for the fishing mortality coefficients in the two periods.

Fortunately, the war-time cessation of fishing for the plaice provides a particular case to which the same treatment can be applied, since for a period of nearly six years (from 1940 to 1945) the fishing intensity in the southern North Sea was practically zero. Of course, no samples could be obtained during this time, but it was found that representatives of the year classes which comprised age-groups V and VI in market samples taken just before the war

TABLE IX.1.

Year-class	
1932	1933
−0·09	0·06
0·22	0·08
0·24	0·06
0·06	0·13
0·01	0·13
0·05	0·26

ESTIMATES OF
NATURAL
MORTALITY
OF PLAICE

The estimates come from sampling before and after the war of 1939–45, during which fishing in the North Sea was very much reduced, and, to a first approximation, can be neglected. The two main year-classes fully represented in both pre-war (1938–39) and post-war (1945–46) samples were those of 1932 and 1933. The table gives a number of partially independent estimates of their average yearly mortality during the war period, as instantaneous coefficients. The negative value of one estimate is due to sampling error, which is high in all the post-war estimates owing to the relatively small number of fish surviving to high ages.

were still present in reasonable numbers as the XIII and XIV groups in the post-war samples. Making allowances for the total mortality between the last pre-war sampling dates and the cessation of fishing (about January 1940) and between the commencement of fishing and the first post-war sampling dates, the estimates of M given in Table IX.1 are obtained. Bearing in mind that there was some fishing by one or two European countries even during the war, a fair estimate of M is about 0·1—certainly the true value is unlikely to be much higher than this. Assuming this value to hold good for the pre-war period also, subtraction from the value of $(F + M)$ previously obtained indicates that the fishing mortality coefficient during the pre-war steady state was between 0·7 and 0·8. For convenience we shall use for computation the actual value of F obtained by subtraction, i.e. 0·73, though of course no reliance can be put on the second decimal place.

Marking experiments. Mortalities can also be determined by means of special marking experiments, the principle being to set up an experimental population of tagged fish about which certain characteristics are known, principally the total number initially liberated and the time that each fish has been at liberty when subsequently recaptured by commercial fishing vessels. Marking experiments have been dealt with in the preceding chapter. They have been used extensively to determine migration paths (e.g. Borley, 1916) and to calculate the total absolute abundance of an animal population (Schaeffer, 1951), but these techniques do not immediately concern us here. Thompson and Herrington (1930) and Graham (1938 ii) have developed approximate methods for computing fishing mortality rates, and Ricker (1948) has also presented an analysis of different possible types of information which can be obtained from marking experiments and of ways of dealing with them. Special theoretical methods must be developed before the results of marking experiments can be used to obtain reliable mortality estimates for the population as a whole.

After a batch of marked fish has been liberated the number remaining at a given time will be governed by most or all of the following factors : (a) the same mortality as that caused by fishing and natural causes in the unmarked population ; (b) any additional deaths due to the process of handling and marking or to subsequent harmful effects of the mark itself ; (c) the detachment of marks from living fish ; (d) any abnormal behaviour of marked fish which renders them more or less likely to be captured and retained by the gear than unmarked fish. The accuracy with which we can estimate their abundance will also depend on whether they are uniformly mixed with the unmarked population and therefore exposed to the same effective overall fishing mortality, since the fishing intensity, and hence the probability of capture, will seldom be the same everywhere in any large area. In addition, any failure of fishermen to notice a marked fish which has been recaptured, or to report its recapture, will introduce errors.

The relative importance of these various factors depends on the circumstances and the exact purpose for which the data are used. Thus some species of fish such as herring are difficult to mark satisfactorily, there being many deaths caused by the actual process of marking, and the methods by which the commercial catches are handled is such that any marked fish in them are not readily detected ; with plaice, on the other hand, neither of these difficulties are important. If work is being carried out in a small lake, rapid uniform mixing may be easy to achieve ; in an area the size of the North Sea this is virtually impossible, and dispersal of marked fish from the marking area is a factor to be reckoned with. Abnormal 'catchability' is probably unimportant in trawl fisheries but it may be critical with those types of gear, such as lines, which depend for their action largely on the behaviour of the fish.   Loss of marks through detachment clearly depends on the type of tag used, but it is rarely negligible.

For setting up a theoretical model of a population of marked fish the important question is whether the various factors (other than fishing and true natural mortality) responsible for the decrease in numbers of the population operate at, or very soon after, marking and liberation, or whether they act in a fairly steady way as long as the fish are at liberty.   These two kinds of losses are called type (1) and type (2) respectively, and it will be seen that marking mortality, for example, involves losses of both types.   The same applies to abnormal behaviour of marked fish, though in this case the losses will be mainly of type (1).   On the other hand, detachment of marks may be primarily of type (2) and incomplete reporting of recaptures entirely so.

Clearly, the existence of these two types of losses must be taken into account in developing methods of analysis which can give unbiased estimates of mortality rates in the unmarked population.   Let us suppose that a certain number, N_0, of fish bearing individually identifiable marks are liberated in a batch and, for the moment, that there are no type (1) losses. The simplest reasonable hypothesis we can make about the type (2) losses is that their total effect can be represented by a constant coefficient, denoted by X.   Assuming that the fishing intensity is constant during the period of the experiment and denoting the fishing mortality coefficient by F as before, we may write as a first approximation

$$\frac{dN}{dt} = (F + X)N$$

to define the rate of decrease in numbers of marked fish.   This equation may be solved in the same way as (6), and remembering that when $t = 0$ the number of marked fish present is N_0, the number of live marked fish remaining at time t is

$$N_t = N_0 e^{-(F+X)t}$$

The number recaptured between the time of liberation and the end of a period of duration τ is given by

$$n_1 = F \int_0^\tau N_t \, dt = \frac{FN_0}{F + X}(1 - e^{-(F+X)\tau})$$

and similarly, those caught during the next period of the same duration τ is

$$n_2 = \frac{FN_0 e^{-(F+X)\tau}}{F + X}(1 - e^{-(F+X)\tau})$$

Dividing n_2 by n_1, gives

$$\frac{n_2}{n_1} = e^{-(F+X)\tau}$$

so that

$$X = \frac{1}{\tau} \log_e \left\{ \frac{n_1}{n_2} \right\} - F$$

Substituting this expression for X in the above equation for n_1, and rearranging gives

$$F = \frac{\dfrac{n_1}{\tau} \log_e \dfrac{n_1}{n_2}}{N_0 \left(1 - \dfrac{n_2}{n_1}\right)}$$

This is an expression for the fishing mortality coefficient which does not contain the " other-loss rate " X, and in which all the parameters are directly known from data ; with F computed in this way, the value of X can be estimated from the formula above.

Although very much simplified assumptions have been made to obtain this result it is worth considering its application before proceeding, to see how errors may arise, since it is very often the case that the information available will not permit the use of a more critical method. The recapture periods may theoretically be of any duration but the factors influencing the choice of the best value of τ to use in practice are several. In the above method it is assumed that F and X are constant, and for this reason τ should be kept as short as is compatible with securing a fair number of recaptures in each period so as to reduce the sampling errors. Thus it may be undesirable to extend the experiment to include more than two recapture periods, as has been suggested by Ricker (1948), though if F and X really are constant it would obviously be an advantage to do so because all or nearly all recaptures could then be used to estimate these parameters. This may be seen from the expression for the number recaptured during the rth period, which is

$$n_r = \frac{FN_0 e^{-(F+X)(r-1)\tau}}{F + X}(1 - e^{-(F+X)\tau}) \qquad . \qquad . \qquad . \qquad . \qquad (25)$$

Taking logarithms of both sides gives an expression which is linear in r (cf. (21)), thus :

$$\log_e n_r = \left[\log_e\left\{\frac{FN_0}{F + X}(1 - e^{-(F+X)\tau})\right\} + (F + X)\tau\right] - (F + X)\tau r \qquad . \qquad (26)$$

Plotting logarithms of the numbers of successive recaptures (n_r) against the number of the period during which they were made (r) would therefore be expected to give a straight line the slope of which defines the value of $(F + X)\tau$ and, since the duration τ is known, of $(F + X)$. Substitution of this in the term in square brackets, which defines the intersect on the y-axis, leads to an estimate of F that takes into account all the available data. It is important to note here, however, that if the points are fitted well by a straight line this does not in itself confirm the validity of the assumption that F and X are constant ; certain kinds of trends—including linear ones—in either or both F and X can result in a series of recaptures the logarithms of which fall so nearly on a straight line that the departures could not be distinguished from sampling variations. Yet if there really is this type of trend in fishing mortality during the experiment, the estimate of F obtained in the above way does not even give an indication of the *average* fishing mortality during this time, but is completely outside the true range of change of mortality which occurred. If the fishing intensity decreases as the experiment proceeds the value of F will be higher than that at the time of liberation, and *vice versa*. On the other hand, if the fishing intensity merely fluctuates without a trend, the effect will be to increase greatly the apparent variation of the data and hence decrease the accuracy with which F can be estimated, but it will not introduce serious bias.

The only really satisfactory method of dealing with the problem of changes in fishing intensity with time is to set up a theoretical model which takes account of those changes ; but in many cases the fishing intensity will be unknown, and in these circumstances it is possible to design the marking experiment in such a way that although the effects of changes in intensity are not eliminated they are at least considerably reduced. This way is to spread liberations over a fairly long period, perhaps a year or more, fish being released at regular intervals in small, roughly equal batches. The length of time during which each recaptured fish had been at liberty is noted, and recaptures are then grouped into equal periods at liberty and summed to give the values of n_r. Thus the recapture periods no longer correspond to chronologically successive time intervals but each refers to a number of overlapping intervals, so that the effect of a change in intensity is spread over several values of n_r, instead of being restricted to one particular value. This is the way in which the plaice marking data used by Graham (1938 ii) were treated, and we shall take them to illustrate the application of (26).

Recaptures were grouped into four periods of time at liberty, each of three months duration ($\tau = 0.25$ years), and proportioned to give the values that would have been expected if exactly one thousand fish had been liberated ($N_0 = 1000$). The resulting values of n_r were as follows :

$$n_1 = 139$$
$$n_2 = 91$$
$$n_3 = 52$$
$$n_4 = 40$$

The natural logarithms of these numbers are shown plotted against r in Fig. IX.4, and it is seen that they fall reasonably close to a straight line having a negative slope 0.44 and an intersect 5.38. Dividing the slope by the above value of τ gives an estimate of $(F + X) = 1.76$, and substituting this in the expression in square brackets of (26) gives the value $F = 0.69$. Finally, X is found by subtraction to be 1.07. The above data were published by Borley (1916) from marking experiments made before 1914, but it is known that the (international) fishing intensity at that time did not differ greatly from that during the pre-war period to which the estimate of $F = 0.73$, obtained previously by other methods, refers. Agreement between the two estimates can therefore be taken as tending to confirm the validity of each. It should be noted also that the value of F obtained by using (26) is very similar to that deduced by Graham using a different method. Graham's procedure is in fact a useful way of obtaining rapidly an approximate estimate of fishing mortality, though it cannot safely be used with much larger values of τ or X than those above.

Another interesting point arising from this example concerns the value of X, which is much higher than the true natural mortality coefficient given previously ($M = 0.1$). The difference must be caused mainly by factors that are peculiar to the marked population. Continuous detachment of marks from living fish, and any additional steady mortality due to the after-effects of the tagging process, for example, would not influence the value of F as. computed from (26) or from the equations based on two recapture periods, but they will be components of the other-loss rate X. In many cases, and certainly in the above example, their combined magnitude exceeds that of the true natural mortality ; for this reason it seems best to regard marking primarily as a means of estimating F, a value of M then being obtained by subtraction from the total mortality coefficient found by analysis of age-composition data. Incomplete reporting by fishermen of the recapture of marked fish will probably change all the n's in roughly the same proportion and thus will not affect the estimate of $(F + X)$, but the value of F subsequently obtained will be too low.

The indications so far are that in choosing recapture periods the value of τ should be kept as small as is consistent with getting a good number of recaptures in each period. On the other hand, if τ is made too small, even though sufficient returns are available, errors caused by assuming that there are no type (1) losses, when in fact there are, will be exaggerated. It is in some circumstances possible to test whether an abnormal mortality occurs

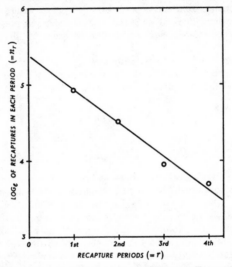

FIG. IX.4.—FISHING MORTALITY FROM MARKING EXPERIMENTS.

Natural logarithms of successive recaptures (n_r) of marked plaice plotted against recapture period (r).

soon after marking by keeping tagged fish in tanks on board ship for some time, and liberating only the most active. Doing this, however, means making further assumptions concerning the similarity in behaviour of fish in the sea and in tanks, and although such an experiment is desirable in practice, it does not necessarily lead to an adequate correction for the effect of such losses in causing F to be underestimated. The existence of type (1) losses of any kind means that although the number of fish actually liberated is exactly known, the number which are alive and still bearing a mark shortly afterwards, and hence the *effective* initial number from which subsequent recaptures are obtained, is unknown. Instead of attempting to estimate these losses directly, another possibility is therefore to use a method for estimating F and X that does not require a knowledge of the initial number liberated, and which is based only on the characteristics of the trend in recaptures. It will be noted, incidentally, that such estimates would not be influenced by the incomplete reporting of recaptures, which, as mentioned above, is unlikely to affect significantly the *ratios* of successive recaptures. It so happens that to use this method the fishing intensity must change during the experiment and the magnitude of this change must be known ; hence in addition to providing a means of eliminating the effects of type (1) losses, the method to be described also deals rigorously with the case of a varying intensity.

Although, in practice, the fishing intensity is likely to change smoothly and without sudden jumps, it is much simpler and for most purposes perfectly satisfactory to assume that changes take place abruptly in finite steps at the beginning of each recapture period. In effect this amounts to supposing that the fishing intensity remains at a constant level during each period which is the average value of the actual fishing intensity during that time. Denoting, then, the fishing intensity during the rth recapture period of duration τ by f_r, the corresponding value of the fishing mortality coefficient can be defined by multiplying by the constant c introduced previously, thus :

$$F_r = cf_r$$

By a direct extension of the methods used to obtain (25) for the case when F is constant, we find

$$n_r = \frac{cf_r N_0 \, e^{-\left\{(r-1)X + \sum\limits_{p=0}^{r-1} cf_p\right\}\tau}}{cf_r + X}(1 - e^{-(cf_r + X)\tau})$$

in which f_0 is taken as zero.

Dividing by the corresponding expression for n_{r+1}, i.e. the number recaptured in the $(r + 1)$th period, gives the relationship

$$\frac{1}{\tau}\left[\log_e\left\{\frac{n_r f_{r+1}}{n_{r+1} f_r}\right\} + \log_e\left\{\frac{(cf_r + X)(1 - e^{-(cf_{r+1}+X)\tau})}{(cf_{r+1} + X)(1 - e^{-(cf_r + X)\tau})}\right\}\right] = cf_r + X \ . \qquad . \ (27)$$

This expression will be seen to be identical in form to (24) for the ratio of the abundances of a year-class in two successive years with the fishing intensity different in each, and solutions for c and X may be obtained by the same iterative method. With c found in this way, the average fishing mortality coefficient in each period can be computed, and this estimate will have been obtained without a knowledge of the number of fish liberated, since this quantity does not appear in (27). Now in the first instance the period of time after liberation during which type (1) losses are operating will be unknown ; probably the best way of using (27) is therefore to make repeated estimates of c and X allowing an increasing interval of time to elapse between the date of liberation and the date on which the first recapture period is taken to start, all recaptures obtained before this date being discarded. In this way it should be possible, by inspection of the points to which the equation is fitted and of the trend in the values of c and X, to find when a time interval has been taken that is long enough for the influence of type (1) losses to be avoided.

This technique is clearly a useful one, but in practice its advantages must to a certain extent be qualified by the greater demands made of the data. The estimation in this way of both F and X depends rather critically on the accuracy with which the changes in fishing intensity can be measured and also on the magnitude of these changes, which should be neither very slight nor too violent, a steady upward or downward trend providing the most satisfactory conditions. It will be remembered that the same limitations apply to the use of (24) for the analysis of age-composition data, but with marking experiments the investigator may well be able to choose conditions in which to conduct the experiment that satisfy these requirements. For example, although the overall annual fishing intensity may be too nearly constant, there will often be certain areas in which a marking experiment could be conducted where the local fishing intensity varies considerably during the course of a year. At the present stage of investigations on the design and analysis of marking experiments it seems that the best procedure is to attempt as far as possible to conduct them in conditions which allow the above method or extensions of it to be used.

Brief mention must be made of the possibility of calculating the natural mortality directly from the loss-rate X by estimating the other components of the latter and subtracting them, leaving M as the residue. This procedure is useful when M is fairly large and provided type (2) losses due to abnormal mortality of marked fish are expected to be relatively small or negligible, since the order of magnitude of the remaining type (2) components of X, i.e. the rate of loss of marks by detachment and the rate at which fish leave the marking area, can in many cases be estimated independently.

An instantaneous coefficient which defines the rate of detachment of marks from living fish can be determined by means of a special experiment in which *two* marks are placed on each fish. The majority of recaptures will probably still bear both marks, since initially all the fish present do so, but if the rate of detachment is at all appreciable some will be recaptured which have lost one mark. A knowledge of the proportion of the latter to the total recaptures in each recapture period provides the basis for estimating the rate at which marks are lost in an ordinary experiment from fish bearing initially a single mark only, which of course could not be studied directly since a fish that has lost the only mark it carried is seldom recognizable as such in the catch. An experiment of this kind using plaice marked in the usual way with ebonite discs joined by a silver wire passing through the body of the fish on one side, suggested that the detachment coefficient was about 0·2 to 0·3. This is not inconsiderable, since it implies that even in the absence of mortality of any kind, a batch of marked fish will be reduced in a year to about 75 per cent of their initial number as the result of marks becoming detached.

Finally there is the question of dispersal of fish from the marking area. If the exploited area is not too big it may be possible to eliminate this factor entirely by liberating marked fish over the whole area in proportion to the population density at each point, so that the marked population is confined within the same limits as the unmarked population and suffers no emigration. An alternative procedure is to restrict liberation to a relatively small marking area, and either to allow dispersal of marked fish from this area to be included in the loss-rate X or to estimate it directly. Theoretical treatment of dispersion can be based on the concept of random movement and can, as we shall show later, describe the local movements of plaice quite well. A rigorous mathematical analysis is highly complex except for the simplest of situations (see Skellam, 1951), and at the present time it is found better to use an approximate theory based on the fact that when movement is random the instantaneous rate of loss of marked fish from an area is approximately proportional to the difference between the densities of marked fish inside and outside the area at that time. The magnitude of the factor of proportionality, termed the *transport coefficient* T, depends not only on the average speed of swimming of the fish in the areas in question (related to environmental conditions such as temperature and the density and distribution of food organisms) but also on the ratio of the perimeter of the area from which the dispersion is occurring to its area. This fact underlies the method used by Jackson (1939) for estimating the rate of migration of tsetse flies from a marking area, and it can readily be adapted to fish marking experiments in the following way. Fish are liberated uniformly over two areas A and B of roughly the same shape but of different sizes ; they can be separate areas or, better, the smaller (A) can be contained within the larger (B). If the residual loss-rate (i.e excluding losses by dispersal) is denoted by X' and assumed to be the same for both areas, and neglecting the fact that a small number of the recaptured fish will probably have earlier left the marking area and then re-entered it, we can write approximately

$$X_A = X' + T_A$$

$$X_B = X' + \frac{b}{a}T_A$$

where T_A is the transport coefficient for area A, and a and b are the perimeter/area ratios for each area respectively. The total other-loss coefficients for each area, X_A and X_B, can be estimated by the methods previously discussed, restricting analysis in each case to recapture from the appropriate areas. The above equations then provide a solution for the transport coefficient T_A, viz.

$$T_A = \frac{X_A - X_B}{1 - \dfrac{b}{a}}$$

The residual loss-rate X' can then be obtained by subtraction. A preliminary analysis of recent marking experiments with North Sea plaice using this and other techniques indicated that the order of magnitude of the transport coefficient was about 10, measured with reference to a marking area about thirty miles square and in time units of a year. Such a value can be interpreted as implying that some 90 per cent of the marked fish initially present in marking area of this size will have crossed its boundary at least once during the following three months.

Mesh Selection and Recruitment

We must now turn to the factors governing the age at which the fishing mortality first operates, i.e. to the process of entry of fish to the exploited phase of the population. The mechanism by which fish become liable to capture varies according to the type of gear used, but nearly always it is the *size* of the fish that determines retention, even when the gear is a baited hook. As will have been gathered from earlier chapters, this effect is quite direct in the case of trawls, seines and in fact all " bag-type " nets where retention or escape is determined by the relation between the maximum girth of the fish and the lumen of the meshes of the net, or often, of a small part of the net. With drift-nets and other types of gill-net the problem is more complex since the fish may be caught by lip, gill-cover, or shoulder and there is selection in the upper as well as the lower size ranges of fish.

The difficulties encountered in determining selectivity of trawl meshes have been adequately described in Chapter VIII, and may be considered to be behind us at this stage.

We are in a position to re-state the theory, from the start, in a short form, suitable for inclusion in the general theory of this chapter. Let us consider what happens to a batch of fish of various sizes after they have entered the mouth of a trawl. They pass down the net and most eventually enter the cod-end, though a few escape before this, either by swimming out of the mouth of the trawl or through the meshes of the " square " and " belly " which are usually larger than those of the cod-end. Although not a great deal is known about this, Todd's work (Ch. VIII) proved that most escape is effected after fish have entered the cod-end, and this can probably be considered to happen by haphazard swimming and wriggling at the meshes. All those fish whose maximum girth is greater than the lumen of the mesh— strictly of the *largest* mesh, since there is always some variation in mesh size within a cod-end —can never escape, except when the net is torn. This is not infrequent on rough grounds, but when it occurs the hole is usually large enough to allow fish of all sizes to escape so that it does not contribute to selection. Those fish whose maximum girth is less than the lumen of the mesh will be able to escape at some time or other provided they are not damaged, the smallest most easily and hence most rapidly. Those whose maximum girth is only *just* less than the mesh opening may have to make many attempts before they are able to squeeze through. Hence, if the size distribution of the fish remaining in the net could be measured after increasing intervals of time had elapsed following entry, it would be expected that the number of fish smaller than the mesh size would decrease, the smallest most rapidly, until eventually the catch would consist only of fish with girths bigger than the mesh. This would be true knife-edge selection but it is never attained in practice for several reasons, the main ones being that the usual duration of a haul (1–3 hours) is not long enough to allow the final state to be reached, that fish are entering throughout the haul and not just at the beginning of it, that some fish are injured by the gear and thus cannot escape although small enough to do so, and finally that as the catch in the cod-end accumulates the net becomes congested and the area through which escape can take place is reduced.

Selection ogive. Now we are interested in determining the variation of fishing mortality with size, and hence with age, over the selection range and so, remembering that F is related to the proportion captured in a given time, it is necessary to find what proportion of all fish of a given size which enter the net are retained during a normal commercial haul. Thus fish of girth greater than the mesh size are exposed to the full fishing mortality rate, whatever its value may be, but fish of a size such that, for example, only half of those entering the net are retained under normal conditions are subject to a fishing mortality of only $F/2$. In practice it is convenient to refer to total length rather than to girth, since the two are directly related and the former is more easily and rapidly measured ; thus the curve obtained when the percentage retained is plotted against length is Buchanan-Wollaston's (1927) *length selection ogive*, and its ordinates define the proportion of the full fishing mortality rate to which fish of each length are exposed. It is now clear that this can be obtained experimentally, if the size distribution of the population entering the net is known ; comparing it with that of the catch will give the proportion retained at each length.

Chapter VIII has mentioned the several practical methods for determining the length selection ogive of a given mesh, these differing mainly in the way in which the size distribution of the population entering the net is measured. The oldest method is to fit the test cod-end with a loose cover of fine-meshed net. The population is estimated from the combined catches in the cod-end and compared with the catch in the cod-end alone to give the selection ogive. This method has the advantage that it greatly reduces the haul-to-haul sampling variation present in the following method, but on the other hand there are two sources of bias that may be present. One is that some fish which have escaped through the cod-end may return to it from the cover before the catch is measured, and the other is that the presence of the cover does to some extent distort the normal fishing characteristics and elective properties of the trawl. Unfortunately, neither of these factors are readily measured in practice.

The method at present used for plaice is to make alternate hauls with the trawl fitted first with the cod-end to be tested and then with a cod-end of a much smaller size, small enough in fact for all fish within the selection range of the test mesh to be retained. Taking ratios of the catches of each length-group in the two nets gives the selection ogive directly,

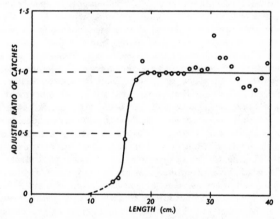

Fig. IX.5.—Selection Ogive of a 70 mm. Trawl Cod-end Mesh for Plaice.

The points are the ratios of the catch of this mesh to that of a much smaller one (about 40 mm. per unit fishing time, adjusted so that the ogive tends to an upper asymptote of unity. The 50 per cent selection point is about 15·5 cm. (accurately, the " 70 " mesh was most probably 72 mm. gauge.)

but allowance has sometimes to be made for a difference in fishing power of the trawls. Chapter VIII mentioned the fact that fishing power with a smaller meshed cod-end is usually lower than that with a larger, due presumably to a slower rate of water flow through the net. The selection ogive for plaice obtained in this way with a cod-end mesh size of 70 mm. on the gauge (Ch. II)—the size in general commercial use just before the war—is shown in Fig. IX.5. A convenient index of the position of the ogive is the 50 per cent retention length, which in the example is seen to be about 15·5 cm. Experiments with larger meshes show that the 50 per cent length and the average mesh size are proportionally related and this is what would be expected since plaice grow more or less isometrically and so length and girth are also proportional. We have therefore a means of predicting the 50 per cent length of plaice for a trawl fitted with a cod-end of any specified mesh size. The 50 per cent length of plaice is $2·2 \times$ gauge mesh.

The final step is conversion of the length ogive into an *age-selection ogive*, and this can conveniently be done by using the fitted von Bertalanffy equation for growth in length to calculate the average age of fish at each length. Incorporation of this retention-age relationship, as it stands, into the population model in the form of a variation of F with age is possible, but it introduces unnecessary complications ; instead, it has been found that a very close approximation is obtained by using the *mean selection age* and assuming that knife-edge selection occurs at that age. When the experimental length ogive is symmetrical, as is roughly the case in Fig. IX.5, the mean selection age corresponds closely to the 50 per cent point. If the selective characteristics of the gear are such that the selection range is above the size at which fish are recruited to the exploited area, then the mean selection age is the required value of $t_{\rho'}$ referring to that gear ; with the relationship between the 50 per cent retention length and mesh size determined, it is thus possible to specify the value of $t_{\rho'}$ for any mesh of such a size that its selection range is completely above the age at recruitment. With smaller meshes, entry to the exploited phase of the population is determined in part or wholly by the process of recruitment into the area, and this was in fact the case when the 70 mm. mesh was in use before the war in the plaice fishery. Before we can arrive at a value of $t_{\rho'}$ for plaice we must therefore investigate the process of recruitment in more detail.

Age at recruitment. In plaice, the geographical separation of the nursery grounds from the main exploited area means that recruitment involves a movement of fish from one to the other. The study of this movement has been described in detail in the previous chapter, and evidence will be presented later to show that it may be considered essentially as a process of random dispersion from the coastal zone. At the moment, however, we need not be concerned with the details, but only with the average age at which the recruits enter the exploited area. Finding this is, in principle, a straightforward matter of sampling to estimate the abundance of those age-groups that are in the process of entering the exploited area. If the cod-end mesh of the commercial gear is small enough and there is no rejection at sea of undersized fish because of their low market value or because of regulation prohibiting their landing, then market-samples of the commercial catch can be used ; otherwise, a research vessel using a small-meshed gear must be used to obtain representative samples of the smallest fish on the commercially fished grounds.

In the plaice fishery, the 70 mm. mesh was sufficiently small to catch and retain most fish on the main grounds, but it is known that there was some rejection at sea of very small fish at certain seasons and in certain places. In the absence of extensive and regular research vessel sampling, the best that can be done is to use market samples obtained before a minimum size limit was imposed by legislation in 1933 to estimate the average proportion of each age-group which were below that limit and therefore rejected at sea after 1933. In this way the market samples for the second part of the pre-war period (1934–38) can be adjusted to allow for the undersized fish that were caught but not landed. This procedure is not entirely satisfactory as there was some rejection of small fish for economic reasons *before* 1933, but it does provide a reasonable indication of the true age-composition of the population on the fishing grounds, and the average figures for the pre-war period are given in row A of Table IX.2. The average length of II-group fish was about 20 cm. and since Fig. IX.5 shows that the gear in use was not selective above a length of about 19 cm. the data can be regarded as unaffected by this factor. It will be seen that the abundance *increases* with age from II through III to age-group IV, indicating that these groups are incompletely

s.f.—d d

68

recruited, which explains why the first three points of Fig. IX.1 fall below the line fitted to age-groups V and above. It shows also that recruitment does not occur when fish reach a certain 'threshold' age but is spread over a number of years, so that we must next find the proportion of each year-class that is recruited into each age-group and for this a further theoretical development is necessary.

TABLE IX.2.

Age-group	II	III	IV	V	VI	VII	VIII	IX	X
A . .	282	1,816	2,728	1,831	786	339	159	70	28
B . .	639	3,068	2,709	376	—	—	—	—	—
C . .	0·094	0·452	0·399	0·055	—	—	—	—	—

CALCULATION OF RECRUITMENT AGE IN PLAICE

Line A gives the average annual age composition of the post-recruit phase of the population, as catch per 100 hours' fishing. Line B shows the number of recruits into each age-group in the same units, these being expressed as proportions of the total recruitment each year in line C.

Since there are no O-group fish in the samples, and an entirely negligible number of I-group, all the II-group fish must be recruits. The data of Table IX.2 refer to indices of mean abundance taken over the year commencing on April 1st, but we now need indices of the actual number entering on the date of recruitment. If for age-group II this is denoted by R_{II}, then from (20) we can write:

$$\bar{N}_{II} = \frac{R_{II}}{F + M}(1 - e^{-(F+M)\phi_{II}})$$

where $\bar{N}_{II}$ is the observed annual mean abundance of all II-group fish and ϕ_{II} is the fraction of the year during which II-group fish are present, i.e. from the date of recruitment to the following March 31st, when they enter age-group III.[1] Hence on rearranging we have

$$R_{II} = \frac{\bar{N}_{II}(F + M)}{1 - e^{-(F+M)\phi_{II}}}$$

In the case of the III-group fish some will be recruits and the rest survivors from the II-group recruited in the previous year. The number of the former will be given by

$$\frac{R_{III}}{F + M}(1 - e^{-(F+M)\phi_{III}})$$

and the latter by

$$\frac{R_{II}e^{-(F+M)\phi_{II}}}{F + M}(1 - e^{-(F+M)})$$

---

[1] It will be noted that when computing N_{II} from, say, monthly values of catch per unit effort of age-group II, values for that part of the year before the recruits enter the fishery are included.

Adding these and rearranging gives

$$R_{\text{III}} = \frac{1}{1 - e^{-(F+M)\phi_{\text{III}}}} \{\bar{N}_{\text{III}}(F + M) - R_{\text{II}}e^{-(F+M)\phi_{\text{II}}}(1 - e^{-(F+M)})\}$$

where $\bar{N}_{\text{III}}$ is the observed annual mean abundance of all III-group fish. Expressions for R_{IV}, R_{V}, etc., involving all the preceding R's, can be obtained in the same way.

In order to use these expressions for computing the recruitment into each age-group, values of the ϕ's are required. These are obtained by noting the date of influx of fish into the low age-groups when samples were taken monthly during the years since the war. This is in the summer, and the actual ages measured from April 1st are found to be

Age-group .	.	.	II	III	IV
Age (yrs.) .	.	.	2·45	3·35	4·33

Not much reliance can or need be placed on the second decimal figure, but the difference between the dates for II- and III-group fish is significant and shows that older fish enter a little earlier in the year than the younger ones. Hence we have

$$\phi_{\text{II}} = 0\cdot55 \text{ yrs.}$$
$$\phi_{\text{III}} = 0\cdot65 \text{ yrs.}$$
$$\phi_{\text{IV}} = 0\cdot67 \text{ yrs.}$$

and it will be assumed that if any recruitment into age-groups V and above does occur it happens on the same average date as that into age-group IV.

When these values of ϕ and those of $\bar{N}$ given in Table IX.2 are used in the above expressions to compute the number of recruits into each age-group, an age is reached above which the result is zero (within the limits of sampling error). By that age, therefore, all fish are recruited, none are left on the nursery grounds and the change in numbers in the exploited phase is due entirely to mortality. In this way it is found that recruitment occurs in age-groups II to V, so that all age-groups from V upwards are fully represented in the exploited area. The indices of R_θ are given in row B of Table IX.2, the values for age-groups VI and above, which do not differ significantly from zero, being omitted. Thus it appears that recruitment is spread over four years of life, and as with gear selection, the *mean* age of recruitment is required ; this is obtained by converting the data of row B to proportions of the total recruitment (row C) and computing the weighted mean age, giving a value of $t_\rho = 3\cdot7(2)$ years.

The remaining problem concerns the combination of the mesh and recruitment ogives when these overlap. We showed above that overlap does not occur to any appreciable extent when a 70 mm. mesh is used, but the problem will arise when we come to use our population model to predict the results of using larger meshes. Both ogives must first be expressed in the same units by calculating the *length* recruitment ogive, and then the product of the ordinates of the two ogives at each length computed. The *resultant ogive* gives the fraction of the full fishing mortality to which all fish of a given length are exposed, since it takes account of the fact that only a fraction of the year-class has, by the time the fish have reached a given length, entered the exploited area, and that of these only a proportion are retained by the commercial gear because of their small size. A resultant ogive can be treated in the same way as before, its mean giving an estimate of the true mean length, $L_{\rho'}$, at which fish enter the exploited phase. The relationship between $L_{\rho'}$ and cod-end mesh size for plaice is shown in Fig. IX.6, and it will be seen that because of the relatively large size at which recruitment occurs the mesh has to be increased considerably—up to as much as 100 mm.— before causing any appreciable increase in $L_{\rho'}$. If the relationship between length and age is taken as constant, this $L_{\rho'}$ curve can be converted into one defining the change of the mean

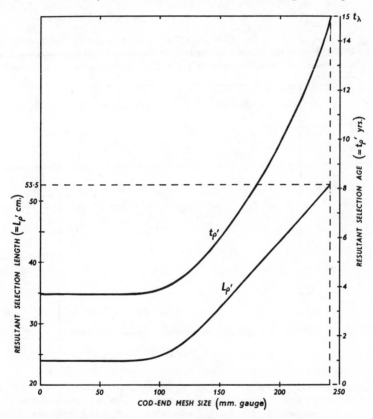

FIG. IX.6.—SELECTION OF PLAICE BY VARIOUS MESH SIZES.

The resultant selection length ($L_{\rho'}$) and age ($t_{\rho'}$) is plotted against cod-end mesh size of trawls, double twine. The curves are terminated at age 15 years and the corresponding length from Fig. IX.2.

age, $t_{\rho'}$, at which fish enter the exploited phase, with mesh size. Using the pre-war growth in length shown in Fig. IX.2 gives the $t_{\rho'}$ curve of Fig. IX.6, though it is important to note that with models in which growth is made to vary with population density no such direct conversion is possible, since increasing mesh size alters the biomass of the population and hence the relation between length and age (see p. 423). Nevertheless, complications of this kind do not affect the length or age at which fish are recruited, and because the selection ogive for a 70 mm. mesh is almost wholly below the recruitment ogive the value of $t_{\rho'}$ for this mesh is virtually the same as t_{ρ} given above, and for the pre-war state we can use the values

$$t_{\rho} = t_{\rho'} = 3 \cdot 7(2) \text{ years}$$

Growth Parameters

Fitting the von Bertalanffy equation. The method by which the primary growth parameters K and W_{∞} and the derived parameter t_0 of von Bertalanffy equation are computed from the weight-at-age data for plaice shown in Fig. IX.3, is best introduced by writing (13)

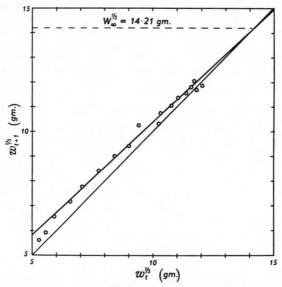

Fig. IX.7.—Fitting von Bertalanffy's Equation to Weight Data.

The method is derived from the form of equation (14) : $w_{t+1}^{1/3}$ is plotted against $w_t^{1/3}$ and the thicker of the two lines shown is fitted. Where the line cuts the bisector (thinner line) there is no change in weight, i.e. $W_{\infty}^{1/3}$ has been reached. The slope of the fitted line is e^{-k}. The growth curve of Fig. IX.3 has been fitted in this way, giving $W_{\infty} = 2867$ gm., $K = 0.095$, and $t_0 = -0.815$ years (see p. 406).

in a form relating the weights at any two ages a year apart, w_t and w_{t+1}, and taking cube-roots, thus obtaining

$$w_{t+1}^{1/3} = W_{\infty}^{1/3}(1 - e^{-K}) + w_t^{1/3}e^{-K}$$

which is an expression linear in $w_t^{1/3}$. Hence if the cube-root of the weight at each age is plotted against the cube-root of the weight when one year older, a straight line should be obtained having a slope e^{-K} and an intersect on the y-axis of $W_{\infty}^{1/3}(1 - e^{-K})$. This is shown in Fig. IX.7. By measuring the slope and intersect an estimate of $W_{\infty}^{1/3}$ can therefore be obtained, but the latter is more easily measured by finding where the line cuts the bisector drawn through the origin, since that is where $w_t^{1/3} = w_{t+1}^{1/3} = W_\infty^{1/3}$. Length, instead of the one-third power of weight, can, of course, be used in a similar way to estimate L_∞ and K : that the lengths of fish at any two successive ages (i.e. l_t and l_{t+1}) are linearly related was, in fact, first deduced by Ford (1933), who used the relation to measure the curvature and asymptotes of herring growth curves. With the plaice data we find that $W_\infty = 2867$ gm., and taking the natural logarithm of the reciprocal of the slope gives the estimate $K = 0.095$. To find t_0, equation (14) can be rearranged to give the expression

$$\log_e\{W_\infty^{1/3} - w_t^{1/3}\} = [\log_e\{W_\infty^{1/3}\} + Kt_0] - Kt$$

which is linear in t. Since W_∞ and K are now known, the intersect of the resulting straight line can be used to compute a value of t_0 directly, and for the plaice data this is $- 0.815$ years.[1] The values of W_∞, K and t_0 obtained in this way give the curve shown in Fig. IX.3 ; it will be seen that apart from the first two points which, as was explained earlier, refer to partially recruited age-groups, the original data are fitted well throughout their range.

Weight and length. Some comment is required at this point concerning the relationship between weight and length in plaice, since as we have seen previously a knowledge of growth in length is needed for several purposes ; moreover, lengths can be measured more rapidly than weights, especially at sea, and much of the data relating to the size distribution of other fish are in terms of length. Although we have assumed hitherto that plaice grow isometrically, it is a fact that the growth of most fish, including plaice, is not quite isometric even when they are adult, but major departures from this type of development are rare. In recent years many authors have used the allometric growth formula

$$w = bl^k$$

where b and k are constants, to express the relationship between length and weight. This usually gives a better fit to data than the simple cubic postulated previously, if only because it is possible to adjust two parameters (b and k) instead of one (q). However, b is no longer equivalent to q dimensionally unless k is exactly 3, and the allometric formula is derived from the differential equation

$$\frac{dw}{dl} = k\frac{w}{l}$$

in which there is no particular reason why k should be a constant. If anything other than a cubic is to be adopted then it must be used also in the derivation of the von Bertanlanffy equation, and for this purpose it is necessary to distinguish between changes in *shape* and changes in *specific gravity*, since the relationship between surface area and length is affected only by the former. There are no fundamental objections to this course being taken, but since the weight-length relationship in plaice is so closely represented by the cubic

$$w = 0.008892l^3$$

as shown in Fig. IX.8, it is not necessary to complicate further the growth formula (and hence the population model). The reader is referred to Hile (1936), Reeve and Huxley (1945) and

---

[1] Treating the data in this way also provides another estimate of K, since this is the slope of the line defined by the above equation. The fact that t_0 happens in this case to be a negative is of no particular significance.

Le Cren (1951) for a further discussion of the problem, which has perhaps taken rather more space in the literature than its importance warrants.

Absolute Number of Recruits and Maximum Age

All the parameters of the population model represented by (16) have now been estimated for the pre-war state of the plaice population except the maximum age t_λ and the annual recruitment R.   The value of the former may to a certain extent be taken arbitrarily since there is no definite evidence of a limit to the life-span in plaice.   As very few fish above

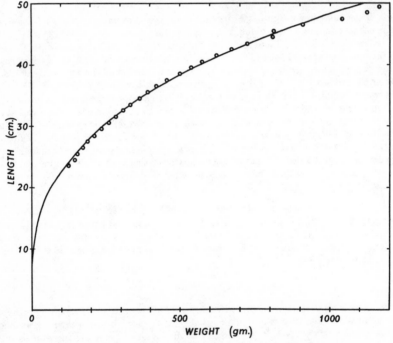

FIG. IX.8.—RELATIONSHIP BETWEEN WEIGHT AND LENGTH IN PLAICE.

The curve is that given by the equation $w = 0.00892l^3$.

15 years are found in market samples it is convenient to take this as the value of t_λ for this species, but it can be shown that the exact value has but little effect on the final conclusions reached by the application of the simple population model to the fishery with which we are concerned here.

In the marine fisheries there is no known method of determining directly the absolute number of recruits entering the exploited area each year, though indices of abundance for each recruit year-class are readily obtained from age-composition data as we have seen. However, a glance at (16) shows that the yield varies directly and proportionally with the annual recruitment, so that the magnitude of the latter does not affect the *shape* of the curve of yield against fishing effort or against mesh size. This makes it possible to deduce most of the important characteristics of an exploited fish population without knowing the actual value of R by plotting the ratio Y_W/R, i.e. the yield *per recruit*, instead of Y_W itself. Strictly speaking, this procedure is valid only if R is constant, but the difficulty is avoided by interpreting R as the mean recruitment, $\bar{R}$, and the yield per recruit as the mean value, $\bar{Y}_W/R$, over a period of time. If the absolute value of $\bar{R}$ is required it can be obtained by treating that parameter instead of the yield as the unknown factor in the yield equation (16) and comparing the mean yield per recruit with the observed mean yield when the population was in a steady state. Thus the average annual catch of plaice by all countries fishing the North Sea during the pre-war period for which values of the mortality and growth parameters have been estimated was $5 \cdot 2 \times 10^4$ tons, or, converting to the appropriate units, $5 \cdot 4 \times 10^{10}$ gm. Calculating the mean yield per recruit from (16) for the same period gives $\bar{Y}_W/R = 194$ gm., so that $\bar{R}$ is of the order

$$\frac{5 \cdot 4 \times 10^{10}}{194} = 2 \cdot 8 \times 10^8$$

i.e. about 300 million, which is not out of keeping with estimates of numbers in the fishable stock derivable from the egg-census (Ch. VI).[1]

PREDICTIONS IN CONTROLLED FISHING

The Variation of Steady Yield with Fishing Intensity and Mesh Size

Yield-intensity and yield-mesh curves. The values of yield per recruit, Y_W/R, calculated from (16) using various values of the fishing mortality coefficient (F) and assuming that the mesh remains at its pre-war size of 70 mm. are shown plotted against F in Fig. IX.9. This will be called a *yield-intensity* curve. All points on it refer, of course, to steady states and it must not be interpreted as showing the immediate effect of a change in fishing intensity (see p. 420). As is required, this curve predicts that no catch would be obtained if there were no fishing, and that with an infinite intensity all fish would be caught as soon as they enter

[1] With the reservation that the data are under revision at the time of writing, the concordance of estimates may be shown :

    (i) Wollaston's under-estimate of the number of ♀ plaice in the Southern Bight in 1914

        (Ch. VI) . . . . $17\frac{1}{2}$ m.

    (ii) Multiply by 3 for fecundity error . . . . 50 m.

    (iii) Add 100 per cent for spawning in Heligoland Bight, etc. . . . . . 100 m.

    (iv) Double for ♂ plus ♀ . . . . 200 m.

    (v) The mature are about $\frac{1}{2}$ of the fishable population, so the fishable population . 400 m.

In the steady state the number of recruits must equal the number that die and are caught, and $M + F = 0 \cdot 83$ (above), whence the number of recruits is obtained from (v).

    (vi) $0 \cdot 83 \times 400$ m. = 320 m.

Number (vi) differs but little from the estimate in the text. Because of uncertainty attaching to steps, especially (iii), not much reliance can be placed on such closeness of agreement—only on the agreement of orders of magnitude.—Ed.

the exploited area, in which case the yield per recruit is equal to the threshold weight at recruitment = 123 gm. We are more concerned with the middle part of the curve than with the extremes, and the most striking feature is the existence of a maximum steady yield corresponding to a fishing mortality coefficient of about 0·22. Remembering that F is proportional to fishing intensity, such a value corresponds to an intensity very much less than the average value, or that during any single year, before (or after) the war. It seems therefore that if the fishing intensity before the war had been considerably smaller, the average annual yield would not have been less, and might perhaps have been rather more, though it must be remembered that this conclusion is so far based on the behaviour of a very much simplified model from which many complicating factors known to exist have been omitted. The ways in which the effects of density dependence can be assessed will be indicated towards

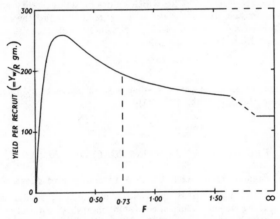

Fig. IX.9.—Annual Steady Yield of Plaice plotted against Fishing Intensity.

Yield of plaice per recruit (Y_W/R) as a function of fishing mortality coefficient (F) with $t_{\rho'} = 3·72$ years, corresponding to a 70 mm. gauge mesh in trawls, double twine. The vertical line at $F = 0·73$ corresponds to the average pre-war fishing intensity.

the end of this chapter. At this stage it is important to note that certain of these relationships tend to cancel each other's effects, and in fact, it is probable that the curve of Fig. IX.9 provides a fairly good basis for deducing the effects on the catch of changes in fishing intensity.

If F is kept constant at the pre-war value of 0·73, equation (16) can be computed for various values of $t_{\rho'}$ to find the effect on yield of increasing the size of mesh. The result—which is called a yield-mesh curve—is shown in Fig. IX.10, the scale of cod-end mesh (taken from the $t_{\rho'}$ curve of Fig. IX.6) being given below that of $t_{\rho'}$. This curve also has a maximum, and the indications are that if the mesh size were to be increased above that in general use an increased yield would be obtained up to a point. However, if the mesh is increased too far the yield falls, because death of so many small and medium-sized fish from natural causes before entering the exploited phase would offset the benefit of allowing individuals to grow to a worthwhile size. In referring to the upper range of the mesh scale it is necessary to bear two points in mind. One, which has already been mentioned in connection with Fig. IX.6, is that this scale (and, of course, the curve itself) assumes that growth remains unaffected by the increase in population density that would be caused by an increase in mesh, a problem which is examined in detail later. The other is that increasing the size of mesh would probably cause some increase in the fishing power of the gear (see Ch. VIII), and in order to keep the fishing intensity constant—as is assumed in Fig. IX.10—it would be necessary, strictly, to make some other compensatory change, e.g. in the size of fleet. For plaice the indications are that the fishing power might perhaps be increased by some 30 per cent over the whole range of mesh shown in Fig. IX.10, so that, without a compensatory change, F would gradually increase with mesh from $0\cdot73$ to a maximum of about $1\cdot0$ at the largest meshes. This could have only a slight effect on the yield-mesh curve (see Fig. IX.11) and in the remainder of this chapter it is ignored.

This is perhaps the best place to mention the effect that the use of different parameter values—particularly of the natural mortality coefficient M and the maximum age t_λ—has on yield curves, either because of a real change in those factors that might occur or because their magnitude may have been incorrectly estimated. In general, if M is increased the maximum of the yield-intensity curve moves to the right and the curve becomes more flat-topped, until at a high enough value the maximum disappears entirely.[1] This means that a higher fishing intensity (or a smaller mesh) would have to be used in order that man as a predator might " compete " more effectively with the other causes of death. Nevertheless, it is found with the plaice that for any practicably acceptable value of M the maximum yield would be obtained with a fishing mortality appreciably less than that caused by the pre-war intensity. In the case of the yield-mesh curve an increase in M causes the maximum to move to the left, and eventually eliminates it—but extremely high values must be reached before the latter happens.

The effect of increasing the arbitrary value of t_λ is to move the maximum of the yield-intensity curve slightly to the left until, when the possible life-span is taken to be of infinite duration, a sharp maximum is obtained at a value of $F = 0\cdot13$. Again, the maximum of the yield-mesh curve moves in the opposite direction, though in neither case is the shift very pronounced and it is clear that these curves are not sensitive to increases in the value of t_λ beyond the one we have used. It will be remembered that the value of t_λ (15 years) was probably underestimated, so that it is not necessary to consider here the effects of taking values lower than this.

In addition to shifting the position of the maximum yield, changes in the values of M and t_λ also alter the level of the yield curves as a whole, but this effect is less important because to convert yield per recruit into absolute yield the former is compared with the observed value of the latter for a known state, and R is computed in the way described previously. Changes in the level of the Y_W/R curves that are roughly proportional for all values of F or $t_{p'}$ therefore leave the absolute curves unaffected, the only result being an inverse change in the estimate of recruitment.

The simple population model represented by (16) and the yield curves of Figs. IX.9 and IX.10 obtained from it have much in common with those of Baranov (1918), who applied a rather similar theory to the North Sea plaice. There are certain peculiarities in the treat-

[1] The word " maximum " is used throughout to denote the peak of a modal curve as opposed to the upper limit approached asymptotically by a curve such as that of Fig. IX.12.

ment by this author that make his model unsuitable as a basis for developing more complex models, but more serious is the fact that he took, in conjunction with an infinite life-span, a simple proportional increase of length with age to represent growth in the population model. Although this growth function—which implies a roughly cubic relationship between weight

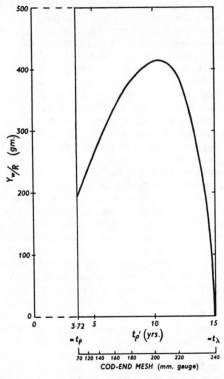

FIG. IX.10.—ANNUAL STEADY YIELD OF PLAICE PLOTTED AGAINST MESH.

The yield per recruit (Y_W/R) is shown as a function of age at entry to the exploited phase ($t_{\rho'}$), with $F = 0.73$. The lower horizontal scale is that of mesh corresponding to the upper scale of $t_{\rho'}$, assuming a constant growth-rate (see, however, Fig. IX.17).

and age—represents the early phases of growth reasonably well it has no inflection, with the consequence that the weight of very old fish is enormously overestimated. As a result, the yield predicted by Baranov's model when the relative abundance of old fish is high, that is, when the fishing mortality is slight or the mesh large, is seriously exaggerated. Nevertheless, essentially correct conclusions were reached as to the way a fish population reacts to exploitation, and he was the first person to attempt a rigorous mathematical formulation of the problem. Indeed, Baranov's contribution was much in advance of contemporary thought and passed unnoticed for many years. More recently, similar theoretical models have been developed by Ricker (1944) and Silliman (1945), but both suffer from the same kind of disadvantage as Baranov's owing to the use of an empirical growth function which is not asymptotic in conjunction with the assumption of an infinite life-span. The simple exponential increase of weight with age used by Ricker has particularly serious effects at low values of F, especially when the natural mortality coefficient is small, and the yield-intensity curves predicted by his model can never have a maximum.

Graham's synthesis. Graham (1935) has tackled the problem in a more general way, not, as here, by an analysis of the separate factors that determine the yield, but by starting with the assumption that the greatest catch will be obtained when the population, as a whole, is growing at its greatest rate. Since the actual growth of a population from an initially very small size is seldom open to direct observation, Graham assumes that growth of a fish population would take the form of the familiar logistic curve which has been observed for human populations (Pearl, 1930) and for certain experimental populations (e.g. Gause, 1934). This leads to the conclusion that the most effective level of fishing is that which maintains the population at an abundance about half that which it would attain if there were no fishing (i.e. in the virgin stock). An advantage of Graham's approach is that it synthesizes the simultaneous effects of all the primary factors and the interaction between them, such as their dependence on density, so that less detailed population studies are needed for its application. On the other hand, the results depend directly on the assumption of symmetry of the population growth curve which cannot easily be verified, and the absence of specific reference to individual factors precludes application of the method to the assessment, for example, of the effects of changes in the size of mesh. Graham's theory has provided a valuable focus for the most recent discussion of the theory of fishing, but it does not lend itself to numerical calculation of the detailed effects of fishing on the stocks. The reader is again referred to the original work of this author (1935, 1939) and to his popular summary of it (1943) for a more complete account.

Thompson's halibut work. W. F. Thompson (1937), whose work produced a worldwide effect in bringing large-scale fishery regulation into the realm of practical politics, used a procedure for assessing the effects of various fishing intensities on the yield in which the survival of a year-class from one year to the next is calculated arithmetically by applying a percentage mortality rate, the yield being computed by multiplying the number of fish caught at each age by the average weight of fish at that age. This method gives a correct general picture of the variation of yield with fishing intensity, and Thompson used it to arrive at the principles on which regulation of the Pacific halibut fishery should be based. The method of regulation was, however, worked out by direct analysis of past commercial statistics of catch and effort, and consisted of setting a limit to the magnitude of the total catch which could be taken each season. The history of the fishery had been well documented from the beginning of the century, and the decline in yield that set in when fishing effort increased during the years immediately prior to the beginning of regulation in 1930 was most marked, and left little doubt as to the direction in which regulation should proceed, namely, reduction of effort. This was achieved indirectly by setting initial catch quotas lower than the previous level of yield, and within a few years the results were spectacular, the catches being maintained and even increased with a greatly reduced effort. Nevertheless, the approach adopted by Thompson is open to the objection that the apparent benefit, or much of it, might in reality have been due to natural changes in the fish population, principally fluctuations in recruitment, rather than to the relaxation of man's activity. This is, in fact, the view taken by Burkenroad (1948), who has criticized Thompson's interpretation of the results of regulation of this fishery. Recently, Thompson (1950) has published a defence of the regulation which, in its turn, has been criticized (Burkenroad, 1950, 1951) and, conditionally, defended (Holt, 1951). The present authors believe that the changes in catch and effort which have occurred since regulation was instituted, together with the previous history of the fishery, are too closely linked to be readily ascribable to natural fluctuations, but commercial statistics of catch and effort alone are scarcely sufficient to give a conclusive answer. The controversy is instructive in that it demonstrates that nothing short of a detailed analytical study of the factors responsible for the behaviour of an exploited population can provide a sound basis for regulation or a means of ascertaining conclusively its true effects. There is, of course, the further difficulty that even if it is accepted that recent trends in the fishery are due to regulation, it does not follow that the present state of the fishery is necessarily the best that could be obtained.

Later on, when discussing the merits of various regulative methods, we shall refer again to the Pacific halibut because until recently it was the only major fishery which is extensively controlled. Before proceeding, however, we must establish more precisely the purposes which the regulation of a fishery is designed to achieve ; in other words, to answer the question posed at the beginning of this chapter—how should a fish stock be exploited to obtain the best results ?

Eumetric Fishing

So far we have investigated the effect on yield of changing the fishing intensity while holding the mesh size constant, and of changing the mesh size with the fishing intensity constant, but in the general case it is necessary to take account of changes in both these characteristics of the fishing activity. This need arises in practice because there is no saving of costs in using a larger cod-end mesh, so that it is far more economical to increase the average age and size of fish caught by reducing the amount of fishing than by increasing the size of mesh. Now if a yield-mesh curve of the kind shown in Fig. IX.10 is drawn for each of a number of different values of F, it is found that as F is increased the mesh required to give the maximum yield, and the magnitude of that yield, also increases. Examples of yield-mesh curves for several different values of F are shown in Fig. IX.11. In a rational fishery, provided there are no special difficulties, a size of mesh would be used that corresponded with the peak of the yield-mesh curve *for the particular fishing intensity* which is generated ; hence the kind of yield-intensity curve that is of the greatest fundamental importance is not that in which the mesh-size is held constant (as in Fig. IX.9), but in which the mesh is adjusted throughout to enable the maximum possible yield to be obtained at each level of fishing intensity. This is called a *eumetric* yield curve, and it may be constructed by plotting the maximum value of Y_W/R from each of a number of yield-mesh curves against the corresponding fishing mortality. Fig. IX.12 shows the eumetric yield curve constructed in this way from the curves of Fig. IX.11. The most important feature of the eumetric yield curve for plaice is that it approaches an asymptote as fishing intensity increases, and has no maximum. Moreover, this general shape does not depend in any way on the particular values of the parameters used in calculating it. Another characteristic is that the biomass remains approximately constant over a very wide range of fishing intensity, since the effects of changes in fishing mortality tend to be counteracted by the corresponding changes in the age at which fish enter the exploited phase. Thus density dependent effects, which are the most likely to invalidate conclusions based on simple population models, are largely minimized with a eumetric yield curve, as will be shown later.

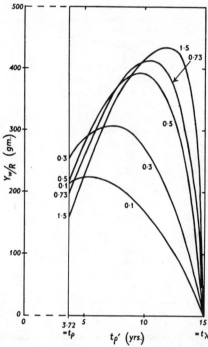

Fig. IX.11.—Annual Steady Yield of Plaice Plotted against Mesh at Various Fishing Intensities.

The yield per recruit (Y_W/R) as a function of $t_{\rho'}$ for various values of F, to show how the height and location of the maximum changes with F. See Fig. IX.10 for mesh scale.

The curve defining the relationship between fishing mortality and $t_{\rho'}$ needed to produce a eumetric yield curve is called a *eumetric fishing* curve ; that for plaice, corresponding to the eumetric yield curve of Fig. IX.12, is shown in Fig. IX.13.

Fishery Regulation

The use of a eumetric yield curve for fishery assessments may now be considered—as a valuable ideal. Clearly, the greatest possible yield can be obtained only with an infinitely high fishing intensity, and hence at a prohibitive cost. With no possibility of obtaining the greatest possible yield, determining the best point on the eumetric yield curve at which to fish becomes essentially an economic problem. Yield and fishing intensity have therefore to be transformed into their economic equivalents of value and running costs—not necessarily in absolute units, but the value of the yield relative to the cost of catching it ; to do this,

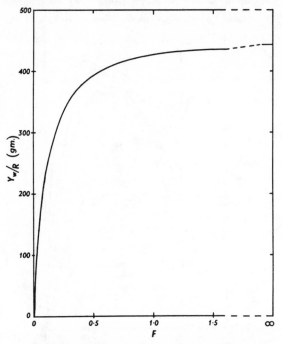

Fig. IX.12.—Eumetric Yield Curve for Plaice.

If a change of mesh accompanies a change in fishing intensity in such a way that each value of F is matched by the mesh that would give the maximum steady yield at that value (Eumetric fishing, Fig. IX.13), the resultant curve of steady annual yield per recruit (Y_W/R) for plaice differs widely from that of Fig. IX.9, particularly in having no maximum.

of course, the yield curve must first be expressed in units of absolute yield and total fishing intensity, though the shape of the curve will not thereby be altered materially.

Deducing the precise relationship between yield and value is difficult, because of the possibility that there may be less consumer demand for a large yield than a small one, with a consequent fall in the price (per lb.) of fish. With a eumetric yield curve, however, the average size of fish caught increases with the magnitude of the yield itself (because the mesh is also increased), thus tending to counteract a fall in price resulting from any slackening of demand that may occur, and it is legitimate as a first approximation to regard the total value of a eumetric yield as proportional to its weight. Fishing intensity can be varied in several ways, such as by altering the size of the fleet, i.e. the total fishing power, the time that vessels are allowed to fish each year, or the efficiency of the gear used. At this stage, however, we are concerned with fishing intensities generated in the most economical way, since a prime motive in any commercial fishery must be to obtain as much income in the form of yield as is possible with a given expenditure in fishing. Graham (1935) has put this in a slightly different way in saying that . . . " the benefit of efficient exploitation lies more in economy of effort than in increase of yield, or preservation of future stocks, though both of these purposes may also be served ". The question of economy of effort is even more vital when deciding the point on a eumetric yield curve that would give the best results, because in this case yield does not rise but falls if the fishing effort is decreased. We shall find shortly that

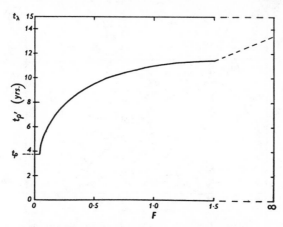

FIG. IX.13.—EUMETRIC FISHING CURVE FOR PLAICE.
The curve defines the relation between F and $t_{\rho'}$ used in constructing Fig. IX.12.

certain other factors may have to be taken into account when regulating a fishery and that these may require some modification of this principle ; but here it implies that changes in fishing intensity must be brought about primarily by changes in total fishing power, and hence that fishing intensity is roughly proportional to the cost of generating it.

Optimum fishing. An hypothetical example of a transformed eumetric yield curve is shown in Fig. IX.14, curve (a). Curves (b) and (c) show the profit (total value minus total cost) and rate of profit (profit per unit cost) as functions of cost, which latter in many situations bears a constant relation to capital outlay. It will be seen that the absolute profit curve has a maximum, and also that at a certain high value of F it crosses the x-axis. Here there is no profit in fishing ; this is, in fact, the state to which an uncontrolled fishery tends, since as Graham (1948) has pointed out, the basic competitive element in fishing causes the intensity of exploitation continually to increase until, to use his words, " average profit =

nil ''.   The aim in a fishing industry that is a self-supporting economic unit might therefore be to fish at such an intensity and with such gear that the steady yield is obtained under conditions fairly close to those for maximum profit, and this will strike a reasonable balance between the benefits to the industry and to the consumer.   On the other hand, the nation's relative need to put financial resources into exploiting the fish population for food or directing them to other channels may to some extent alter the situation and merit the exertion of a somewhat higher fishing intensity.   Again, it may be desirable for social reasons to maintain

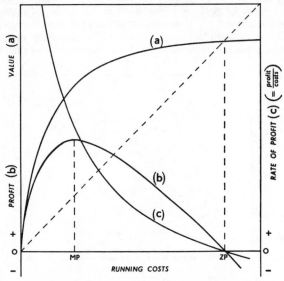

FIG. IX.14.—AN EXERCISE IN BIONOMICS.

Curve (a) is a eumetric curve in which steady yield instead of being expressed in weight is now in money value, which is plotted against the economic equivalent of intensity of fishing, namely, running costs.   From those are derived (b) the annual profit and (c) the profit expressed as a rate on running costs, which in many situations bear a constant relation to capital outlay.

MP = maximum profit point ; ZP = zero profit point.

a rather larger industry and fleet than would be necessary to obtain the maximum profit. Thus it is better to regard curve (b) as showing the maximum *potential* profit, since when all relevant factors are taken into account the best conditions may not be those in which the whole of the maximum potential profit is actually realized, but in which some is absorbed in obtaining rather more yield or maintaining a larger fleet.   The best fishing intensity to use in a particular case, which we shall call the *optimum fishing*, depends on these and other criteria which it would be outside the scope of this book to discuss further, but it can be specified if the relevant information is available and if it is known on what basis the fishing industry is to operate.

S.F.—E E

This use of the term optimum differs from that of Hjort, Jahn and Ottestad (1933), Graham (1935, 1939 and 1948) and Baerends (1947), who all define it as the maximum of a curve of equilibrium yield varying with fishing intensity with a constant mesh in use, but we have seen that the existence of such a maximum is not an inevitable characteristic of the dynamics of a fishery, especially when variation of mesh size is considered. Other authors, notably Herrington (1943), Nesbit (1943), Ricker (1945) and Baerends (1947), have realized the complimentary nature of fishing intensity and gear selectivity, and the first three define the " optimum catch " as, in effect, that which would be obtained using an infinite fishing intensity and catching all fish of each year-class as soon as the latter reached its greatest total weight. Thus their optimum is identical to what we have termed the greatest possible catch, but the latter name is more appropriate because this catch is specifically not the " best " but only the greatest. Another term occurring frequently in the literature is " overfishing ", the meaning of which has already been discussed in earlier chapters (pp. 50 and 262). It is generally meant to indicate a state in which too high a fishing intensity is being used, with the corollary that the remedy should be to decrease it. However, it might be better in some cases to increase the size of mesh, so a wider interpretation of the word is to regard it as indicating a biological and economic maladjustment of the fishing activity which requires regulative action to reduce the fishing intensity or increase the size of mesh or perhaps both. With this definition, overfishing is any combination of fishing intensity and mesh size lying below the eumetric fishing curve (see Fig. IX.13).

So much, then, for the fundamental objectives that fishery regulation should be designed to achieve. We have established them for a simple case of a fishery based on a single species, but the same principles apply to the more complex situations that are met in practice. In the North Sea, for example, there are, besides plaice, a number of different species of commercial importance caught by the same general type of demersal gear, and the distributions of some of these fish overlap so much that they have to be regarded as constituting a single composite fishery. Any regulation applied to them will affect each somewhat differently and must be a compromise between their particular requirements, so that assessments must be made on the basis of a combined yield curve and the economic transformation must take account of their different market values. An important point here is that whereas changes in fishing intensity will not cause any drastic alteration in the contribution of the various species to the combined yield, this will not necessarily be true with changes in mesh size. A mesh large enough to obtain the best yield from plaice or cod, for example, would probably result in virtually complete loss of the smaller fish such as whiting or sole ; but some of the latter are the basis of special fisheries, while the maintenance of variety is important to the consumer and cannot be sacrificed for quantity alone. Nevertheless, the problems raised by a composite fishery are not insuperable and the concepts of eumetric and optimum fishing apply to a fish conmunity just as they do to a single population.

Methods of regulation. Practical problems of other kinds arise when regulation of a fishery is actually attempted. Some of these are administrative rather than technical, but they are worthy of brief mention to show the background against which the ideal requirements of fishery control must be viewed. One of these concerns the *method* by which a fishery may be regulated, and here again it is useful to take the case of the North Sea demersal fisheries for illustration. A summary of methods that have been discussed for the

North Sea is given in the Final Report of the Standing Advisory Committee of the International Conference on Overfishing, April 1947, and is as follows :
1. Minimum size of mesh to be used.
2. Minimum legal sizes at which various species of fish may be landed.
3. Reduction of the power of fishing fleets.
4. Reduction of the total annual catch by setting a limit.
5. Control of the building of fleets.
6. Control of the amount of fishing activity, i.e. the amount of time vessels spend fishing.
7. Restriction of fishing operations to a specified fishing season.
8. Closure to fishing of certain parts of the fishing area.

Despite their apparent diversity, these all reduce in effect either to decreasing the fishing mortality or increasing the age at entry to the exploited phase, which we have seen to be the fundamental factors that must be adjusted in order to obtain the best results. Some, however, are more direct in their action than others, the setting of a limit to the annual catch and closure of part of the area to fishing being examples of indirect methods of restricting fishing mortality. Catch limitation is indirect because the size of the catch does not bear a precise relationship to the fishing mortality owing to the influence of natural factors, notably fluctuations in recruitment. Its use will also result in secondary effects such as a shortening of the fishing season, as has happened in the Pacific halibut fishery where this method of regulation has been adopted—unless, of course, further legislation is brought in to prevent them.

There is, however, one consideration of particular importance in deciding the best method of regulating to achieve optimum fishing as previously defined. Since an essential feature of this is to maintain, within limits, an agreed upon margin of profit—not necessarily the maximum—it follows that regulation must include control of total fishing power. Otherwise, improvements made by other methods, whether they involve control of fishing time, closure of areas, or any other form of regulation—including mesh regulation—will be largely nullified by the inevitable increase in fishing power that the initially more profitable fishing will encourage. Nevertheless, the best method or combination of methods to adopt in the circumstances is to some extent a matter of expediency and depends partly on factors such as the economic structure of the fishing industry concerned.

In the North Sea it is, indeed, fortunate, that several different methods of regulation are available, because it is possible for each of the various countries concerned in fishing the area to adopt that which is most suited to the particular structure and economy of its industry, and yet for each to make a contribution to the conservation of the area as a whole. This raises the need for assessing the equivalence, in terms of regulative value, of the various methods, notably that between increases in size of mesh and reduction in fishing intensity. This can be done roughly from a diagram such as Fig. IX.11 which shows the effect on yield of changes in both F and $t_{\rho'}$, or more precisely by setting up a theoretical model in which the fishing mortality coefficient is partitioned into components representing the different fleets concerned, each component of F being associated with a value of $t_{\rho'}$ appropriate to the selective properties of the gear used by the fleet in question. Such a model can be used to assess the effect on the total yield, and on the yield to each fleet, of changes in the fishing intensity of some fleets and in the size of mesh used by others.

The transitional period. The immediate effects of applying a regulative measure to a fishery must also be considered and given due weight. It will be remembered that the yield-intensity and yield-mesh curves of Figs. IX.9 and IX.10 refer to steady states, so that they cannot be used to predict the yields to be expected during the transitional phase following a change in F or $t_{\rho'}$, that is, during the interval between the old steady state and the new one. Theoretical models can, however, be developed for this purpose and they differ from

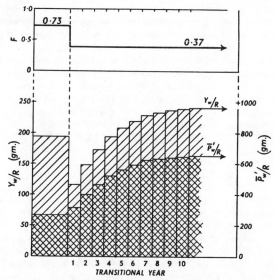

Fig. IX.15.—Immediate Changes on Reduction of Fishing.

The diagram shows the changes that would be expected, relative to what would have been obtained had there been no change in intensity of fishing and if fluctuations are disregarded. The transitional period in plaice lasts about 10 years, approximating to the fishable life-span of this species. The intensity of fishing is here supposed to have been suddenly reduced to half its pre-war level (from $F = 0.73$ to $F = 0.37$). The annual yield in successive years of the transitional period is shown by medium-shaded histograms, on which are superimposed those (darkly shaded) representing the catch per unit effort. It is seen that the latter begins to rise above its former value even during the first year, whereas gain in yield is not shown until the fifth year.

the type we have been using hitherto in that the annual yield from each of the year-classes present in the population is evaluated separately and then summed to give the total yield in each year of the transitional phase. An example, based on plaice, of the transitional yields following a sudden change in F from the pre-war value of 0.73 to 0.37, that is, reduction to half, is shown by the open histogram of Fig. IX.15, the corresponding changes in the catch per unit effort being indicated by the shaded histogram. It will be seen that immediately after the decrease in fishing mortality there is a drop in yield which is roughly proportional to that in F, but that it then gradually builds up to the new steady value ; the duration of the transitional phase is equal to the fishable life-span, since after this time has elapsed all year-classes present in the exploited phase of the original population have been replaced. The catch per unit effort, on the other hand, increases throughout the transitional period. A sudden reduction in fishing intensity to half its former value is of course very drastic, and in practice reductions would probably have to be made more gradually ; in this case the initial drop in yield would be less but it would take longer to reach the final steady state.

The transitional yields following a sudden increase in the size of mesh follow the same pattern as that shown in Fig. IX.15, though the catch per unit effort remains proportional to the yield throughout and therefore does not change in the same way as after a reduction in fishing intensity.

This discussion has been taken far enough to show that theoretical population models can be of great value in designing a programme of fishery regulation as well as in establishing the fundamental objective. Actual regulation would probably be experimental in nature, proceeding in relatively small steps ; and with a properly designed research programme, the information gained after each step had been taken would increase the accuracy with which the next could be predicted from the theoretical model.

MORE COMPLEX THEORY

To conclude this account of the theory of fishing we shall consider briefly two aspects of the dynamics of demersal fisheries that have not been taken into account in developing the simple population models at the beginning of the chapter. These are the dependence of recruitment, growth and mortality rates on population density, and the variation of the latter and of the fishing intensity from one part of the exploited area to another. A discussion of these factors has been left until last because in some cases the lack of critical data necessarily means that assessment of their effects must be speculative, and, in fact, study of them, both theoretical and practical, is an important feature of present-day research on the dynamics of fish populations.

The Variation of Parameters with Population Density

Growth. Investigation of the dependence of fish growth on population density involves two complementary methods. On the one hand, empirical regressions and correlations may be sought ; on the other, assumptions may be made about the causes of the dependence on density, appropriate mathematical models set up, and their properties examined. There is sufficient information about the North Sea plaice for both techniques to be used, so that the results obtained by the two methods can be compared.

The variation in abundance of plaice during the pre-war period was too small to have had a detectable effect on growth, but the cessation of fishing during the 1939–45 war caused the population to increase greatly (Margetts and Holt, 1948) ; the average biomass for the whole war period has been estimated as about three times the pre-war level. Investigations in 1946 on the weight-at-age of old fish which had lived through the war showed that their growth had been appreciably slower than normal, with an average value of L_∞ of 60·6 cm. compared with the pre-war average of 68·5 cm.[1] Thus we have two levels of growth and biomass which as a first approximation can be joined by a straight line to predict the growth (in terms of L_∞) corresponding to any intermediate level of biomass. If the latter is expressed in terms of biomass per recruit, $\overline{P}_W/R$, we can write

$$L_\infty = a + b\left(\frac{\overline{P}_W}{R}\right) \qquad . \qquad . \qquad . \qquad . \qquad (28)$$

For pre-war conditions we have $L_\infty = 68·5$ and $\overline{P}_W/R = 266$ gm. (calculated from (18) using the appropriate growth and mortality estimates), so that (28) becomes

$$68·5 = a + b \times 266$$

while for the war period we have

$$60·6 = a + 3b \times 266$$

Solving simultaneously gives the estimates $a = 72·47$ cm. and $b = -0·0149$. A theoretical model incorporating this relationship between growth and density can now be constructed by substituting for $\overline{P}_W/R$ from (18) in (28) ; the simplest form this takes is when $t_{p'} = t_p$, as is the case for plaice when a 70 mm. mesh is used (see p. 404). Making this simplification in (18) and remembering that $L_\infty = \left(\dfrac{W_\infty}{q}\right)^{1/3}$, we find that

$$\left(\frac{W_\infty}{q}\right)^{1/3} = a + b\left\{W_\infty \sum_{n=0}^{3} \frac{\Omega_n e^{-nK(t_p - t_0)}}{F + M + nK}(1 - e^{-(F+M+nK)\lambda})\right\} . \qquad . \qquad (29)$$

In order to use this equation to estimate the growth rate—in terms of W_∞—corresponding to any specified value of F, it is necessary to restrict the change in growth to the post-recruit population, because the density of the pre-recruit phase is not altered by changing F. In other words, fish must still reach their (constant) weight at recruitment, W_p, at the same age, t_p, as before, and this can be made to happen by adjusting the origin (t_0) of the new growth curve, i.e. that defined by the new value of W_∞. The required value of W_∞ is calculated from the relationship between these two parameters expressed by (14), i.e.

$$W_p = W_\infty(1 - e^{-K(t_p - t_0)})$$

in which W_p and t_p are known constants defining the weight and age at recruitment.

With t_0 related to W_∞ in this way, it will be seen that the latter is the only unknown parameter in (29) ; for any specified value of F the required estimate of W_∞ is, in fact, that which satisfies both sides of the equation.[2] Pairs of values of F and W_∞ can then be calculated and used in the simple yield equation (16) to construct a yield-intensity curve for plaice with a 70 mm. mesh that incorporates this, assumed linear, relationship between growth and

---

[1] It will be remembered from the discussion at the beginning of the chapter of the significance of the parameters K and W_∞ (or L_∞) in the von Bertalanffy equation that changes in growth due to variations in the supply of food (which is probably the main effect of density) are best represented in terms of the latter parameter. The change in L_∞ noted here may not seem great, but it implies a war-time value of W_∞ of 1982 gm. which is 69 per cent of the pre-war average and outside the normal range of fluctuation of this parameter during the pre-war period.

[2] This is the basis of methods actually used to estimate W_∞. One of these is to take a series of values of W_∞ and to use each in turn to compute both sides of (29). Values of the latter are then plotted separately against those of W_∞ used to compute them, and at the intersection of the two resulting lines both sides of the equation have the same numerical value. The value of W_∞ at this intersection is therefore the estimate required, and can be read off the graph.

density, and it is shown in Fig. IX.16 as curve (b) compared with curve (a) for constant growth. It will be seen that an effect of the density dependence of growth is to depress the maximum and move it to the right, though it is still well below the level of fishing mortality sustained before the war. In other words, allowing the density to increase by reducing fishing intensity is still advantageous, though not quite to the extent that would be predicted from the simple model, and a smaller reduction in intensity would be necessary to reach the maximum yield.

A yield-mesh curve incorporating a density dependent growth can be constructed in a similar way, starting with the relationship between $L_{\rho'}$ and cod-end mesh shown in Fig. IX.6. The resulting curve plotted against mesh size with F constant at 0·73 is shown

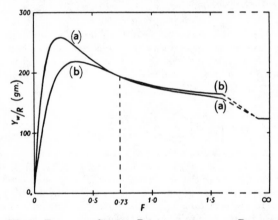

FIG. IX.16.—EFFECT OF GROWTH RATE CHANGING WITH DENSITY : 1.

As in Fig. IX.9, the yield of plaice per recruit (Y_W/R) is shown as a function of fishing intensity (F) with mesh constant at 70 mm., but in calculating curve (b) the growth-rate was reduced progressively as the density of stock increase, and *vice versa*. The difference from curve (a), constant growth-rate, is appreciable, but not radical.

in Fig. IX.17 [curve (b)] together with that calculated with constant growth parameters for comparison [curve (a)] ; the latter is that of Fig. IX.10 plotted against mesh size instead of $t_{\rho'}$, using the relationship between these two factors defined by the $t_{\rho'}$ curve of Fig. IX.6. Curve (b) still reaches a maximum value, though it is considerably lower than before and occurs at a smaller mesh size. These differences are due to the fact that enlarging the size of mesh increases the duration of the pre-exploited phase [i.e. the value of ρ in the yield equation (16)] and hence increases also the total biomass ; this, in turn, reduces the rate of growth and tends to prolong still further the duration of the pre-exploited phase. Thus increasing mesh size has a kind of 'servo-effect', and this is taken into account by the theoretical model used to construct curve (b) of Fig. IX.17. Comparing Figs. IX.16 and IX.17 shows, incidentally, that the yield-mesh curve for plaice is more sensitive to the density dependence of growth that is the yield-intensity curve, and it is clear that an accurate prediction of the effects of a large change in mesh size in this species could not be made from the simple model with constant growth parameters.

Leaving the empirical solution we may now describe the analytical approach, and the two results can be compared.

In the sea, direct space and crowding effects can hardly have much importance in affecting the growth-rate ; yet, nevertheless, density can have marked and significant effects

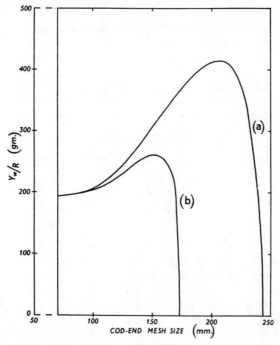

FIG. IX.17.—EFFECT OF GROWTH RATE CHANGING WITH DENSITY : 2.

As in Fig. IX.10, the yield of plaice per recruit (Y_W/R) is shown as a function of mesh size with fishing intensity constant at $F = 0.73$, but in calculating curve (b) the growth rate was assumed to be reduced progressively as the density of stock increases. The differences from the use of constant growth rate, curve (a), are considerable. Not only is the benefit from increase of mesh estimated to be less, but the maximum yield is reached at a considerably smaller mesh size.

(Ch. VII, p. 318). In all probability the effect of density on growth is due almost entirely to competition for a limited food supply and a theoretical model can best be constructed on this basis. The first stage is to derive an expression for the total amount of food consumed annually by a population, in terms of its growth and mortality and taking into account the food required for maintenance purposes. It was mentioned in the preceding chapter that

Dawes (1930, 1931) determined experimentally the requirements of plaice, using *Mytilus edulis* for food, and analysis of his data shows that the rate at which a plaice consumes food for maintenance is proportional to the two-thirds power of its weight, that is, we can write

$$\frac{d\chi_A}{dt} = \zeta w^{2/3}$$

where $\frac{d\chi_A}{dt}$ denotes the rate of consumption of food for maintenance, and ζ is the maintenance food coefficient which is found to have the value 20·5 per year. The rate of consumption of maintenance food by all the fish, N_t, of a year-class that survive to age t is therefore given by

$$\frac{dX_A}{dt} = \zeta w_t^{2/3} N_t$$

Hence the total amount of food consumed for maintenance by the year-class throughout its post-recruit life, which is the same as the annual maintenance food consumption of the population in a steady state, is given by

$$X_A = \zeta \int_{t_\rho}^{t_\lambda} w_t^{2/3} N_t \, dt$$

Substituting for w_t from (14) and for N_t from (3) and (7) and integrating, gives

$$X_A = \zeta R W_\infty^{2/3} \sum_{n=0}^{3} \Omega_n' e^{-nK(t_\rho - t_0)} \left\{ \frac{1 - e^{-(M+nK)\rho}}{M + nK} + \frac{e^{-(M+nK)\rho}(1 - e^{-(F+M+nK)\lambda})}{F + M + nK} \right\}$$

where
$$\Omega_0' = 1$$
$$\Omega_1' = -2$$
$$\Omega_2' = 1$$
$$\Omega_3' = 0$$

It will be noted that this equation is similar in many respects to (18) for the annual mean biomass, and a convenient way of writing it is

$$X_A = \zeta \overline{P}_{W^{2/3}} \qquad . \qquad . \qquad . \qquad . \qquad . \qquad . \qquad (30)$$

Dealing now with the fraction of the food consumed that is utilized for growth, it is first necessary to determine the total growth increment produced by the population each year, which we shall call the *annual production, A.P.* The term " production " is used here in the sense of Tansley (1929) and the " corrected production " of Lindemann (1942), and it is also equivalent to the integrated form of the " gross production rate " of Clarke (1946) and Ricker (1946) ; the reader is referred to Macfadyen (1948) for a review of the definitions of the terms " production ", " productivity " and " yield " which have appeared in the litera-ture. If the rate of growth of an individual at age t is denoted by $\left(\frac{dw}{dt}\right)_t$ the total amount of growth produced by a year-class throughout its post-recruit life, which for a population in a steady state is the same as the annual production, can be written as

$$\text{A.P.} = \int_{t_\rho}^{t_\lambda} \left(\frac{dw}{dt}\right)_t . N_t \, dt \qquad . \qquad . \qquad . \qquad . \qquad (31)$$

The most convenient form in which to express $\left(\frac{dw}{dt}\right)_t$ is deduced from the original differential form of the von Bertalanffy equation (10), and after substituting for H, D, and s this becomes

$$\frac{dw}{dt} = 3K(W_\infty^{1/3}w^{2/3} - w)$$

Hence (31) can be written as

$$\text{A.P.} = 3K\int_{t_\rho}^{t_\lambda} (W_\infty^{1/3}w_t^{2/3} - w_t)N_t\,dt \qquad . \qquad . \qquad . \qquad . \qquad (32)$$

However, from the way (30) is derived it can be seen that

$$\int_{t_\rho}^{t_\lambda} w_t^{2/3}.N_t\,dt = \overline{P}_{W^{2/3}}$$

and since also

$$\int_{t}^{t_\lambda} w_t.N_t\,dt = \overline{P}_W$$

(32) becomes

$$\text{A.P.} = 3K(W_\infty^{1/3}\overline{P}_{W^{2/3}} - \overline{P}_W)$$

This equation defines the total growth increment produced annually by a population in terms of the growth rate (i.e. the value of W_∞) and the fishing and natural mortality rates which are contained in the expressions for $\overline{P}_{W^{2/3}}$ and $\overline{P}_W$. Now the amount of food, X_B, which the population must eat to give this annual production may be represented as

$$X_B = \frac{\text{A.P.}}{\varepsilon} = \frac{3K}{\varepsilon}(W_\infty^{1/3}\overline{P}_{W^{2/3}} - \overline{P}_W) \qquad . \qquad . \qquad . \qquad (33)$$

where ε is the efficiency of utilization of food for growth. For a particular type of food, ε is therefore the ratio of a given amount of growth to the amount of food required to produce it, and from Dawes' data an approximate value of it for plaice feeding on *Mytilus* is 0·20. This figure means that of the food eaten in excess of that required for maintenance purposes, about one-fifth by weight is eventually converted into body material.

Finally, the total amount of food consumed annually by a population in a steady state, for both maintenance and growth, can now be obtained by summing (30) and (33), giving

$$X = X_A + X_B = \left(\zeta + \frac{3KW_\infty^{1/3}}{\varepsilon}\right)\overline{P}_{W^{2/3}} - \frac{3K}{\varepsilon}\overline{P}_W \qquad . \qquad . \qquad (34)$$

This equation can be used to compute a value of the total annual food consumption of a population, X, if values of the coefficients ζ and ε, and also of the mortality and growth parameters, are known, as is the case for North Sea plaice during the pre-war period. The value of X per recruit is found in this way to be 1122 gm. (in terms of *Mytilus* flesh), which is equivalent to an absolute value of $3·14 \times 10^{11}$ gm., obtained by multiplying by the estimate of $\overline{R}$ given previously. By making certain assumptions as to how the total food consumption varies with stock abundance, some appreciation of the effects on growth of fish of competition for food may be made. One approach that recent investigations by the authors have found to be fruitful is to regard X as the ' yield ' obtained by fish grazing the food population, so that it can be expressed by an equation similar to that describing predation by man on the fish. In the simplest case this is

$$X = r\omega(1 - e^{-G\lambda'}) \qquad . \qquad . \qquad . \qquad . \qquad . \qquad (35)$$

where r is the annual recruitment into the food population, ω is the average weight of the food organisms, λ' is their life-span and G is the grazing mortality coefficient due to predation by the fish in question. Now this equation contains, in effect, only two unknown quantities—the products $r\omega$ and $G\lambda'$—so that if two levels of growth and mortality in the fish population are observed, two values of X can be computed from (34) and hence used to estimate $r\omega$ and $G\lambda'$ from (35) by simultaneous equations. Equating (34) and (35) we have

$$r\omega(1 - e^{-G\lambda'}) = \left(\zeta + \frac{3KW_\infty^{1/3}}{\varepsilon}\right)\overline{P}_W^{2/3} - \frac{3K}{\varepsilon}\overline{P}_W \qquad . \qquad . \qquad . \qquad (36)$$

so that for any specified value of F this equation contains only one unknown parameter, namely, the corresponding value of W_∞, for which a solution can be found by methods similar to those described in connection with (29).

Certain additional complications have to be introduced into (36) before it can be used in this way, but they cannot be discussed in detail here. The grazing mortality G is of course not constant, and depends on the magnitude of the fish population and on the size composition of fish comprising it. The utilization coefficients ζ and ε are also variables and depend on the rate of food consumption ; the higher this is the lower the efficiency of utilization of food for both maintenance and growth. But, although there are certain additional parameters not represented in (36), these can all be determined from data and the principle by which solutions for the unknown growth parameter W_∞ are found remains the same.

With pairs of values of F and W_∞ computed as above a yield-intensity curve can be calculated from the simple yield equation (16) as in the empirical method outlined above. Taking the pre-war and war-time levels of growth and density to estimate $r\omega$ and $G\lambda'$, this procedure gives a result so similar to curve (b) of Fig. IX.16 that the two cannot usefully be contrasted graphically. It is therefore tabulated in column A of Table IX.3, with values for curve (b) in column B for comparison.

TABLE IX.3.

F	Y_W/R (gms.)		YIELD PER RECRUIT OF PLAICE AS A FUNCTION OF F
	(A)	(B)	
0·01	18	13	The figures in column (A) are obtained by postulating a linear relationship between L_∞ and biomass per recruit (P_W/R); those in column (B) are derived from a more complex model taking into account the grazing of the fish population on its food. Both are based on associated changes in growth and density during a war period.
0·05	80	64	
0·1	136	121	
0·2	197	195	
0·3	217	218	
0·4	217	218	
0·5	213	213	
0·73	194	194	
1·0	179	179	
1·5	165	162	

It is seen that the two sets of figures are virtually identical for all values of F above 0·2, but diverge at lower values. The biological implications of the two methods of relating growth to density are seen better by plotting against biomass per recruit the values of L_∞ predicted by

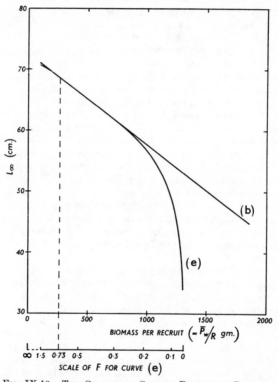

FIG. IX.18.—THE CHANGE OF GROWTH RATE WITH DENSITY.

The relations illustrated are between the growth parameter (L_∞) and the biomass per recruit ($\bar{P}_W/R$) of plaice. Curve (b) is the empirical linear relationship, curve (e) is that predicted from analysis of grazing by the fish population on its food. The changes in biomass are those expected from changes in intensity of fishing. For example, the scale of F corresponding to the biomass changes used for curve (e) is shown below the figure.

each, as in Fig. IX.18. The straight line (b) is the empirical linear relationship, and this is almost exactly coincident with the curve (e) predicted by the more complex model over a range of biomass up to some four times the pre-war level (indicated by the vertical broken line) ; the corresponding scale of F for curve (e) is also shown. A linear relationship between L_∞ and biomass is thus in harmony with what would be expected from a theoretical analysis of the interaction between a fish population and its food, except when biomass is extremely large. With this qualification, it would therefore seem to be a satisfactory method of representing the variation of growth with density in theoretical models, at least until further research shows that a more detailed theory is necessary.

Natural mortality. Variation of the natural mortality rate with density cannot be treated as fully as that of growth, and it is still necessary to rely on empirical methods. In the derivation of the logistic law of population growth (Lotka 1925, Pearl 1930, Kostitzin 1939) it is sometimes assumed that the exponential coefficient of natural mortality itself varies linearly with population density in such a way that the percentage mortality increases as density increases, though strictly speaking, it is only necessary to postulate that the *net* rate of increase or decrease, i.e. the difference between the mortality and reproduction coefficients, is linearly dependent on density. Such a device describes satisfactorily one of the factors that cause a population to tend towards a limiting abundance instead of increasing indefinitely, and it seems that a density-dependent mechanism of some kind must contribute to the maintenance of equilibria in the majority of natural populations. A full analysis of the available vital statistics of plaice or of any other marine species has not been made from this point of view, but a complication here is that in an exploited population trends in natural mortality are to some extent masked by the presence of a much larger fishing mortality. Hence, even if a trend could be detected, it is unlikely that the data would be good enough to distinguish between a linear and any other function. If, however, the natural mortality rate is dependent on density—and it is almost certain to be so to some degree and over some range of abundance—evidence from experimental populations suggests that a straight line relationship between M and the population numbers, $\bar{P}_N$, would probably give a good approximation, so that it can be used to gain some appreciation of the effect of such dependence on yield curves.

Taking the value of M for the pre-war plaice population to be $0 \cdot 1$, as we have determined it above, gives one point on the line relating M to density. Now the greatest slope the line can have is that which results in the coefficient of natural mortality being zero when the density is zero, since any steeper slope would cause the natural mortality to become negative at a finite density, which is inadmissible. Fixing the coordinates of the line in this way therefore gives the maximum change of M with density that is possible using a linear relationship and with M having the value $0 \cdot 1$ at $F = 0 \cdot 73$ (i.e. the pre-war value of F). Values of M corresponding to other fishing intensities can then be calculated directly, and these are given in Table IX.4. They can be used in the simple yield equation (16) to construct a yield-intensity curve that takes into account what is probably a fairly extreme variation of M with density. This is shown in Fig. IX.19, curve (c), with the simple yield-intensity curve from Fig. IX.9 for comparison [curve (a)], and it will be seen that the effect is to lower the maximum value and move it to the right. These changes are very similar to those caused by the dependence of growth on density, shown by curve (b), and the same conclusions are reached, notably that a considerable reduction in fishing intensity from the pre-war level could be made before any decrease in yield resulted, and this of course means that the catch per unit effort could be increased correspondingly.

Growth and natural mortality together. So far we have dealt separately with the density-dependence of growth and natural mortality, but in practice the two may well occur together—indeed could hardly fail to do so—and it is necessary to obtain some idea of their combined effect. Since we have taken M to be related to the total population *numbers*, $\overline{P}_N$, the variation of M with F is not influenced by changes in growth provided fish enter the exploited phase as soon as they are recruited to the exploited area, as is the case with the plaice when a 70 mm. mesh is used. The dependence on density of both growth

TABLE IX.4.

F	M
0·0	0·200
0·05	0·18
0·1	0·17
0·2	0·15
0·3	0·13
0·4	0·12
0·5	0·11
0·73	0·100
1·0	0·09
1·5	0·08

DENSITY-DEPENDENT NATURAL
MORTALITY

Pairs of values of natural and fishing mortality coefficients used in computing the yield-intensity curves (c) and (d) of Fig. IX.19 (plaice). The values of M increase linearly with population numbers, $\overline{P}_N$.

and natural mortality can therefore be taken into account simultaneously by computing a value of W_∞ for each value of F by either of the methods discussed previously, but using the appropriate value of M in each case from Table IX.4. In this way it is possible to obtain

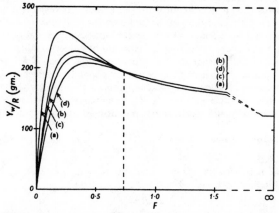

FIG. IX.19.—EFFECTS OF PARAMETERS CHANGING WITH STOCK DENSITY.

In addition to curve (a), from Fig. IX.9, calculated using constant parameters, are plotted curves of yield of plaice per recruit (Y_W/R) making growth (b), natural mortality (c), and both (d) dependent upon density. The differences are not appreciable within the limits likely to be acceptable as a reduction from the pre-war intensity shown by the ordinate at $F = 0.73$.

a value of both W_∞ and M to be used in (16) for each value of F, and the resulting yield-intensity curve (d) is shown in Fig. IX.19. Now although when dealt with separately the density dependence of growth and natural mortality causes a similar kind of change in the shape of the yield curve, their combined effect is not simply additive. For example, as F is decreased the value of M increases, thus causing a lower population abundance than if M remained constant ; but although this tends to decrease the yield, nevertheless the lower density will allow a higher growth rate than if growth alone varied with density, which tends to counteract the effect of the increased M. In fact, the combined effect turns out to be a further reduction of the maximum yield and increase in the value of F at which it occurs, but to a lesser extent than would result from direct addition of the individual effects. This is seen from Fig. IX.19, and serves to demonstrate the power of the analytical method, since there is no other way by which the resultant effect of an interaction as complex as this could be assessed quantitatively.

Recruitment. Density may also affect recruitment. Here the relationship is of a rather different character. The number of recruits, R, which enter the exploited area in a given year are the survivors from the total number of eggs, E, laid at some previous spawning season by the adult population, supposing here for convenience that the whole of a year-class is recruited in a single year, so that all the recruits of a given year are the progeny of the same adult stock. The mechanism of the dependence of recruitment on population density must therefore be analysed in two stages, the first being to express the total egg production in terms of the characteristics of the adult population which spawned them, and the second being to relate the number of surviving recruits to the number of eggs of which they are the progeny. This problem has already been stated qualitatively in Chapter VI, where Simpson has followed the formulation of Kesteven (1947). It remains to make a quantitative model for the problem, and Simpson's studies on the fecundity of fish have shown that the number of eggs laid by a female at each spawning season is roughly proportional to its weight (see Chapter VII also). Denoting this constant of proportionality by ξ, and assuming that the sex-ratio s remains unchanged, the annual egg production E will be proportional to the annual mean biomass of the mature section of the population. Denoting the average age at which fish first become mature by t_η, and supposing for simplicity that t_η is greater than $t_{\rho'}$, the methods used to deduce (17) for $\bar{P}_W$ can be employed to give an expression for E in terms of growth, mortality and, particularly, recruitment, in the adult population. It is :

$$E = s\xi R W_\infty e^{-M\rho - (F+M)(t_\eta - t_{\rho'})} \sum_{n=0}^{3} \frac{\Omega_n e^{-nK(t_\eta - t_s)}}{F + M + nK}(1 - e^{-(F+M+nK)(t_\lambda - t_\eta)}) \quad . \quad (37)$$

In what follows we need also the *egg production per recruit*, γ, which may be obtained from (37) simply by dividing both sides by R. Thus we have

$$E/R = \gamma \quad . \quad . \quad . \quad . \quad . \quad (38)$$

where γ includes the whole of the right hand side of (37) except the quantity R.

In the second stage of the analysis, lack of detailed information concerning the factors influencing survival of eggs, larvae and young fish, must make any attempt at a mathematical formulation of their effects to some extent speculative. Nor is it possible to deduce an empirical relation between eggs and recruits from existing data with any certainty, because of the enormous year-to-year fluctuations in larval survival which usually mask the effects of any changes in egg production which may have occurred. Nevertheless, the fact that the long-term trends in recruitment that might have been expected as exploitation of stocks has intensified have not been found, suggests that some kind of compensatory mechanism may be present, with the effective mortality rate during the pre-recruit phase decreasing as the number of young decreases. The simplest of these is the one used previously to relate the adult mortality rate to density, and in the present case this amounts to postulating a linear relationship between the effective pre-recruit mortality rate and the number of young fish comprising the year-class in question. The term " effective " is used here because one of the outstanding features of the mortality during early life in most fishes is that it fluctuates enormously from one stage to the next (see Chapter VI), but as long as the natural mortality rate in at least one stage varies with density it will have the required compensatory effect. A possible mechanism would be competition for food among the young larvae at a critical stage such as that when the yolk-sac is first exhausted, because it is known that shortage of food can readily cause the death of young larvae. In these circumstances it can be shown that the relationship between a given number of eggs and the number of recruits surviving to age t_ρ takes the form

$$1/R = \alpha + \beta/E \qquad . \qquad . \qquad . \qquad . \qquad . \qquad (39)$$

where α and β are density-independent parameters related to the coefficients of larval mortality, and which vary only with environmental conditions and with the duration of the pre-recruit phase. The curve given by (39) starts at the origin and proceeds towards an asymptotic value of R as E tends to infinity, the value of which is $1/\alpha$. The curvature depends on the value of β, and examples of the relationship for certain pairs of values of α and β are shown in Fig. IX.20. The curves are similar in shape to those suggested by Tester (1948) and Kesteven (1947), and it is interesting to find that postulating a larval mortality rate which, at one or more stages in the early life-history, is linearly dependent on larval density provides a quantitative expression of these authors' ideas.

In plaice there are insufficient data to give any indication of a relationship between egg production and recruitment, though it seems likely that the former is large enough for recruitment to be near the asymptotic level and therefore effectively independent of fecundity. Data for haddock in the North Sea cover more years and a wider range of egg production, and here there is some evidence of a trend, though it is not conclusive, and the fluctuations are too great to allow (39) to be fitted accurately. The best procedure at the present time is, therefore, to examine the effects of various values of α and β such as those shown in Fig. IX.20, which are based on fecundity data and age-compositions of the haddock given by Raitt (1933, 1939), and see whether any generalization can be made.

The first requirement is to deduce an equation for yield in which recruitment is expressed in terms of egg production as in (39). Now both equations (38) and (39) contain E and R, so that by eliminating E between them we find

$$R = \frac{1}{\alpha}\left(1 - \frac{\beta}{\gamma}\right)$$

Substituting this expression for R in equation (16) for yield gives

$$Y_W = \frac{FW_\infty}{\alpha}\left(1 - \frac{\beta}{\gamma}\right)e^{-M\rho}\sum_{n=0}^{3}\frac{\Omega_n e^{-nK(t_{\rho'}-t_0)}}{F + M + nK}(1 - e^{-(F+M+nK)\lambda}) \qquad . \qquad . \qquad (40)$$

in which it will be remembered that γ is itself given in terms of the growth, mortality and fecundity of the population as in (37). This may be termed a *self-regenerating* population model, because recruitment does not appear as an arbitrary number of fish but is expressed in terms of the biomass of the adult population of the previous generation. It will be noted also that (40) gives the actual catch, since the concept of ' yield per recruit ' is no longer useful when recruitment is made to vary with density. Using this equation, with constant growth and mortality parameters determined for the North Sea haddock and the values of α and β producing the curves shown in Fig. IX.20, gives the corresponding yield-intensity curves of Fig. IX.21. Noteworthy features of these are that the magnitude of the absolute

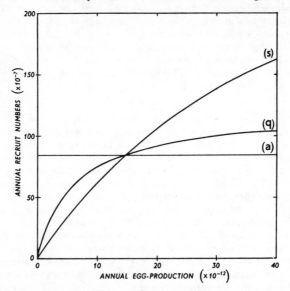

FIG. IX.20.—THEORY OF RELATION BETWEEN EGG-PRODUCTION AND RECRUITMENT.

This figure shows the three curves relating egg-production and recruitment that were used in constructing the two figures that follow. Parameters (see text) :

$$\text{Curve (s)} \quad \alpha = 0\cdot 29 \times 10^{-9}$$
$$\beta = 13{,}000$$
$$\text{Curve (q)} \quad \alpha = 0\cdot 83 \times 10^{-9}$$
$$\beta = 5{,}000$$

Curve (a) Recruitment constant.

yield varies enormously with F, although the value of F defining the position of the maximum is virtually unaffected. It is also seen that at a certain high fishing intensity, depending on the particular values of α and β, the population falls to zero. When depletion is too great insufficient survivors are left to maintain egg production and recruitment drops ; this results in an even smaller stock and eventually the process culminates in extinction, although such an end-point is most unlikely to be reached in a commercial fishery because fishing would have become unprofitable long before. Some authors have defined overfishing as excessive depletion of this kind, but we have seen that a stock can be overfished simply through

S.F.—F F

insufficient numbers being left to grow to a reasonable size, regardless of any relationship between egg production and recruitment.

Recruitment and growth together. Now the curves shown in Fig. IX.21 have been calculated from (40) in which all factors other than recruitment are assumed to be independent of density, but it is clear that the predicted changes in density can be so great

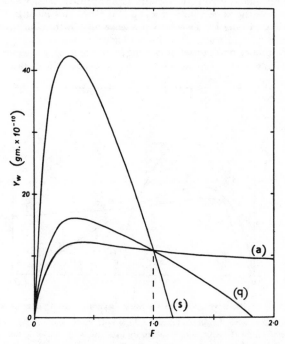

FIG. IX.21.—EFFECT OF RECRUITMENT VARYING WITH DENSITY.

Curves (s) and (q) show the effect on the relation between yield of haddock and fishing intensity of making recruitment increase with egg-production according to curves (s) and (q) of Fig. IX.20. In curve (a) recruitment is held constant for comparison. The large changes in yield, especially curve (s), are considerably reduced when the variation of growth with density is also introduced (Fig. IX.22).

that in practice some kind of compensatory mechanisms would come into play. The growth rate of the adult population would almost certainly be affected, and probably the natural mortality also. To obtain a more realistic assessment it is therefore important to examine the behaviour of a self-generating population model in which other parameters vary simultaneously in this way. Not enough is known for this purpose about natural mortality in haddock, but a relationship between growth and density has been detected (Raitt, 1939) and can be incorporated in a self-generating population model by methods similar to those described previously for plaice. The details of the model are complex, but in principle it amounts to expressing the parameter W_∞ in terms of a self-regenerating equation for population numbers P_N, the solution for W_∞ for a given value of F being obtained by an iterative procedure similar to that described above in connection with the variation of growth with density. Fig. IX.22 shows examples of yield-intensity curves calculated from a model of this kind based on haddock, with both recruitment and growth density dependent. Curves (q) and (s) incorporate the same values of α and β as those illustrated in Fig. IX.21 and it will be seen how introducing the density dependence of growth reduces the height of the maxima and also results in them occurring at higher values of F.

Density dependence and eumetric fishing. Before leaving the question of the density dependence of parameters we must look briefly at its effect on eumetric yield curves, as

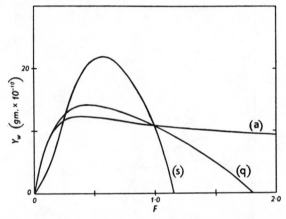

Fig. IX.22.—Effect of Recruitment and Growth changing simultaneously with Density.

Comparing this figure with the previous one, making both recruitment and growth density dependent does not materially alter the conclusions, except that the (unlikely) curve (s) here shows expectation of substantially less benefit from reduction of fishing. The same effect is shown in much less degree by curve (q).

opposed to simpler ones. Curve P(b) of Fig. IX.23 is a eumetric yield curve for plaice incorporating the same linear relationship between L_∞ and biomass as in curves (b) of Figs. IX.16 and IX.17. It is the same shape as curve P(a) calculated with constant growth parameters but is considerably lower; this might indeed have been anticipated from Fig. IX.17, where the maximum of curve (b) with growth density dependent is lower than that of curve (a). The reason why curve P(b) is below P(a) is that the growth parameters used to construct P(a) were estimated for the pre-war period when the stock was depleted by fishing and the growth rate correspondingly high. Eumetric fishing, however, allows a stock abundance in the order of four times the pre-war level, with a correspondingly lower growth rate, and hence a smaller yield than that predicted by keeping growth constant.

Eumetric yield curves in which both growth and recruitment are density dependent can be constructed using the methods described above for haddock. Taking, for example, the relationship between recruitment and egg-production defined by curve (q) of Fig. IX.20, the resulting eumetric yield curve is shown by H(q) of Fig. IX.23. The close similarity between this and curve H(a) calculated with constant parameters is at once apparent : what in fact has happened is that the effect of this particular degree of change of recruitment with egg-production in tending to increase yield has almost exactly counterbalanced the depressing

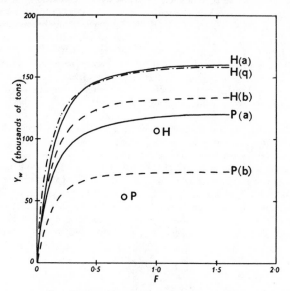

FIG. IX.23.—SOME EUMETRIC YIELD CURVES.

In " eumetric " fishing the mesh used would be of the best size for the intensity of fishing, that is, as F increases along the x-axis of this figure, the mesh is increased according to the maxima in a family of curves such as those shown in Fig. IX.11. Although fully eumetric fishing may be out of reach in practice, it represents, as it were, completion of the theory of fishing. It also provides a rational basis for adjustments within the limits of the practicable. All the curves show the possibility of much better fishing than that represented by the points (H) haddock and (P) plaice for pre-war intensities. (Curves : P, plaice ; H, haddock ; (a) constant parameters ; (b) density-dependent growth ; (q) density-dependent growth and recruitment.)

effect of the density dependence of growth alone [shown by curve H(b)]. This is, of course, a coincidence, and postulating a more pronounced change of recruitment with egg-production would result in a eumetric yield curve lying above H(a), and *vice versa*. What is important is that the general shape of eumetric yield curves predicted by theoretical models incorporating one or more density dependent mechanisms is little different from those calculated with constant parameters, thus confirming their value as a basis for establishing the principles of fishery regulation.

Spatial Variation in Density of Fish and Intensity of Fishing

The last problem to be discussed concerns the spatial variation of factors, that of the fishing intensity being the only one we mention here.   We have hitherto spoken of an exploited area and of a fishing mortality coefficient applying to the fish population living in it, but this is a simplification of the real situation.   A fish population is seldom, if ever, spread uniformly throughout its range.   The density of the North Sea plaice population, for example, varies greatly from one region to another, though some fish are always to be found everywhere within the normal limits of the range of the population ; with a highly migratory fish such as herring, on the other hand, the population may often be concentrated at any one time into a relatively small part of its range.   As would be expected, the fishing intensity varies accordingly, and tends to be greatest where fish are most dense.   As a consequence, the magnitude of the instantaneous fishing mortality coefficient itself varies spatially, so that the first question arising in these circumstances is how to define and compute from data a single index of fishing intensity and a fishing mortality coefficient, for the population and exploited area as a whole.

It will be clear that the total catch obtained by a fleet, and hence the total fishing mortality generated by it, depends on how the vessels are distributed relative to the concentrations of fish ; the more perfectly the highest concentrations of vessels coincide with the highest density of fish the greater will be the mortality generated.   What is needed is therefore a weighted index of fishing intensity that takes into account the relative distribution of fish and effort, and it has been found that this can be computed quite easily if the necessary statistics are available.   Let us suppose that the whole area is divided into Z sub-areas and that commercial statistics give the catch and effort in each.   The required index, which is called the effective overall fishing intensity $\bar{f}$, is then given by the expression :

$$\bar{f} = \frac{\sum_{i=1}^{Z} f_i (\bar{P}_N)_i}{\sum_{i=1}^{Z} (\bar{P}_N)_i} \qquad . \qquad . \qquad . \qquad . \qquad . \qquad (41)$$

where f_i and $(\bar{P}_N)_i$ denote the fishing intensity and the number-density in the ith sub-area. The overall intensity, $\bar{f}$, is therefore a weighted mean of the intensities in each sub-area, the weighting factors being the number-density in each.   The latter are obtained from the ratio of the catch to the fishing intensity in each sub-area, and for these estimates to be accurate it is of course necessary that the size of each sub-area should be as small as the accuracy of commercial statistics will permit.

Corresponding to $\bar{f}$ is the effective overall fishing mortality coefficient, $\tilde{F}$, the two being related by the constant c introduced earlier. $\tilde{F}$ is the coefficient that would be estimated from random samples of the catch and it could be used instead of F in all the population models developed in this chapter, but it will be seen that it can only be calculated from (41) by means of past data of catch and effort.   The more important use of population models is in the prediction of future events, but (41) can be used for this purpose only with considerable reserve, because if there is a change in the total effort—or its distribution—the densities in each sub-area will also change.

The solution to this problem lies in a knowledge of the laws of local movements and interchange of fish within the exploited area, a phase of the problem which is only just beginning to be studied. It was met first when discussing the analysis of marking experiments, and there the concept of a transport coefficient was introduced to define the rate at which fish disperse from a specified area. Now the theoretical basis for this approach lies in the assumption that there is a random element in the movement of fish, and that the pattern of dispersion obeys, fundamentally, the laws of physical diffusion that describe the conduction of heat or the diffusion of gases. Some evidence that this is so in plaice comes from a study of the migration of young fish from the nursery grounds on the Dutch, German and Danish coasts into deeper water, which we have mentioned as the cause of recruitment in this species. Data were provided by regular sampling by a research vessel on a line of stations—the Leman-Haaks line—extending from the coast nearly to the middle of the North Sea, which has been described in detail in Chapter VIII. If dispersal of young fish from the coastal region was a diffusion effect, it would be expected that the distribution of each year-class at right angles to the coast would be that of one-half of a normal curve. An indication that this is so is shown in Fig. IX.24, where data referring to age-group III over a number of years are plotted in this way, allowance having been made for local irregularities, and a normal curve fitted to them. This simple mechanism of uniform dispersal cannot, however, account for the uneven and patchy distribution of adult fish over the area as a whole, and the hypothesis of random movement has been extended to cover this by supposing that the speed of movement of fish varies from one part of the area to another. This treatment implies that fish tend to accumulate where their speed of movement is least, which might be where the abundance of food is greatest and hence where the distance travelled by fish in moving from one food patch to the next is small. The treatment of such a situation by means of rigorous diffusion theory would be most difficult, but by postulating a transport coefficient to define the rate of movement of fish out of each sub-area the problem becomes more amenable to analysis, though it is still complex. In this way a family of differential equations may be set up to represent the rate of movement of fish into and from each sub-area, the solutions of which define the number of fish in each sub-area in the steady state generated by any specified distribution of fishing effort. It is then possible to deduce the yield from each sub-area and hence that from the whole population.

Searching efficiency. Population models of this kind have an application to several problems of importance in the study of a fishery. For example, they can be used to take some account of the effects of ' cover ', that is, of areas of bottom too rough for trawling but which are nevertheless inhabited by fish and, similarly, of the value of closing part of the exploited area to fishing as a regulative measure ; in both these cases the assessment is made by putting to zero the fishing mortality coefficient in those sub-areas where no fishing occurs. But perhaps the most interesting of future applications of these models will be to the analysis of what may be called the searching efficiency of a fleet. We mentioned above that the catch obtained by a fleet of a given size depends partly on how the vessels are distributed in relation to the centres of aggregation of the fish population, and in practice vessels spend an appreciable amount of their time in searching for the highest concentrations of fish, especially in fisheries based on highly migratory species. Much of this searching is done by the ships individually, their skippers using past experience of the likely whereabouts of fish and contemporary catches as guides, but there is also an element of co-operation, with information on the magnitude and position of catches being exchanged between vessels by wireless.

The degree of co-operation varies greatly, but in some fleets it is developed to the extent of the majority of the fleet working as a group and spending no time searching, while the remainder act purely as fish-searchers. Now it is clear in a general sense that the greater the degree of co-operation the more effectively the fleet as a whole can distribute itself over the fish population and hence the greater the total catch it can obtain ; but it is possible to

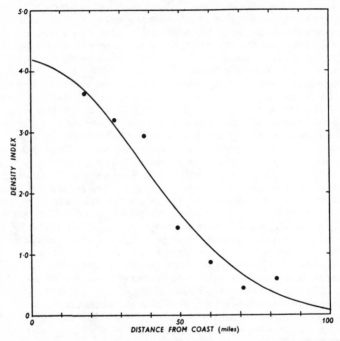

Fig. IX.24.—Random Dispersion of Young Plaice from Nursery Grounds.

Despite vagaries from year to year and locality to locality, the data summed over a number of years of observations on the Leman-Haaks Line (Fig. VIII.4) show that densities of, for example, age-group III, when plotted against distance from the coast, fit reasonably closely to one half of the Normal Curve— that is, to random dispersion.

go further than this, and define the particular distribution the fleet must take up in order to obtain the greatest catch that is possible in the circumstances. Thus for a given expenditure of effort, that is, of total time spent actively fishing by a fleet, the vessels should be distributed so that two criteria are satisfied ; (a) that all vessels are fishing at any moment on the same density of fish, i.e. that the catch per unit fishing time should be the same in all parts of the fished area, and (b) that no part of the unfished area should have a higher density than the fished area. This is called the *limiting distribution of effort*, but in practice it would not be quite the best distribution for a fleet to adopt because of the greater time that is spent in steaming to and from the more distant grounds. Thus by somewhat reducing the average distance from port at which each vessel worked—in fact, until the catch per trip of each vessel was the same, instead of the catch per unit fishing time—the saving in steaming time would more than offset the slightly less effective distribution of effort that would result. The tactics that a fleet should adopt in order to obtain the greatest catch, which may be called the *optimum tactics*, is therefore a compromise between the requirements for the limiting distribution of effort and the need to avoid expenditure of an excessive amount of time in steaming. Evaluation of the optimum tactics in a given case is clearly not a simple task, but neither are the difficulties insuperable. Basically, a knowledge of the rates of movement of fish in various parts of the exploited area is required, to be incorporated into a theoretical model of the kind mentioned above ; for a fleet of a given size and with a given system of communication it should then be possible to employ mathematical methods similar to those described by Morse and Kimball (1951) to determine the best tactics to adopt in a given set of circumstances, having in mind the criteria set out above. No analysis of this kind has yet been attempted for any fishery, and it may turn out that in some cases the present system of fishing is very nearly the best that could be adopted. Where this should not be found to be so, however, the potential economic benefits might be considerable, because it would mean that the efficiency of the fleet could be increased without additional expenditure ; in this case, however, the individual vessels would no longer be economically independent—the fishing unit would be the fleet instead of a single ship.

CONCLUSION

We have now taken our discussion of the theory of fishing to the limits of present research, and indeed have allowed ourselves to speculate beyond them in the last few pages. We began by pointing out that fishing was basically akin to hunting, in that fishing activity contains fundamental elements of both intra- and inter-specific competition, and also because a fish population is a natural prey which cannot be cultivated. These are the two main reasons why, in the past, uncontrolled fishing has almost inevitably led to excessive depletion of the stocks, bringing with it economic distress to the fishery industries. We then showed that, while still retaining the competitive element, the number of competitors (i.e. the amount of fishing) and the properties of the fishing gear could be adjusted to obtain the best results in terms of a balance between the magnitude of the yield and the economic conditions under which it is obtained. Finally, we have found that this ' best ' is limited, ultimately, by the competitive element itself, and that a truly rational exploitation of a fish resource requires a high degree of co-operation between fishing units and the adoption of a conscious and balanced fishing attack ; in other words, man's predation must become social rather than individual. When it is remembered that no major fishery—nor probably any but a very few minor ones—is fished at the present time in a way that satisfies either of these criteria, our findings are encouraging from every point of view. They are encouraging to the fishermen and the fishing industries—because of the brighter social and economic prospects which properly regulated fishing can offer ; to the consumer—because of the cheaper and more

CONCLUSION

abundant supply of high-grade protein food which is potentially available ; and to the fisheries research worker—because his assistance will be vital to bring these changes into effect. The lesson from it all is that in order to get the best possible results from the exploitation of these natural resources, man must be prepared to seek a full understanding not only of the characteristics of the fish populations as prey but of his own behaviour as predator.

R. J. H. B.
S. J. H.

References

Baerends,G.P., De rationale exploitatie van den zeevischstand, in het bijzonder van den vischstand van de Noordzee, *U.S. Fish and Wildlife Service, Special Scientific Rep., Fisheries,* **13**, 1947.

Baranov,T.I., The biological basis of fisheries, *U.S.S.R. Rep. Div. Fish. Management and Sci. Study of the Fishery Industry,* **1**, No. 1, 1918.

von Bertalanffy,L., Mathematische und physiologische Gesetzlichkeiten des Wachstums bei Wassertieren, *Archiv. f. Entwicklungsmech.,* **CXXXI**, 613, 1934.

von Bertalanffy,L., Inquiries on growth laws, *Human Biology,* **10**, no. 2, 181, 1938.

von Bertalanffy,L., Problems of organic growth, *Nature,* **CLXIII**, 156, 1949.

von Bertalanffy,L., The theory of open systems in physics and biology, *Science,* **III**, 23, 1950.

Beverton,R.J.H., and Holt,S.J., The dynamics of exploited fish populations, *Fish. Invest.,* ser. *II* (in the press).

Bidder,G.P., The mortality of plaice, *Nature,* **CXV**, 495, 1925.

Borley,J.O., Review of plaice marking experiments, *Fish. Invest.,* ser. *II*, no. 3, 1916.

Buchanan-Wollaston,H.J., On the selective action of a trawl net, *Cons. Int. J., II*, no. 3, 343, 1927.

Burkenroad,M.D., Fluctuation in abundance of Pacific halibut, *Bingham Ocean. Coll. Bull.,* **XI**, 81, 1948.

Burkenroad,M.D., Population dynamics in a regulated marine fishery, *Texas J. Sci.,* **2**, no. 3, 438, 1950.

Burkenroad,M.D., Some principles of marine fishery biology, *Publ. Inst. Mar. Sci.,* **2**, (1), 177, 1951.

Clarke,G.L., Dynamics of production in a marine area, *Ecol. Monogr.,* **XVI**, 322, 1946.

Dawes,B., Growth and maintenance in the plaice, *J.M.B.A.,* **XVII**, 103, 1930; 877, 1931.

Deming,W.E., Statistical adjustment of data, New York, 1948.

Ford,E., Herring Investigations, 1924-1933, *J.M.B.A.,* **XIX**, 303, 1933.

Gause,G.F., The struggle for existence, Baltimore, 1934.

Glaser,O., Growth, time and form, *Biol. Rev.,* **XIII**, no. 1, 20, 1938.

Goodeve, Sir Charles, Operational Research, *Nature,* **CLXI**, 377, 1948.

Graham,M., Modern theory of exploiting a fishery, *Cons. Int. J.,* **X**, 264, 1935.

Graham,M., The sigmoid curve, *Cons. Int. Rapp.,* **CX**, 15, 1939.

Graham,M., The fish gate, London, 1943.

Graham,M., Rational fishing of the cod in the North Sea, Buckland Lect., London, 1948.

Graham,M., Rates of mortality from marking experiments, *Cons. Int. J.,* **XIII**, 76, 1938 ii.

Herrington,W.C., Limiting factors for fish populations. Some theories and an example, *Bingham Ocean. Coll. Bull.,* **XI**, 229, 1948.

Hile,R., Age and growth of the cisco, *Bull. U.S. Bur. Fish.*, **XLVIII**, 211, 1936.

Hjort,J., Jahn,G., and Ottestad,P., The optimum catch, *Hvalråd Skr.*, no. 7, 92, 1933.

Holt,S.J., Review of Thompson, 1950, *Cons. Int. J.*, **XVII**, 320, 1951.

Jackson,C.H.N., The analysis of an animal population, *J. Anim. Ecol.*, **VIII**, no. 1, p. 238, 1939.

Jensen,A.J.C., The relation between the size of mesh and the length of fish released, *Cons. Int. Rapp.*, **CXXV**, 65, 1949.

Kesteven,G.L., Population studies in fisheries biology, *Nature*, **CLIX**, 10, 1947.

Kostitzin,V.A., Mathematical biology, London, 1939.

Le Cren,E.D., The length-weight relationship in the perch, *J. Anim. Ecol.*, **XX**, 201, 1951.

Lindemann,R.L., The trophic-dynamic aspect of ecology, *Ecology*, **XXIII**, no. 4, 399, 1942.

Lotka,A.J., The elements of physical biology, Baltimore, 1925.

MacFayden,A., The meaning of productivity in biological systems, *J. Anim. Ecol.*, **XVII**, no. 1, 75, 1948.

Morse,P.M., and Kimball,G.E., Methods of operations research, New York and London, 1953.

Nesbit,R.A., Biological and economic problems of fishery management, *U.S. Fish and Wildlife Service, Spec. Sci. Rep.*, no. 18, 23, 1943.

Pearl,R., The biology of population growth, New York, 1930.

Rafferty,J.A., Mathematical models in biological theory, *American Scientist*, **38**, 549, 1950.

Raitt,D.S., The fecundity of the haddock, *Sci. Invest. Fish. Scot.*, 1932, no. 1, 1933.

Raitt,D.S., The rate of mortality of the haddock in the North Sea stock. 1919-1938, *Cons. Int. Rapp.*, **CX**, no. 6, 65, 1939.

Reeve,E.C.R., and Huxley,J.S., Some problems in the study of allometric growth, *Essays on growth and form presented to D'Arcy Wentworth Thompson*, 121, Oxford, 1945.

Ricker,W.E., Further notes on fishing mortality and effort, *Copeia*, no. 1, 23, 1944.

Ricker,W.E., A method of estimating minimum size limits for obtaining maximum yield, *Copeia*, no. 2, 84, 1945.

Ricker,W.E., Production and utilization of fish populations, *Ecol. Monogr.*, **XVI**, 373, 1946.

Ricker,W.E., Methods of estimating vital statistics of fish populations, *Indiana Univ. Pub. Sci. Serv.*, **15**, 1948.

Robertson,T.B., The chemical basis of growth and senescence, Philadelphia, 1923.

Rounsefell,G.A., Methods of estimating total runs and escapements of salmon, *Biometrics*, **V**, no. 2, 115, 1949.

Russell,E.S., The overfishing problem, Cambridge, 1942,

Schaeffer,M.B., Estimation of size of animal populations by marking experiments, *U.S. Fish. Bull.*, **LII**, no. 69, 189, 1951.

Silliman,R.P., A method of computing mortalities and replacements of *Sardinops caerulea*, *U.S. Fish and Wildlife Service, Spec. Sci. Rep.*, no. 24, 1943.

Silliman,R.P., Mortality rates from length frequencies of the pilchard, *Copeia*, **4**, 1945.

Skellam,J.G., Random dispersal in theoretical populations, *Biometrika*, **XXXVIII**, 196, 1951.

Tansley,A.G., Succession, the concept and its values, *Proc. Int. Congr. Plant Sci., Ithaca*, 1926, **I**, 677, 1929.

Tester,A.L., The efficacy of catch limitations in regulating the British Columbia herring fishery, *Trans. Roy. Soc, Canada*, **XLII**, ser. 3, sect. 5, 135, 1948.

Thompson,W.F., The effect of fishing on the stocks of halibut in the Pacific, *Publ. Fish. Res. Inst. Univ. Washington*, 1950.

Thompson,W.F. and Herrington,W.C., Life History of the Pacific Halibut (1) Marking experiments, *Rep. Int. Fish. Commn*, **2**, 1930.

Some aspects of the dynamics of populations important to the management of the commercial marine fisheries

M.B.Schaefer

Bulletin, Inter-American Tropical Tuna Commission, **1,** 27 – 56, 1954. Pages 27 – 38 are reproduced here.

Editor's introductory notes

The first eleven pages of Schaefer (1954) are presented together with the full (1957) paper. The first paper states the development from the logistic equation and estimates an MSY from the curve of catch on stock. The second describes the use of a linear inverse dependence of catch per unit of effort on fishing effort to develop a curve of catch upon fishing effort. It was applied to the stock of yellowfin tuna in the Eastern Tropical Pacific with an estimated MSY and errors attached.

The great advantage of Schaefer's method is that only simple statistics are needed, catch, fishing effort (time spent fishing) and catch per unit of fishing effort. Given an adequate range of estimates an MSY can be determined and stocks managed with the Schaefer method are unlikely to suffer from recuitment over-fishing under the best conditions. The Schaefer model and its successors can be applied readily to populations which cannot be aged. A disadvantage of the model is that estimates can only be improved with the slow accretion of observations year by year. A more serious difficulty is that the catchability coefficient may vary inversely with abundance and this potential bias should be eliminated or the estimate of MSY will be biased.

The logistic model is the usual one employed by ecologists both in their general and theoretical studies. Schaefer developed its application to fisheries explicity and successfully.

Some aspects of the dynamics of populations important to the management of the commercial marine fisheries

M.B. Schaefer

A population of oceanic fish under exploitation by a fishery may be influenced by a great number of elements in the complex ecological system of which it forms a part. Of these, however, only one, predation by man, is capable of being controlled or modified to any significant degree by man's actions. Any management or control of the fishery, to the extent this may be possible at all, must, therefore, be effected through control of the activities of the fishermen. It seems important to elucidate some of the basic principles of the effect of fishing on a fish population and, conversely, the effect of the fish population on the amount of fishing, in order to understand in what circumstances and in what manner such control of the activities of the fishermen can influence the fish population and the yield obtained therefrom.

Management of a fishery has as its purpose the modification or limitation of the activities of the fishermen in order to realize a change in the fish population, or the catch, or both, which in some manner is preferable to that which would obtain if the fishermen were allowed to operate without these modifications or limitations. What may be "preferable" involves in the general case a great many economic and sociological matters difficult or impossible to treat objectively, and not susceptible to quantitative reasoning. We must, therefore, confine our attention to a less general case, but one most often met in practice, where the purpose of management is to obtain a larger average total catch per unit of time than would be obtained without management. An important special case of this is management directed toward obtaining the maximum average total catch per unit of time, which is often referred to, somewhat ambiguously, as the "optimum catch."

The Inter-American Tropical Tuna Commission has the task, specified by the Convention under which it is organized, of gathering and interpreting factual information to facilitate maintaining the populations of the tropical tunas and of the tuna-bait fishes at levels which will permit maximum sustained catches year after year. Information respecting these populations is not at the present time adequate for this purpose. An analysis of the fundamental relationships between population size, intensity of fishing, and catch is a valuable, if not indispensable, basis upon which to plan the efficient collection and interpretation of the information required to accomplish the purposes of the Commission.

The staff of the Commission has directed a large share of its attention since its inception to the collection and compilation of reliable data respecting the total catch and catch per unit of fishing effort of each tuna species over the period of growth and development of the fishery in the Eastern Tropical Pacific. This task is nearing completion. The next step in the investigation is to employ these data together with such ancillary vital statistics as may be required and may be obtainable, to the estimation of the level of maximum sustained yield of each tuna stock and the determination of the present condition of the fishery with relation thereto. This step requires the employment of a suitable mathematical model describing the effect of fishing on the tuna stocks. Models which have been applied in the past to other fisheries are not satisfactory for this purpose. It has, therefore, proven necessary to undertake the investigations reported in this paper directed toward the development of a suitable model, and of methods of its application to fisheries data, which can be applied to the data of the tuna fishery. These studies, although of a theoretical nature, are of the most direct practical importance to the objectives of the Commission, since they are fundamental to the interpretation of the catch data and related information being collected by the staff.

It is well known that in dealing with oceanic fisheries we have to do with very complex ecological systems and that, therefore, the effects of the amount of fishing on the size of the fish population and on the catch is difficult to estimate. Some recent and current controversies bear adequate witness to this. The very complexity of these systems tends, however, to divert attention from consideration of the fundamental laws of population growth which make it possible for a species to survive increases in predator populations, and, by the same token, make possible that extensive predation by man which is commercial fishing.

In this investigation it will be attempted to indicate the manner in which the fundamental laws of population growth operate in the case of a commercial fishery, and so, perhaps, clarify some of the important considerations basic to the management of the oceanic fisheries. These will be shown by means of mathematical models. Certain parts of these models or very similar ones have been employed in predator-prey investigations of other organisms (Gause 1934, Lotka 1925), and there have been limited attempts to apply somewhat similar techniques to the fisheries, as will be noted subsequently. There is rather good reason to believe that the models sufficiently describe reality to be useful in furthering our understanding.

In pursuing the investigation we wish to elucidate the dynamics of a population of oceanic fish not related to environmental variations, that is the dynamics of the "mean" population under average environmental conditions. We shall, therefore, consider the situation in which all the factors of the environment are constant except predation by man, i.e. the amount of fishing. In application, the effect of variation due to environmental changes is treated as a random variable, independent of size of population.

The law of population growth in populations which tend to stability

Populations of organisms living in a constant environment with a limited food supply may be of one of two kinds. In one kind of population, exemplified in particular by some insects, different stages of which are in competition with each other for the means of life, the number of adults fluctuates periodically and continuously (Nicholson 1949, 1950). In the other type, the population tends to stability, so that for a particular set of environmental conditions the population has at each size a definite potential rate of increase which is dependent only on the existing size of the population. A great many populations from the yeasts and protozoa to man have ben shown to be of this sort. Most, at least, of the populations of fishes are believed to be of this kind.*

The general law of population growth for such a population, P, may be expressed as

$$\frac{dP}{dt} = f(P) \quad\text{..................................... (1)}$$

where $f(P)$ is continuous, positive and single valued between $P = O$ and $P = L$, the maximum population which the living space and food supply can support, and zero at these limiting values of P. We will call $f(P)$ the *natural rate of increase.*

A particular function which has been shown to fit experimental data as well as data from populations in nature for a good many organisms is the Verhulst-Pearl logistic

$$\frac{dP}{dt} = k_1 P(L - P) \quad\text{..................................(2)}$$

$$\text{where } k_1 \text{ is a constant}$$

In this case, of course, $f(P)$ is a parabola with its axis along $P = L/2$. It is shown graphically in Figure 1. Integrating, we may obtain P as a function of t, which is a sigmoid curve with an upper asymptote at $P = L$ and an inflection point at the value of P for which $\frac{dP}{dt}$ is a maximum, i.e. at $P = L/2$.

This law has been employed to describe the growth of a considerable variety of organisms, for example yeasts (Gause 1934, p. 78, Pearl 1925, p. 9), protozoa (Gause 1934, p. 36, p. 93 et. seq.), fruitflies (Pearl 1925, p. 11), and humans (Pearl 1925).

Büchman (1938) has considered the general dynamics of commercial fish populations based on this relationship, as has Graham (1939). Graham (1935) employed this growth law in an analysis of the effect of World War I on the abundance and landings of demersal fishes from the North Sea, and Baerends (1947) has made similar analyses.

It is possible that for fish populations the special case of (1) represented by the logistic (2) is not, in general, an exact representation of the

---

*One notable exception may be some species of Pacific Salmon, which tend to periodic fluctuations characteristic of the species. These may be found to be the result of the direct or indirect competition of different year classes for the means of subsistence. This has been little investigated, however.

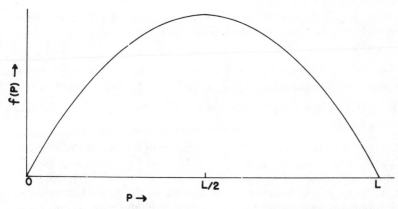

Figure 1. Natural rate of increase of a population which grows according to the Verhulst-Pearl logistic.

population growth law. In particular, the relationship (2) is a parabola, symmetrical with respect to its axis, from which it follows that the maximum natural rate of increase occurs at a value of P half way between zero and the maximum population value, L. There is reason to believe that in at least some populations of fishes, the curve is actually somewhat asymmetrical, with the maximum value of $\frac{dP}{dt}$ at a value of P less than $L/2$. Experimental data have also shown this is sometimes the case for other organisms, for example the yeast data of Gause (1934, p. 68).

Effects of Fishing

A fishery, that is removal of fish from the stock by man, has the effect of subtracting from that increase in stock which would occur at the existing level of population if no fishing were taking place. In other words, the rate of change in the stock will be less than the natural rate of increase by an amount equal to the rate of catching of the fish. That is

$$\frac{dP}{dt} = f(P) - P\,\phi(F) \quad\text{............................}(3)$$

where $P\,\phi(F)$ is the *rate of catching,* depending on the size of the fish stock and some positive single valued function of the number of units of fishing effort, F.

It is obvious from (3) that whenever the rate of catching is less than the natural rate of increase, the population will increase. Conversely, when the rate of catching exceeds the natural rate of increase, the population shrinks in size. When the rate of catching is exactly equal to the natural rate of increase, $\frac{dP}{dt} = O$, the population remains unchanged, and the fishery is said to be in equilibrium for that level of population and fishing effort. The annual catch made under such a condition of equilibrium has been called the stabilized catch, the equilibrium yield, and other things. We shall call it the *equilibrium catch.*

From the equation (3) and the general form of $f(P)$ certain conclusions of importance in fishery research and management may be immediately perceived:

(1) As the fishery increases in intensity (as F increases), the stock P, decreases. This decrease in population is a necessary consequence of increasing fishing intensity, and, thus, is an inevitable result of the development of a fishery.

(2) The stock, and the corresponding equilibrium catch, can be held constant, by regulating the amount of fishing, at any value less than $P = L$. Stability or instability of the population and catch over a given period of time has, therefore, no necessary relationship to the level of abundance, but merely reflects whether the rate of catching is changing or is constant.

Catch per unit of effort

Let it be assumed that the fishery operates on the stock in such a manner that one unit of fishing effort produces the same relative effect on the stock, that is it catches the same percentage of the stock, regardless of the time or place it is applied. Then

$$\phi(F) = k_2 F \qquad \text{where } k_2 \text{ is a constant}$$
$$\text{and } P \phi(F) = k_2 P F \ \dots\dots\dots\dots\dots\dots\dots\dots\dots\dots\dots\dots(4)$$

Under these circumstances, the rate of catching per unit of fishing effort is

$$\frac{k_2 P F}{F} = k_2 P \ \dots\dots\dots\dots\dots\dots\dots\dots\dots\dots(5)$$

and is, thus, proportional to the stock. The average catch per unit of effort during a given period of time will be proportional to the average size of the fish stock encountered by the fishery during the period. The average catch per unit of effort per year, or some other short time period, has been extensively used by fishery scientists to measure changes in the size of fish populations.

Maximum equilibrium catch

As has been shown above, when the rate of catching is just equal to the natural rate of increase, the stock will remain unchanged and the catch obtained will, of course, be stabilized also. The size or sizes of stock at which the equilibrium catch may be maximized are levels of maximum equilibrium catch. In general, it is supposed that a fish population has a growth law at least similar to (2) in that there is but a single maximum. In this case, there is but a single size of population at which the equilibrium catch may be maximized. This size of population has been referred to as the optimum stock and the corresponding rate of catching as the optimum catch. I prefer the expression *maximum equilibrium catch* as being more descriptive of exactly what is meant.

Determination of the status of the fish population and estimation of equilibrium yields

In the practical consideration of management of a fishery we are interested in finding out whether the fish population has been driven below the point at which maximum equilibrium catch may be obtained. If so, curtailment of the intensity of fishing will result in increased average catches. It is also of interest to estimate, if possible, the maximum equilibrium catch and the size of population at which it may be obtained.

The investigation of these matters involves, essentially, estimating the equilibrium catch at various levels of population. This may be accomplished by application of equations (3) and (4) where the assumptions underlying these equations are sufficiently nearly realized.
From (3) and (4) we have

$$\frac{dP}{dt} = f(P) - k_2 PF \dotfill (6)$$

Integrating over the year, we obtain

$$\int_{P_0}^{P_1} dP = \int_0^1 f(P)dt - \int_0^1 k_2 PF dt \dotfill (7)$$

where $P = P_0$ at $t = t_0$
and $P = P_1$ at $t = t_1$
from which

$$P_1 - P_0 = \triangle P = \overline{f(P)} - k_2 F_t \overline{P} \dotfill (8)$$

Where $\overline{f(P)}$ is the annual natural rate of increase and, hence, the annual equilibrium catch corresponding to the mean stock $\overline{P}$ encountered by the fishery during the year*. F_t is the total fishing intensity for the year, $F_t = \int_0^1 F dt$. $k_2 F_t \overline{P}$ is, of course, the total catch during the year.

The average catch per unit of effort is

$$U = \frac{k_2 F_t \overline{P}}{F_t} = k_2 \overline{P} \dotfill (9)$$

If we have adequate statistical records of the fishery we know the amount of effort, the catch, and the catch per unit of effort year-by-year. If we can evaluate k_2 in (9) we shall be able to compute $\overline{P}$ for each year from the catch statistical data. From values of $\overline{P}$ we may estimate P_1 and P_0

---

*$\overline{P}$ is the average of P taken with respect to the units of effort applied during the year.

That is, $\overline{P} = \dfrac{\displaystyle\int_0^1 PF dt}{\displaystyle\int_0^1 F dt}$

approximately by interpolating between values of $\overline{P}$ for successive years. Given $P_1 - P_0$ and the catch, we can estimate $\overline{f(P)}$, the annual equilibrium catch corresponding to $\overline{P}$ during each year of the series.

One estimate of k_2 is provided by data from tagging experiment, since $F_t k_2$ is simply the instantaneous rate of fishing mortality, that is $f = 1 - e^{-k_2 F_t}$, where f is the annual fishing mortality rate, which may be determined from the recovery rates of marked fish. Other means also exist, of course, for estimating k_2.

An application to the Halibut fishery of the North Pacific

The manner in which this procedure may be applied can be illustrated by the example of the fishery for Pacific Halibut, using for our example the population of Area 2 (the region south of Cape Spencer). Statistics of catch and catch per unit are given by Thompson and Bell (1934) and by Thompson (1950). Revised, and presumably more accurate, values have been furnished recently by Bell to Dr. R. VanCleve (MS) from which I have taken the values employed here; see the first three columns of Table 1.

Tagging experiments (Thompson and Harrington, 1930) conducted in Area 2 indicated an annual fishing mortality rate of approximately 40% in 1926. Subsequently Thompson and Bell (1934) found that 47% was perhaps more realistic. Using 47% as the annual fishing mortality rate in 1926, we have

$$e^{-F_t k_2} = 0.53, \text{ and } F_t = 494{,}078 \text{ skates (Thompson 1950, table 2)}$$

Then $F_t k_2 = 0.635$

and, $1/k_2 = \dfrac{494}{635} \times 10^6 = 778 \times 10^{3*}$

Multiplying the values of U for each year by $1/k_2$ we obtain estimates of $\overline{P}$ (Table 1, column 4). Interpolating between successive values of $\overline{P}$, we obtain estimates of the stock at the beginning of each year (column 5). Differences of the values for successive years indicate the increase or decrease of the stock which resulted from the catch taken during the year ($\triangle P$, column 6). In accordance with (8) we add $\triangle P$ to the annual catch to obtain $\overline{f(P)}$, the annual equilibrium catch corresponding to $\overline{P}$ (column 7).

We now have estimates of the stock and equilibrium catch obtainable from that stock for the series of years from 1916 to 1946. Plotting $\overline{f(P)}$

---

*By considering changes in catch per unit of effort and total catch over the period 1926 to 1933, during which period the stock fell and then returned again to the original level, Thompson (1950) arrived at a value for $1/k_2$ of 335×10^3. This corresponds to a fishing mortality rate of about 77% in 1926, which seems unreasonably high from the tagging results, age composition data, and other information respecting this fishery. An indication of why his analysis gives this result will be given later (p. 37).

against $\overline{P}$ we would expect the points to fall on a curve (a parabola, if $f(P)$ is the logistic) in the absence of other influences. Actually, due to unknown effects of variable environmental factors, measurement errors, and other unaccounted-for sources of variation, the points will tend to scatter about an average curve. By observing the trend of the plot of $\overline{f(P)}$ against $\overline{P}$, we may ascertain, however, how the equilibrium catch for this population varies, on the average, with the size of the population. This has been done in Figure 2, where the small, solid points represent the annual values from Table 1. The centers of the crosses represent the mean values calculated for each 10 units of U.

It is quite obvious that the equilibrium catch increases, on the average, up to a catch per unit of effort of about 80 pounds per skate, at least, corresponding to a mean population of some 62,000,000 pounds. Data beyond this population level are not available (the single point for 1916 at 114 lbs. per skate is not deemed adequate for extending the relationship). Certainly it appears that, contrary to the contention of Burkenroad (1951, 1953), this halibut population was driven below its point of maximum equilibrium catch, and the curtailment of fishing had a beneficial effect on the subsequent catches.

It is unfortunate that reliable data are not available for earlier years when the population was, presumably, larger, which would enable us to estimate equilibrium catches for higher population values and so find out where the maximum occurs. It appears that it might be desirable, if possible, in order to find this out, to curtail fishing to permit higher levels of population to be reached.

This example points out clearly the desirability of obtaining adequate statistical data on a fishery during its early stages so that the maximum equilibrium catch may be estimated, approaching it from those population levels which are too high to give the maximum equilibrium catch. It is practically difficult, once the maximum has been passed, to drive the stock back up past the point of maximum return for purposes of investigation, since the immediate economic welfare of the industry must always be considered in practical regulations.

In the analysis of the halibut data thus far, we have not specified the form of $f(P)$ beyond the general restrictions on (1). As a matter of illustrating methodology, it is of interest to see what results are obtained if we specify that the curve be the logistic (page 29), so that

$$\overline{f(P)} = k_1 \, \overline{P} \, (L - \overline{P}) \text{ or,}$$

since $\overline{P} = \dfrac{U}{k_2}$,

$$\overline{f(P)} = \frac{k_1}{k_2^2} U \, (L_u - U),$$

where $L_u = k_2 L$

Fitting a curve of this form to the mean values (crosses) of Figure 2, (with

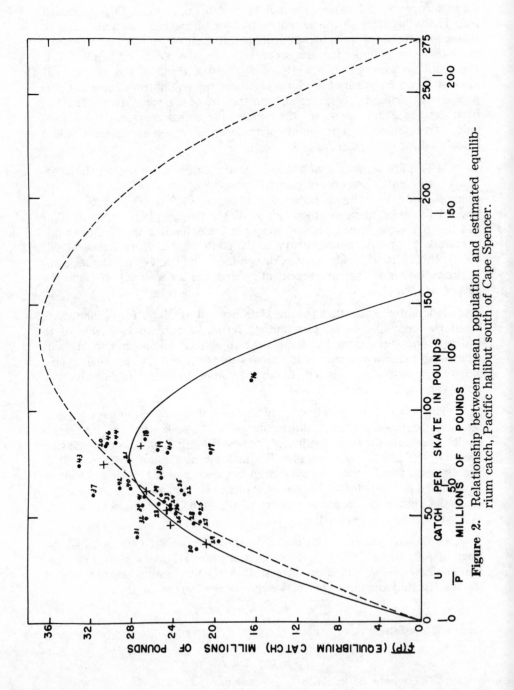

Figure 2. Relationship between mean population and estimated equilibrium catch, Pacific halibut south of Cape Spencer.

TABLE 1. Estimation of Equilibrium Catches for the Population of Pacific Halibut of the Region South of Cape Spencer (I. F. C. Area 2)

Year	Catch in 1000's of pounds $(k_2 F_t \overline{P})$	Catch per unit of effort in pounds per skate $(U = k_2 \overline{P})$	$\overline{P}$	P_0	$\triangle P$	$\overline{f(P)}$
1915	44,023	117.5	91,415			
1916	30,278	114.1	88,770	90,092	−14,082	16,196
1917	30,803	81.3	63,251	76,010	−10,542	20,261
1918	26,270	87.0	67,686	65,468	+ 195	26,465
1919	26,602	81.8	63,640	65,663	− 1,323	25,279
1920	32,358	83.6	65,041	64,340	− 2,100	30,258
1921	36,572	76.4	59,439	62,240	− 8,364	28,208
1922	30,482	62.1	48,314	53,876	− 7,663	22,819
1923	28,008	56.7	44,113	46,213	− 2,645	25,363
1924	26,155	55.3	43,023	43,568	− 2,101	24,054
1925	22,637	51.3	39,911	41,467	− 1,360	21,277
1926	24,711	51.7	40,223	40,107	− 974	23,737
1927	22,934	48.9	38,044	39,133	− 1,722	21,212
1928	25,416	47.3	36,799	37,411	− 3,530	21,886
1929	24,565	39.8	30,964	33,881	− 4,900	19,665
1930	21,387	34.7	26,997	28,981	+ 272	21,659
1931	21,627	40.5	31,509	29,253	+ 5,718	27,345
1932	21,988	49.4	38,433	34,971	+ 4,279	26,267
1933	22,530	51.5	40,067	39,250	+ 2,217	24,747
1934	22,638	55.1	42,868	41,467	+ 4,240	26,878
1935	22,817	62.4	48,547	45,707	− 311	22,506
1936	24,911	54.3	41,245	45,396	− 778	24,133
1937	26,024	60.4	46,991	44,618	+ 5,640	31,664
1938	24,975	68.8	53,526	50,258	+ 39	25,014
1939	27,354	60.5	47,069	50,297	− 2,372	24,982
1940	27,615	62.7	48,781	47,925	+ 233	27,848
1941	26,007	61.1	47,536	48,158	+ 622	26,629
1942	24,321	64.3	50,025	48,780	+ 4,707	29,028
1943	25,311	73.2	56,950	53,487	+ 7,819	33,130
1944	26,517	84.4	65,633	61,306	+ 2,840	29,357
1945	24,378	80.5	62,629	64,146	+ 39	24,417
1946	29,678	84.5	65,741	64,185	+ 2,100	31,778
1947	28,652	85.9	66,830	66,285		

$\overline{P}$ and $\overline{f(P)}$ in thousands of pounds, U in pounds per skate) under the criterion of least squares, we obtain

$$\frac{k_1}{k_2{}^2} = 4.64 \qquad\qquad L_u = 156.1$$

This curve is plotted as the solid line in Figure 2*. It may be seen that it has a maximum value of 28.25 million pounds for the equilibrium catch at $k_2 \overline{P} = 78.05$ pounds per skate.

This curve depends, of course, only on the points to which it is fitted, and may be rather different beyond those points from the curve which would be obtained if we had some values of $\overline{f(P)}$ for higher population levels. The calculated maximum population, corresponding to 156.1 pounds

*See footnote, Page 37.

per skate, is much less than is shown by the available data of catch per skate for the early years of the fishery. From the few data available it is indicated by the International Fisheries Commission (Thompson and Bell 1934, table 1) that in the early 1900's the catch per skate was as high as 270 or 280 pounds. This is not, however, necessarily inconsistent with our results, since in the early days of the fishery the vessels may have been operating on local concentrations of halibut more abundant than the average for the entire area fished in later years. Thompson (1950, p. 2) states of the records on which these values are based: "It is my opinion, from personal experience, that such records showed a higher catch per set than the present comprehensive methods of collecting would have shown."

On the other hand, if we assume that the data from the 1900's are representative of the population in an almost unfished condition, so that the maximum population which the area will support corresponds to a catch per skate of, say, 275 pounds, we may fit a logistic to the available points, as before, but with the further restriction that $L_u = 275$. This results in a value of $k_1/k_2^2 = 1.95$. This curve is plotted as a broken line in Figure 2. It will be seen that now the estimated maximum equilibrium catch is 36.9 million pounds at a population corresponding to 137.5 pounds per skate. This result is not entirely unreasonable in the light of the total catches of 50 to 60 million pounds per year which were actually obtained by the fishery at its peak of production. (Thompson and Bell, table 1).

It is, it seems, not possible from the data to estimate precisely the population level giving the maximum equilibrium yield. We can, however, state with some certainty that it is at least as high as about 62 million pounds, corresponding to a catch per skate in the neighborhood of 80 pounds, and that at lower values the stock is overfished. This limited conclusion is, however, of very great interest in view of current controversy over the effect of regulation on the halibut stocks.

The nature of the growth of the amount of fishing

The intensity of fishing also may be expected to increase or decrease according to some regular law in response to economic factors. In general, as in any business, new investment of capital and effort will be attracted

---

*It may now be indicated why Thompson's method of determining $1/k_2$ gives a value higher than that from the tagging data. He assumed that the equilibrium catch for the years 1926 to 1933 was a constant. Actually the equilibrium catch was not constant over this period . The deviations of actual catches for this series of years from the equilibrium catches estimated from the logistic with the constants indicated are, on the average, greater than the deviations from the average of $\overline{f(P)}$ over the same period of years. As may be observed from Thompson's formulae on p. 20 of his paper, this will result in a higher value of his "K", which is the same as our $1/k_2$.

to come into the fishery as long as the expected return is equal to or greater than that from alternative enterprises in which the investment might be made. Put in another form, we may state this according to the theory of the "marginal" factor, according to which the cost of the last unit of fishing effort applied will, in general, be equal to the return from that unit.

Under the economic system in effect in most parts of the world, in which the above type of law holds true, as the fishery proves profitable, vessels and fishermen are attracted to it, increasing the rate of catching. This, of course, results in a decrease in the population of fish, lowering the return to each unit of fishing effort, and making the fishery less attractive to new investment. Ultimately, as the fishery grows, that level of fish population will be reached at which the return per unit of effort is so low that the cost of the next unit will be greater than the return from it. If the population falls below this level, vessels will tend to leave the fishery. This may be formulated

$$\frac{dF}{dt} = \psi(F, P - b) \dots\dots\dots\dots(10)$$

where ψ is positive when $P > b$ and negative when $P < b$, F being, as before, the number of units of fishing effort, and b the critical level of fish population at which further investment in fishing becomes unprofitable.

To arrive at a particular function to describe the change of the intensity of fishing with the size of the population, we may consider that the incentive for new investment is proportional to the return to be expected, in which case there will be a linear relation between the percentage rate of change of fishing intensity and the difference between the level of fish population and its economically critical level, b. This function will, then, be

$$\frac{dF}{dt} = k_3 F(P - b) \dots\dots\dots\dots(11)$$

where k_3 is a constant.

It may be noted that this is the law of growth of predator populations which has beeen arrived at in various predator-prey studies, for example Lotka (1925, p. 88), Volterra and d'Ancona (1935).

A study of the dynamics of fishery for yellowfin tuna in the Eastern Tropical Pacific Ocean

M.B.Schaefer

Bulletin, Inter-American Tropical Tuna Commission, **2**, 247-285, 1957.

A STUDY OF THE DYNAMICS OF THE FISHERY FOR YELLOWFIN TUNA IN THE EASTERN TROPICAL PACIFIC OCEAN

by

Milner B. Schaefer

INTRODUCTION

The central problem of the staff of the Inter-American Tropical Tuna Commission is the determination of the effects of fishing on the populations of tunas in the Eastern Tropical Pacific Ocean, in order to estimate the relationship between fishing intensity and the sustainable yield of each tuna species at different levels of fishing intensity. On this basis we may estimate the maximum sustainable yield, and the fishing intensity corresponding thereto, and also determine the current condition of the fishery with respect to the condition of maximum harvest.

One method of arriving at such estimates is the analysis of statistics of fishing intensity and resulting catch over a series of years encompassing different levels of fishing intensity. Data of this sort have been compiled for both yellowfin tuna and skipjack for the period 1934 through 1954, and have recently been published by Shimada and Schaefer (1956). It was concluded by these authors that the intensity of fishing during this series of years had been sufficiently great to have a pronounced effect on the abundance of yellowfin tuna, and that during the years of most intense fishing the level of maximum sustainable catch had been approached.

In order to provide more exact estimates of the relation of fishing effort to population abundance and catch, the data of Shimada and Schaefer, plus the corresponding data for the year 1955, will be examined here in the light of some theory of population dynamics, the basic principles of which have been previously discussed by Schaefer (1954). Some extension of the theory is also involved.

ACKNOWLEDGMENTS

The basic data for this study were obtained from the detailed records of fishing operations and results provided through the cooperation of vessel masters and other members of the tuna fishing industry. The indispensable cooperation of these men is again gratefully acknowledged.

BASIC DATA

In this study we shall consider as a single, independent biological unit the yellowfin tuna supporting the fishery off the west coast of the Americas, extending from Baja California to northern Peru. That these fish are distinct from the populations further to the westward in the Central Pacific and Southeast Polynesia is evidenced by morphometric data (Godsil and Greenhood 1951, Royce 1953, Schaefer 1952, 1955), by tagging results (Shimada and Schaefer 1956, and unpublished data) and by the changes of the population in response to changes in fishing effort (Shimada and Schaefer 1956, Broadhead 1957). Whether the population *within* the region of the American fishery is homogenous is, however, not yet demonstrated. The ensuing analysis is, to this extent, provisional.

Data on total catch are collected for the entire region of the fishery. Abundance is measured by the catch-per-unit-of-fishing-effort of a very large sample of the fleet of tuna clippers, which fish by the live-bait method throughout the whole range, the unit of measurement being the catch-per-day's-fishing of a standard size-category of vessel, having a fish-carrying capacity of 200 to 300 tons of frozen tuna. Total fishing intensity, in terms of equivalents of fishing days of a standard vessel, is calculated for the entire fishery by dividing the catch-per-standard-day's-fishing into the total catch by all vessels of all types. The details of the collection, compilation, and standardization of the data have been given by Shimada and Schaefer (1956) and need not be repeated here. In Table 1 the data for the years 1934 through 1954 have been recapitulated from Table 19 of that publication. The data for 1955, which have since been compiled in exactly the same manner, have been added to the table.

MATHEMATICAL MODEL

Description of the model

The mathematical model employed is essentially the same as that discussed by Schaefer (1954), with some modifications in notation. The theory is, however, also extended in application to provide estimates of all the essential constants from the catch data alone, without recourse to tagging data for estimating fishing mortality which was required in the earlier paper.

The basic concept is that to every value of magnitude of population of tuna of commercial sizes there corresponds, on the average, a certain ability of the population to increase in weight *(the rate of natural increase)*, for all population magnitudes between zero and the maximum population which this sea area will support under average environmental conditions. The rate of natural increase is, of course, zero for zero population, and is also zero, on the average, when the limiting population size has been reached. Now if, at any level of population, the fishery takes exactly the natural increase, there will be no change in the population. If the fishery takes more than the natural increase, the population will diminish by the amount the natural increase, the population will be augmented. The annual catch of the excess catch. Similarly, if the fishery takes less than the amount of which corresponds to the annual rate of natural increase has been termed the equilibrium catch, because the harvest of the fishery is in equilibrium with the growth potential of the population. This may be formulated as follows:

$$\Delta P = C_e - C \qquad \qquad (1)$$

where

ΔP is the change in total weight of the population of commercial sizes of fish during a year

C_e is the equilibrium catch (equivalent to the annual rate of natural increase)

C is the catch during the year

For the general relationship between population and equilibrium catch (or annual rate of natural increase) we have

$$C_e = k_i \bar{P} \, f(\bar{P}) + \eta \bar{P} \qquad \qquad (2)$$

where,

$\bar{P}$ is the mean population during the year

k_i is a constant

η is a random variable

$f(\bar{P})$ is some single valued function of $\bar{P}$ which is zero at the limiting population magnitude and increases with diminishing population. This is a "negative feedback" term which is necessary to describe the self-regulating property by which the rate of increase is regulated appropriately to the size of a restricted environment (Hutchinson, 1954).

Now the simplest assumption we can make about the form of $f(\bar{P})$ (Moran 1954) is that it is linear with $\bar{P}$, i.e.

$$f(\bar{P}) = L - \bar{P} \qquad \qquad (3)$$

where,

L is the limiting population magnitude

So that

$$C_e = k_i \bar{P} \, (L - \bar{P}) + \eta \bar{P} \qquad \qquad (4)$$

It will be shown subsequently that this assumption appears to correspond adequately to the data for the yellowfin tuna over the range of population sizes so far observed. The simple linear relationships for the "feedback" term may not prove adequate, of course, over a wider range of population sizes, but there seems no reason to employ a more complex model than is required by the available data.

We now make the usual assumption that the catch-per-unit-of-effort is proportional to the magnitude of the population, i. e.:

$$U = k_2 P \qquad . \qquad . \qquad . \qquad . \qquad . \qquad . \qquad . \qquad . \qquad (5)$$

where U is the catch-per-unit-of-effort

k_2 is a constant (the instantaneous rate of fishing mortality per unit of fishing effort)

From this it immediately follows that

$$\bar{U} = k_2 \bar{P} \qquad . \qquad . \qquad . \qquad . \qquad . \qquad . \qquad . \qquad . \qquad (6)$$

and

$$\Delta U = k_2 \Delta P \qquad . \qquad . \qquad . \qquad . \qquad . \qquad . \qquad . \qquad . \qquad (7)$$

where $\bar{U}$ is the mean catch-per-unit-of-effort during a year

ΔU is change in catch-per-unit-of-effort during a year

Substituting in (1) and (4) and accumulating some constants for simplicity of notation, we obtain

$$C_e = C + 1/k_2 \, \Delta U \qquad . \qquad . \qquad . \qquad . \qquad . \qquad . \qquad (8)$$

and

$$C_e = a\bar{U} \, (M - \bar{U}) + \epsilon \bar{U} \qquad . \qquad . \qquad . \qquad . \qquad . \qquad (9)$$

where a and M are constants and ϵ is a random variable, leading to

$$1/k_2 \frac{\Delta U}{\bar{U}} = a \, (M - \bar{U}) - \frac{C}{\bar{U}} + \epsilon \qquad . \qquad . \qquad . \qquad (10)$$

When $\Delta U = 0$ and $\epsilon = 0$ we have

$$\frac{C_e}{\bar{U}} = a(M - \bar{U}) \qquad . \qquad . \qquad . \qquad . \qquad . \qquad . \qquad (11)$$

which has been termed the "line of equilibrium conditions", by Schaefer (1954). This describes the *average* relationship between fishing effort ($F = C/\bar{U}$) and population abundance ($\bar{U}$) when the annual catch is in equilibrium with the annual rate of natural increase.

ΔP may be estimated for a given year, i, as was done by Schaefer (1954), by

$$\Delta P_i = \frac{\bar{P}_{i+1} - \bar{P}_{i-1}}{2}$$

Correspondingly,

$$\Delta U_i = \frac{\bar{U}_{i+1} - \bar{U}_{i-1}}{2} \qquad . \qquad . \qquad . \qquad . \qquad . \qquad . \qquad (12)$$

Equation (10), with ΔU estimated by (12), expresses a theoretical relationship among population abundance, measured by the average catch per day's fishing $(\bar{U})$, the total effort $(C/\bar{U} = F)$, and the change in population abundance (ΔU) during a year. To this model will be applied the data of Table 1 to provide estimates of the parameters for the yellowfin tuna fishery in the Eastern Pacific.

Some implicit assumptions

The simple law relating rate of natural increase of population to population size, given by (2), implies certain assumptions about fish populations which are to some degree unrealistic and which, if the degree of departure from reality be sufficiently great, limit its usefulness. Although it is believed that in the case of the fishery for the yellowfin tuna this is not a serious source of error, as will be brought out below, it is of value to consider these implicit assumptions in order to indicate some of the limitations on the general applicability of the model, and the need for some caution in its employment. Some of these matters have also been discussed by Gulland (1955) and Watt (1956).

There are two implicit assumptions about the fish population itself:

1)   That the rate of natural increase responds immediately to changes in population density. That is, delayed effects of changes in population density on rate of natural increase, such as the effects of the time lag between spawning and recruitment of resulting progeny into the catchable stock, are ignored.

2)   That the rate of natural increase at a given weight of population is independent of the age composition of the population.

Neither of these conditions is exactly fulfilled by populations of fishes, or other multicellular organisms. Certainly the effects of intraspecies competition on individuals between the egg and the age of entry into the catchable stock involve some time lag. There also may be some delayed effects of population density on mortality and growth among individuals of catchable sizes. With respect to the second assumption, it is known that the factors of fecundity, growth, and mortality are all to some degree age-specific as well as density-dependent, so that the age structure of the population as well as the total bio-mass enter into the potential for population growth.

It should be noted that if equation (2) is applied to data for steady states only, these assumptions are not necessary. If data were used for only those periods when the fishing effort and population were in steady-state equilibrium, the age structure of the population would be uniquely determined by its biomass. Also, since in the steady state the population during any year is, on the average, of the same size and age structure as during any subsequent year, the effect of time-lag on recruitment, et cetera, is eliminated. Actual fisheries are, however, seldom in steady states, so we must deal with transient states, in which these effects appear.

It is important in this connection to note that the changes in fishing effort in commercial fisheries are often gradual, so that the displacement from steady state is not large. In the case of the yellowfin tuna, it may be seen from Table 1 that there have been fairly large changes in fishing effort and population abundance over a few years, especially at the beginning of World War II, immediately after the war, and again in the last years of the series. This fishery, however, depends on only a few age classes, of which the youngest is believed to be one year old. Consequently, the renewal of the population is very rapid in comparison with many fisheries of high latitudes which exploit a larger number of age groups, and where the age of entry into the fishery is several years. Therefore, even in the presence of moderately large changes in fishing effort, the displacement from the steady state is not great for the yellowfin tuna of the Eastern Pacific.

The facts that the yellowfin tuna enter the fishery at only one year of age, reach sexual maturity very early (some are probably mature at one year and most at two years of age), and have a very short average life span, imply that the time-lag between spawning and recruitment is small. Likewise, effects of changes in age composition, due to changes in fishing effort, will be of lesser importance than for species of greater life span.

The rate of natural increase of a fish population is determined not only by the magnitude and age structure of that population, but also by many environmental factors. Variation in these factors will cause departures from the rate of increase which would occur under average environmental conditions. In applying the model to the data from the fishery, we assume that the effects of variable environmental factors on recruitment, survival and growth are random, or at least that they are not correlated with population changes due to changes in fishing effort, so that they may be averaged out. In other words, we assume that the mathematical expectation of ϵ in equation (10), (which includes the effects of environmental variations as well as various types of measurement errors) is zero.

Variations in the fishes' environment may also affect their behavior so as to make them more or less accessible to capture. It is, again, implicitly assumed in the application that effects of environmental factors on catchability of the population are random, or at least uncorrelated with changes in fishing effort, so that the mathematical expectation of k_2 remains the same for different levels of fishing effort.

It may happen, of course, that the important factors of the environment may change in a cyclic fashion, that the cycle is long with respect to the series of data at hand, and that the fishing effort is, fortuitously or otherwise, correlated with the environmental changes. In this event we should be led erroneously to ascribing changes in the population to changes in fishing effort when such is not, in fact, true. In the case of the yellowfin tuna, while this is not impossible, it seems unlikely in view of the fact that we have had sizable changes in fishing effort in the 1940's, as a result of the effects of the war, and again in 1954 and 1955 as a result of competition with foreign imports. It is improbable that the associated changes in population abundance are, in both cases, due to environmental variations which, accidentally, occurred at these same times.

Finally, in applying the model to the data, it is necessary that the changes in population associated with changes in fishing effort be sufficiently great in relation to the variations due to other causes to permit a usefully reliable determination of the parameters in equation (10). In other words, in the terminology of information theory, (Wiener 1948, page 78) the message signal must be sufficiently large with respect to the background noise to yield a useful amount of information.

ESTIMATION OF PARAMETERS

From equation (10) we have, for any year, i,

$$1/k_2 \frac{\Delta U}{\bar{\bar{U}}_i} = a(M - \bar{U}_i) - \frac{C_i}{\bar{\bar{U}}_i} + \epsilon \qquad . \qquad . \qquad (13)$$

ΔU_i being estimated by (12)

Summing over n years, we have for the expected sum

$$1/k_2 \sum_{i=1}^{n} \frac{\Delta U_i}{\bar{\bar{U}}_i} = \sum_{i=1}^{n} a(M - \bar{U}_i) - \sum_{i=1}^{n} \frac{C_i}{\bar{\bar{U}}_i} . \qquad (14)$$

since the expectation of $\sum_\epsilon$ is zero.

This is equivalent to

$$1/k_2 \sum_{i=1}^{n} \frac{\Delta U}{\bar{\bar{U}}_i} = naM - a \sum_{i=1}^{n} \bar{U}_i - \sum_{i=1}^{n} \frac{C_i}{\bar{\bar{U}}_i} . \qquad (15)$$

In Table 1 are tabulated the values of C, $\bar{U}$, $C/\bar{U}$, ΔU, and $\Delta U/\bar{U}$ for each year of our series. The series of values of $\bar{U}$ and $C/\bar{U}$ are also shown graphically in Figure 1, and the series of values of C and $C/\bar{U}$ are shown in Figure 2.

Since positive values of ΔU are, on the average, associated with catch less than the equilibrium catch for the corresponding value of $\bar{U}$, and conversely, we may also write, corresponding to (14) the equation:

$$1/k_2 \sum_i \left|\frac{\Delta U_i}{\overline{U}_i}\right| = \sum_i \underbrace{\left[a(M - \overline{U}_i) - \frac{C_i}{\overline{U}_i}\right]}_{\Delta U_i > 0} - \sum_i \underbrace{\left[a(M - \overline{U}_i) - \frac{C_i}{\overline{U}_i}\right]}_{\Delta U_i < 0}$$

. . . (16)

That is when we sum the absolute values of $\dfrac{\Delta U_i}{\overline{U}_i}$ on the left hand side, on

the right hand side we $add \left[a(M - \overline{U}_i) - \dfrac{C_i}{\overline{U}_i}\right]$ for years when ΔU_i is

positive and *subtract* this quantity when ΔU_i is negative.

In order to estimate the parameters of the model for the yellowfin tuna

data of Table 1, we have employed equations (15) and (16) in a method of successive approximations.

Computing the sums for (15) for the two periods 1935-1944 and 1945-1954 we obtain:

$$\left.\begin{array}{l} 1/k_2\ (-.11520) = 10\ aM - a(100{,}289) - 77{,}290 \\ 1/k_2\ (-.37665) = 10\ aM - a(\ 75{,}700) - 234{,}242 \end{array}\right\} \quad \cdot \quad \cdot \quad (17)$$

Since in each period the *net* change in $\Delta U/\overline{U}$ was small, we have, as a first approximation, neglected the terms on the left side of these equations and solved for a and M, obtaining

$$a = 6.38301$$
$$M = 11{,}239$$

With these values of a and M, we now compute the terms of equation (16), obtaining, for a first approximation to $1/k_2$,

$$1/k_2\ (2.11843) = 53{,}940$$
$$1/k_2 = 25{,}462$$

Substituting this value of $1/k_2$ into (17), we obtain a second approximation to a and M

$$a = 6.11229$$
$$M = 11{,}246$$

With the new values of a and M we again compute the terms of equation (16) and solve for $1/k_2$, obtaining, as the second approximation:

$$1/k_2 = 26{,}277$$

Continuing this procedure, the series of successive approximations tabulated in Table 2 are obtained.

133

It may be seen that the approximations converge very rapidly. We obtain as our best estimates of the values of the parameters:

$$a = 6.10366$$
$$M = 11,246$$
$$1/k_2 = 26,275$$

For the line of equilibrium conditions, equation (11) relating fishing effort to average population density when the catch is in equilibrium with the rate of natural increase, we have

$$F = \frac{C_e}{\bar{U}} = 6.10366\ (11,246 - \bar{U}) \qquad . \qquad . \qquad . \qquad (18)$$

It immediately follows that

$$C_e = 6.10366\ (11,246 - \bar{U})\ \bar{U} \qquad . \qquad . \qquad . \qquad (19)$$

expressing the relationship between population density and average equilibrium catch.

Also,

$$C_e = \frac{F}{6.10366}\ (6.10366 \times 11,246 - F)$$

or $\quad C_e = .16384\ (68,641 - F)\ F \qquad . \qquad . \qquad . \qquad (20)$

expressing the relationship between fishing effort and average equilibrium catch.

These functional relationships are shown graphically in Figure 1 by the broken line labelled "estimated line of equilibrium". On this graph, the abscissae indicate values of F or $C/\bar{U}$, while the ordinates indicate values of $\bar{U}$. The equilateral hyperbolae indicate values of total catch, corresponding to the products of the ordinates and abscissae; the values of total catch for which the hyperbolae have been drawn are labelled at the ends of these hyperbolae at the top of the graph. Values of average equilibrium catch are given by the intersections of the broken line with the hyperbolae for the corresponding values of F or $\bar{U}$.

Since the values of C for which the hyperbolae have been computed do not form a linear series, it is difficult to interpolate by eye between them. In order, therefore, to permit the relation between fishing effort and average equilibrium catch to be more easily visualized, there has been plotted, as a broken line, in Figure 2 the relationship (20), together with the observed values of fishing effort and total catch for the series of years 1934-1955.

It may be seen from the equations (18) to (20), and from Figures 1 and 2, that this application of the model to the yellowfin tuna data indicates that the maximum average annual equilibrium catch is about 193 million pounds, and corresponds to an average annual fishing intensity of 34,300 days and to an average catch-per-unit-of-effort of 5623 pounds per standard-day's-fishing.

Effects of variation in $1/k_2$

The constant k_2, estimated above, is the instantaneous rate of fishing mortality per unit of fishing effort. With $1/k_2 = 26,275$, we have $k_2 = 38.059 \times 10^{-6}$. For a fishing effort of 25,000 days, which is about the level obtaining in recent years, this would indicate an instantaneous fishing mortality rate of 0.9515, corresponding to an annual fishing mortality rate of 0.614.

There arises the question of the precision with which the constant $1/k_2$ has been estimated, and what effect this has on the determination of the other parameters of the equation of the model. In order to examine this, we have *assumed* a series of values of $1/k_2$, corresponding to annual fishing mortality rates (at 25,000 days' fishing) ranging from 0.300 to 0.900, and have computed the other constants of the model. In Table 3 are shown the assumed values of the constant $1/k_2$, and the corresponding rates of instantaneous and annual fishing mortality at 25,000 units of effort.

The method of computation is similar to that given by Schaefer (1954). For a given value of $1/k_2$ we compute, for each year, the value of $\Delta P = 1/k_2 \Delta U$. From equation (1) we then obtain C_e for each year, by

$$C_e = C + \Delta P$$

Dividing C_e by $\bar{U}$ for each year, we obtain the series of values of $C_e/\bar{U}$. To the series of values of $C_e/\bar{U}$ and $\bar{U}$ for the several years, we fit an equation of the form

$$F = \frac{C_e}{\bar{U}} = a(M - \bar{U})$$

or

$$F = \beta\bar{U} + a \ . \ . \qquad . \qquad . \ . \qquad . \qquad . \qquad . \qquad (21)$$

where

$$\beta = -a$$
$$a = aM$$

This may be illustrated for the case $1/k_2 = 26,275$ (the value of this constant estimated in the preceding section). In Table 4 are shown the calculations of C_e and $C_e/\bar{U}$ for each year. The values of $\bar{U}$ and $C_e/\bar{U}$ are also plotted in Figure 3.

In fitting an equation of the form of (21) to these data, it should be borne in mind that we wish to obtain estimates of the parameters of the functional, or structural, relationship between the two variables, where both are subject to random errors. This, as shown by Wald (1940), Bartlett (1949), and others is not the same problem as the problem of predicting one variable by means of the other. In the latter case, it is appropriate to employ the mean square regression of one variable on the other, but it is not necessarily appropriate for finding the structural relationship.

For estimating the parameters of the functional relationship, Wald (1955) proposed a method which consists of dividing the data into two groups and finding the slope of the line passing through the mean coordinates of the two groups. The functional relationship is taken as the line with this slope which passes through the mean coordinates of all the points.

Bartlett (1949) presented a modification of Wald's method, having the advantage of greater accuracy. This consists of dividing the n points into three groups, the equal numbers k in the two extreme groups being chosen to be as near $n/3$ as possible. The join of the mean coordinates for the two extreme groups is used to determine the slope. The functional relationship is the straight line with this slope passing through the mean coordinates of all the points.

We have followed Bartlett's procedure. The two extreme groups of seven points each were taken as the points having the seven highest and the seven lowest values of $C_e/\bar{U}$. The equation for the resulting line has estimated values of the two parameters:

$$a^* = 68,505 \qquad \beta^* = -6.0884$$

so that $M^* = 11,252$

For this assumed value of $1/k_2$ we have, then, for the functional relationship:

$$-\frac{C_e}{\bar{U}} = 6.0884 \ (11,252 - \bar{U}) \quad . \quad . \quad . \quad . \quad . \quad (22)$$

or

$$\bar{U} = .16425 \ (68,505 - \frac{C_e}{\bar{U}})$$

This line is plotted on Figure 3.

It may be seen, by comparison with (18), that this relationship is nearly identical with that obtained in the preceding section.

Assuming the values of $1/k_2$ tabulated in Table 3, each in turn, we have followed the same procedure as shown by the above example in computing estimates of a^* and β^* for the corresponding functional relationships. The values of these estimated parameters are tabulated in Table 5. The lines representing these relationships are shown in Figure 4, each line being labelled with the value of f (annual fishing mortality rate at 25,000 units of effort) corresponding to the assumed value of $1/k_2$. The lines for $f = 0.614$ and $f = 0.800$ are not shown, because they both fall between the lines for $f = 0.600$ and $f = 0.700$, which are themselves very near to each other.

In the last two columns of Table 5 are also shown the value of maximum average equilibrium catch, and the fishing intensity corresponding to maximum average equilibrium catch, for each of the assumed values of fishing mortality rate.

It may be seen from Figure 4 and Table 5 that, over the range of values of $1/k_2$ corresponding to annual fishing mortality rates (at 25,000 units of effort) between about 0.6 and 0.8, there is little variation in the values of the parameters of the line of equilibrium conditions, and all provide rather similar estimates of the level of maximum average equilibrium catch, and of the level of fishing effort corresponding thereto. It would seem, therefore, that over this range of values of $1/k_2$ there is little variation in the computed values of the other parameters. With assumed values of annual fishing mortality rate below 0.6, however, the values of the parameters of the line of equilibrium conditions change quite rapidly.

The degree of scatter of the points about the line of equilibrium conditions in each case, may be judged from the "variance" $(s^2_{n-3}\beta)$ computed from the formulae of Bartlett on page 209 of his paper. These values are tabulated in the sixth column of Table 5. Now, it is reasonable to expect that the scatter of points $(C_e/\bar{U}, \bar{U})$ will be smallest when the value of $1/k_2$ is near the true value and will become increasingly large as our assumption departs from reality. It may be seen from the tabulated values that the smallest variance was obtained with $f = 0.8$, and did not change greatly over the range $f = 0.6$ to $f = 0.8$. With values of f below 0.6, the calculated variance is very greatly increased, and it also increases substantially at $f = 0.9$.

It appears reasonable to conclude that (1) the annual fishing mortality rate (at 25,000 units of effort) probably lies somewhere between 0.6 and 0.8 and (2) over this range of values, the values of the other parameters are not much affected by errors in the determination of $1/k_2$.

Our estimate, then, of $1/k_2 = 26,275$ should be regarded as being not precise, but this lack of precision makes practically no difference to the estimate of the parameters of the line of equilibrium conditions.

LINEARITY OF THE RELATIONSHIP BETWEEN $C_e/\bar{U}$ AND $\bar{U}$

We have (page 249) in our mathematical model assumed a linear relationship between fishing effort and population abundance, on the average, for steady-state equilibrium.

In order to verify the validity of this assumption over the range of values of fishing effort observed, we have considered the series of values of $\bar{U}$ and $C_e/\bar{U}$, taking $1/k_2 = 26,275$ (i. e. the points of Figure 3). Applying the "t" test given by Bartlett (1949, page 211) it was found that the points are adequately fitted by a linear functional relationship. There seems to be, then, no reason to employ a more complex model, at least over the range of values of fishing effort and abundance so far observed.

ESTIMATES OF AVERAGE POPULATION ABUNDANCE AND
EQUILIBRIUM CATCH CORRESPONDING TO DIFFERENT
VALUES OF FISHING EFFORT, AND CONFIDENCE
LIMITS ON THE ESTIMATES

In the foregoing we have been concerned with determining from the data of Table 1 the values of the constants in equation (11) which best represent the functional relationship among fishing effort, average abundance, and average catch, when the annual catch is in equilibrium with the rate of natural increase. A somewhat different problem is: for specified values of fishing effort to predict the average abundance and average catch under equilibrium condition.

For the latter problem, as noted by Wald (1940) and Bartlett (1949), it is appropriate to employ the least squares regression of $\bar{U}$ on $C_e/\bar{U}$. We have, therefore, fitted to the points of Figure 3 (Table 4) the mean square linear regression, obtaining

$$\bar{U} = 0.15670 \ (71{,}081 - C_e/\bar{U}) \ . \qquad . \qquad . \qquad . \qquad (23)$$

which differs a little from the estimated functional relationship. This line is also shown on Figure 3.

The standard error of estimate (standard deviation from regression) is 1,477 pounds.

Confidence limits on the estimate of the mean value of $\bar{U}$, given the fishing effort $(C_e/\bar{U})$, have been calculated, for the 90% confidence level, according to the procedure of Cramér (1946, page 550). The confidence limits are shown also on Figure 3.

For any chosen value of fishing effort, the expected equilibrium catch, with corresponding confidence limits, may be read off Figure 3 by means of the equilateral hyperbolae, which indicate the products of the values of the ordinates and abscissae. Since, however, the values of total catch shown by the hyperbolae of Figure 3 do not form a linear series, it is difficult to interpolate by eye between them. We have, therefore, also prepared Figure 5, which shows the values of estimated equilibrium catch, and the corresponding 90% confidence limits, for values of fishing effort on the horizontal axis. On the same Figure are shown the values of the actual fishing effort and catch for each year of our series.

It may be seen from Figure 5, that the estimated maximum average equilibrium catch of 198 million pounds corresponds to a fishing intensity of 35,540 standard days' fishing. The maxima of the lower and upper 90% confidence limits are about 176 million pounds and 221 million pounds, respectively, and occur at slightly lower and higher values of fishing effort.

PRESENT STATUS OF THE FISHERY

It may be seen from the foregoing analyses that we may expect the maximum average equilibrium catch of yellowfin tuna to be obtained with about 35,000 standard days' fishing effort, and to be in the vicinity of 195 million pounds. This amount of fishing effort was approached in 1950, 1952, and 1953. In 1954 and 1955, however, the fishing effort decreased markedly, due to economic circumstances. Data are not yet complete for 1956, but from information so far tabulated it appears that the effort was in the vicinity of 25,000 standard days. With the fleet now in being, it is unlikely that, even with full operation, the intensity can reach 35,000 standard days during the next year.

It would appear, therefore, that the current level of fishing effort is somewhat below that corresponding to maximum average equilibrium catch, and is likely to remain so in the immediate future.

It should be noted that the fitted lines drawn in Figures 1, 2, 3, and 5 are extrapolated somewhat beyond the range of the data. In order to verify the validity of these extrapolations, i.e. the applicability of the model developed herein at higher levels of fishing effort, it will be necessary to have observations of fishing effort, and resulting catch, at levels higher than those yet observed.

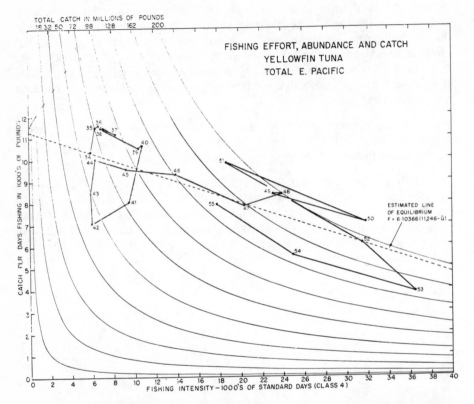

Figure 1. Relationships among fishing intensity, abundance and total catch. Points connected by solid line indicate actual values for each year 1934-1955. Broken line is estimated functional relationship between fishing intensity and average abundance under equilibrium conditions.

Figure 1. Relaciones entre la intensidad de la pesca, la abundancia y la pesca total. La línea continua une los valores realmente observados en cada año durante el período 1934-955. La línea a trazos representa la estimación de la relación funcional entre la intensidad de la pesca y la abundancia medio, en condiciones de equilibrio.

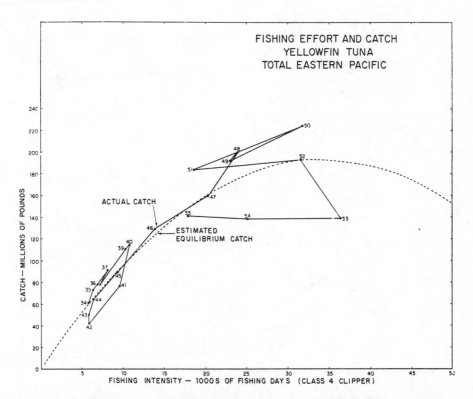

Figure 2. Relationship between fishing intensity and total catch. Points connected by solid line indicate actual values for each year 1934-1955. Broken line is estimated functional relationship between fishing intensity and average equilibrium catch.

Figura 2. Relación entre la intensidad de la pesca y la pesca total. La línea continua une los valores realmente observados en cada año de 1934 a 1955. La línea a trazos representa la relación funcional entre la intensidad de la pesca y la pesca media de equilibrio.

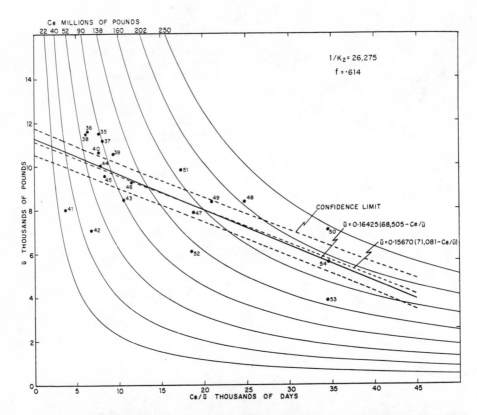

Figure 3. Relationships among fishing intensity, abundance, and equilibrium catch, assuming $1/k_2 = 26,275$. Points indicate values computed for each year 1935-1954. Solid line is estimated functional relationship between fishing intensity and average abundance under equilibrium conditions. Broken lines indicate the values of average abundance estimated for given values of fishing intensity, with the 90% confidence limits on the estimates.

Figura 3. Relaciones entre la intensidad de la pesca, la abundancia, y la pesca de equilibrio, suponiendo $1/k_2 = 26,275$. Los puntos corresponden a los valores computados para cada año de la serie 1935-1954. La línea continua representa la estimación de la relación funcional entre la intensidad de la pesca y la abundancia media en las condiciones de equilibrio. Las líneas a trazos representan los valores de la abundancia media estimados para determinados valores de la intensidad de la pesca, con los límites de confianza del 90 por ciento para las estimaciones.

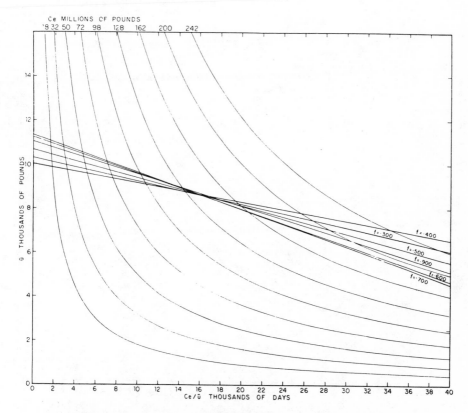

Figure 4. Lines of equilibrium conditions calculated for different assumed values of annual rate of fishing mortality (at 25,000 standard days' fishing). Lines for $f = 0.614$ and $f = 0.800$ not shown (see Table 5); both fall between lines for $f = 0.600$ and $f = 0.700$.

Figura 4. Líneas de condiciones de equilibrio calculadas para diferentes valores supuestos de tasas de mortalidad anual por pesca (al nivel de los 25,000 días de pesca estándar). Las líneas para $f = 0.614$ y para $f = 0.800$ no han sido representadas (Véase la Tabla 5); una y otra caerían entre las líneas para $f = 0.600$ y $f = 0.700$.

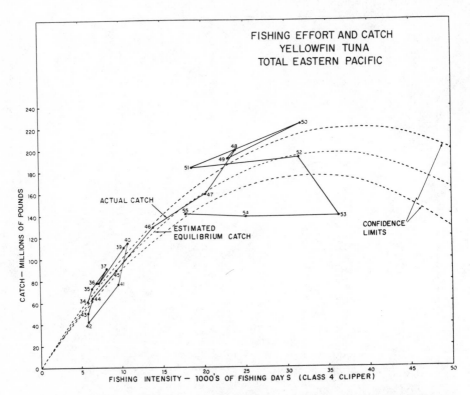

Figure 5. Average equilibrium catch predicted for given values of fishing intensity,
with 90% confidence limits (broken lines), together with actual values of
effort and catch for the years 1934-1955 (points connected by solid line).

Figura 5. Promedios de pescas en equilibrio pronosticadas para predeterminados
valores de la intensidad de la pesca, con límites de confianza del 90 por
ciento (líneas a trazos), además de los valores actuales del esfuerzo y de
la pesca para el período 1934-1955 (línea continua).

TABLE 1. Total Eastern Pacific catch of yellowfin tuna, catch per standard day's fishing, and calculated relative fishing intensity.

TABLA 1. Pesca total de atún aleta amarilla en el Pacífico Oriental, pesca por día de pesca estándar, y valores calculados de la intensidad relativa de la pesca.

Year	Eastern Pacific total catch (in 1000's of pounds) C	Catch per standard day's fishing (in pounds/class 4 day) $\bar{U}$	Calculated intensity (in class 4 days) $F = C/\bar{U}$	ΔU	$\dfrac{\Delta U}{\bar{U}}$
Año	Pesca total Pacífico Oriental (en miles de libras) C	Pesca por día estándar de actividad (libras/día/ clase 4) $\bar{U}$	Intensidad calculada (días de la clase 4) $F = C/\bar{U}$	ΔU	$\dfrac{\Delta U}{\bar{U}}$
1934	60,913	10,361	5,879		
1935	72,294	11,484	6,295	+ 605.0	+.05268
1936	78,353	11,571	6,771	− 184.0	−.01590
1937	91,522	11,116	8,233	− 54.0	−.00486
1938	78,288	11,463	6,830	− 294.0	−.02565
1939	110,417	10,528	10,488	− 427.0	−.04056
1940	114,590	10,609	10,801	−1255.0	−.11830
1941	76,841	8,018	9,584	−1784.5	−.22256
1942	41,965	7,040	5,961	+ 211.5	+.03004
1943	50,058	8,441	5,930	+1489.5	+.17646
1944	64,094	10,019	6,397	+ 535.5	+.05345
1945	89,194	9,512	9,377	− 363.5	−.03821
1946	129,701	9,292	13,958	− 827.5	−.08906
1947	160,134	7,857	20,381	− 469.5	−.05976
1948	200,340	8,353	23,984	+ 253.0	+.03029
1949	192,458	8,363	23,013	− 648.0	−.07748
1950	224,810	7,057	31,856	+ 723.0	+.10245
1951	183,685	9,809	18,726	− 480.0	−.04893
1952	192,234	6,097	31,529	−2997.5	−.49164
1953	138,918	3,814	36,423	− 275.5	−.07223
1954	138,623	5,546	24,995	+2040.5	+.36792
1955	140,581	7,895	17,806		
Sums—Sumas 1935-1944		100,289	77,290		−.11520
Sums—Sumas 1945-1954		75,700	234,242		−.37665

TABLE 2. Successive approximations for parameters.

TABLA 2. Aproximaciones sucesivas para los parámetros.

a	M	$1/k_2$
6.38301	11,239	
		25,462
6.11229	11,246	
		26,277
6.10362	11,246	
		26,275
6.10366	11,246	

TABLE 3. Table of values of $1/k_2$ and corresponding values of instantaneous rate of fishing mortality and annual rate of fishing mortality at $F = 25,000$ days.

TABLA 3. Tabla de valores de $1/k_2$, las correspondientes tasas instantáneas de mortalidad por pesca y las tasas anuales de mortalidad por pesca, para $F = 25,000$ aías.

$1/k_2$	Instantaneous fishing mortality rate k_2F	Annual fishing mortality rate $1-e^{-k_2F}$
$1/k_2$	Tasa instantánea de mortalidad por pesca k_2F	Tasa anual de mortalidad por pesca $1-e^{-k_2F}$
70,093	.35667	.300
48,940	.51083	.400
36,067	.69315	.500
27,284	.91629	.600
26,275	**.95147**	**.614**
20,765	1.20397	.700
15,533	1.60944	.800
10,857	2.30259	.900

TABLE 4. Estimation of equilibrium catches for $1/k_2 = 26,275$.

TABLA 4. Estimación de las pescas de equilibrio para $1/k_2 = 26,275$.

Year Año	C	$\overline{U}$	ΔP $(=1/k_2 \Delta U)$	C_e $(=C+\Delta P)$	$C_e/\overline{U}$
1935	72,294	11,484	+15,896	88,190	7679
1936	78,353	11,571	− 4,835	73 518	6354
1937	91,522	11,116	− 1,419	90,103	8106
1938	78,288	11,463	− 7,725	70,563	6156
1939	110,417	10,528	−11,219	99,198	9422
1940	114,590	10,609	−32,975	81,615	7693
1941	76,841	8,018	−46,888	29,953	3736
1942	41,965	7,040	+ 5,557	47,522	6750
1943	50,058	8,441	+39,137	89,195	10567
1944	64,094	10,019	+14,070	78,164	7802
1945	89,194	9,512	− 9,551	79,643	8373
1946	129,701	9,292	−21,743	107,958	11618
1947	160,134	7,857	−12,336	147,798	18811
1948	200,340	8,353	+ 6,648	206,988	24780
1949	192,458	8,363	−17,026	175,432	20977
1950	224,810	7,057	+18,997	243,807	34548
1951	183,685	9,809	−12,612	171,073	17440
1952	192,234	6,097	−78,759	113,475	18612
1953	138,918	3,814	− 7,238	131,680	34525
1954	138,623	5,546	+53,614	192,237	34662

TABLE 5. Estimates of parameters for various values of $1/k_2$.

TABLA 5. Estimación de los parámetros para varios valores de $1/k_2$.

f (Annual fishing mortality rate at 25,000 days fishing) / f (Tasa de mortalidad anual por pesca al nivel de 25,000 dias de actividad)	$1/k_2$	β^*	α^*	$M^* = \dfrac{\alpha^*}{\beta^2}$	Variance[1] $s^2_{n-3}(\beta^*)$ / Variancia[1] $s^2_{n-3}(\beta^*)$	C_e Maximum (10^3 pounds) / C_e Maximum 10^3 libras	F at C_e Maximum / F al nivel del Maximum de C_e
.300	70,093	− 9.8400	100,441	10,207	$422,777 \times 10^3$	256,290	50,220
.400	48,940	−11.3632	114,366	10,065	$633,525 \times 10^3$	287,785	57,183
.500	36,067	− 7.8155	83,462	10,679	$340,230 \times 10^3$	222,822	41,731
.600	27,284	− 6.0903	68,497	11,247	$146,959 \times 10^3$	192,598	34,248
.614	**26,275**	**− 6.0884**	**68,505**	**11,252**	**$145,908 \times 10^3$**	**192,709**	**34,252**
.700	26,765	− 5.9568	67,481	11,328	$134,781 \times 10^3$	191,999	33,740
.800	15,533	− 5.9708	67,735	11,344	$109,814 \times 10^3$	192,030	33,867
.900	10,857	− 6.7646	74,835	11,063	$168,147 \times 10^3$	206,980	37,417

[1] See Bartlett (1949) page 209.

[1] Ver Bartlett (1949) página 209.

LITERATURE CITED — BIBLIOGRAFIA CITADA

Bartlett, M. S.

    1949 Fitting a straight line when both variables are subject to error.
Biometrics, Vol. 5, No. 3, pp. 207-212.

Broadhead, Gordon C.

    1957 Changes in the size structure of the yellowfin tuna population
of the Tropical Eastern Pacific Ocean from 1947-1955.
Inter-Amer. Trop. Tuna Comm., Bull., Vol. II, No. 1, pp. 1-20.

Cramér, H.

    1946 Mathematical methods of statistics.
Princeton University Press, Princeton, N. J., 575 pp.

Godsil, H. C., and E. C. Greenhood

    1951 A comparison of the populations of yellowfin tuna, *Neothunnus
macropterus,* from the Eastern and Central Pacific.
Cal. Div. Fish and Game, Fish. Bull. No. 82, 33 pp.

Gulland, J. A.

    1955 Review of Schaefer (1954), Bull. Inter-Amer. Trop. Tuna Comm.
Vol. 1. No. 2.
Cons. Perm. Int. Expl. Mer., Journal, Vol. 20, No. 3, pp. 330-332.

Hutchinson, G. E.

    1954 Theoretical notes on oscillatory populations.
Jour. Wildlife Management, Vol. 18, No. 1, pp. 107-103.

Moran, P. A. P.

    1954 The logic of the mathematical theory of animal populations.
Jour. Wildlife Management, Vol. 18, No. 1, pp. 60-66.

Royce, W. F.

    1953 Preliminary report on a comparison of the stocks of yellowfin
tuna.
Proc. Indo-Pacific Fish. Council, 1952, Section II, pp. 130-145.

Schaefer, M. B.

  1952 Comparison of yellowfin tuna of Hawaiian waters and of the American West Coast.
  U. S. Fish and Wildlife Service, Fish. Bull. No. 72, Vol. 52, pp. 353-373.

  1954 Some aspects of the dynamics of populations important to the management of the commercial marine fisheries.
  Inter-Amer. Trop. Tuna Comm., Bull., Vol. 1, No. 2, pp. 26-56.

  1955 Morphometric comparison of yellowfin tuna from Southeast Polynesia, Central America, and Hawaii.
  Inter-Amer. Trop. Tuna Comm., Bull., Vol. 1, No. 4, pp. 90-136.

Shimada, B. M. and M. B. Schaefer

  1956 A study of changes in fishing effort, abundance, and yield for yellowfin and skipjack tuna in the Eastern Tropical Pacific Ocean.
  Inter-Amer. Trop. Tuna Comm., Bull., Vol. 1, No. 7, pp. 350-469.

Wald, A.

  1940 The fitting of straight lines if both variables are subject to error.
  Ann. Math. Stat., Vol. II, pp. 284-300.

Watt, K. E. F.

  1956 The choice and solution of mathematical models for predicting and maximizing the yield of a fishery.
  Jour. Fish. Res. Bd. Canada, Vol. 13, No. 5, pp. 613-645.

Wiener, N.

  1948 Cybernetics, or control and communication in the animal and the machine.
  John Wiley and Sons, New York, 194 pp.

Stock and recruitment

W.E.Ricker

Journal of the Fisheries Research Board of Canada, **11,** 559 – 623, 1954.

Editor's introductory notes

Petersen (1894) thought that recruitment overfishing was the more important form because once fishermen realized the economic benefits of only catching the larger fish, they would no longer take the smaller ones. Russell, Thompson and Graham had all been aware of the possibility that the magnitude of recruitment might be reduced as stock declined under the pressure of fishing. None really believed it. Ricker's paper is important for two reasons, first that the possibility is shown and secondly that the first theory of the generation of recruitment was written, a consequence of cannibalism. Ricker made a very important assumption, that no compensatory mortality occurs among the mature stock: 'opportunities for compensatory effects are so much greater during the small vulnerable stages of a fish's life, that restriction of compensation to those stages seems likely to have wide applicability as a useful approximation'.

The paper consists of three parts: (a) the form of curve needed to stabilize a population (which look a little extreme thirty years later) and possible time sequences in stabilization; (b) eye fitted curves to the meagre quantity of data then available and (c) a brief model of stabilization by cannibalism.

The derivation of the Ricker curve, an appendix to Ricker (1958) is given here in addition to Ricker's (1954) paper. The compensatory mortality is generated no longer by cannibalism but by the aggregation of predators; in particular Ricker believed that the brown trout in a lake aggregated on the patches of sockeye salmon alevins as they emerged from their parent stream. There are two forms to the equation: (a) the general form $R = \alpha P \exp - \beta P$, where R is recruitment in numbers, P is stock in eggs, α is the coefficient density independent survival and β is the coefficient of density dependent mortality and (b) the form about the replacement line, which is particularly useful for the Pacific salmon stocks, $R/R_r = W \exp a(1 - W)$, where $R_r = P_r$, recruitment and stock at replacement, $W = P/P_r$, $a = P_r/P_m$, where P_m is the stock at which recruitment is maximal. Both have been used extensively in fisheries science since their development.

References

Petersen,C.G.J.: (1894), 'On the biology of our flatfishes and on the decrease of our flatfish fisheries,' *Rep IV Dan. Biol. Sta.,* 146pp.
Ricker,W.E.: (1958), 'Handbook of computations for biological statistics of fish populations,' *Fish. Res. Bd. Can. Bull.,* **119**, 300pp.

Stock and Recruitment[1]

By W. E. Ricker
Pacific Biological Station, Nanaimo, B.C.

ABSTRACT

Plotting net reproduction (reproductive potential of the *adults* obtained) against the density of stock which produced them, for a number of fish and invertebrate populations, gives a domed curve whose apex lies above the line representing replacement reproduction. At stock densities beyond the apex, reproduction declines either gradually or abruptly. This decline gives a population a tendency to oscillate in numbers; however, the oscillations are damped, not permanent, unless reproduction decreases quite rapidly *and* there is not too much mixing of generations in the breeding population. Removal of part of the adult stock reduces the amplitude of oscillations that may be in progress and, up to a point, *increases* reproduction.

CONTENTS

INTRODUCTION
General 560
Theory of population regulation 560
Age incidence of compensatory mortality 561
Kinds of population control mechanisms 562
TYPES OF REPRODUCTION CURVES 563
REPRODUCTION IN THE ABSENCE OF DENSITY-INDEPENDENT VARIABILITY 567
VARIATIONS IN REPRODUCTION PRODUCED BY FACTORS INDEPENDENT
OF DENSITY 572
COMBINATIONS OF COMPENSATORY AND NON-COMPENSATORY MORTALITY 579
EFFECTS OF REMOVAL OF MATURE STOCK 580
EXAMPLES OF REPRODUCTION CURVES 586
Pacific herring (*Clupea pallasi*) 587
Pink salmon (*Oncorhynchus gorbuscha*) 588
Coho salmon (*Oncorhynchus kisutch*) 590
Sockeye salmon (*Oncorhynchus nerka*) 590
Haddock (*Melanogrammus aeglifinus*) 593
Fruit fly (*Drosophila melanogaster*) 595
Water-flea (*Daphnia magna*) 596
Starfish (*Asterias forbesi*) 601
Sinuous reproduction curves 602
Discussion 604
OTHER REPRODUCTION SITUATIONS 606
Larger immature fish taken by the fishery 606
Exploitation during the compensatory phase 607
Compensation by immature members of the population 608
TOWARD A THEORY OF RECRUITMENT 609
Theory of predation 609

[1]Received for publication June 1, 1953; as revised, April 30, 1954. Portions of this paper appeared in the *Journal of Wildlife Management* for January, 1954, as a contribution to a symposium on cycles in animal populations.

J. FISH. RES. BD. CANADA, 11(5), 1954.

Cannibalism   610
Predation by other organisms   613
Other compensatory agents   617
Fitting the curve $z = we^{1-w}$ 618
SUMMARY   619
ACKNOWLEDGMENTS   621
REFERENCES   621

INTRODUCTION

GENERAL

There exists today a considerable body of knowledge which goes by the name of "the theory of fishing" or "the modern theory of fishing"—the work of a succession of the most distinguished fishery biologists of our time. It is concerned mainly with predicting what catch can be obtained from a given number of young fish recruited to a fishery, if their initial size and the growth and natural mortality rates prevailing are known. That is, methods have been developed for computing the effects of different rates of exploitation, of changes in rate of exploitation from year to year, of different minimum size limits, etc., upon the yield obtained. Not only that, but much progress has been made in developing methods of determining the actual magnitudes of the population statistics required to make these calculations.

Valuable as the above contributions have been, they comprise only half of the biological information needed to assess the effects of fishing and an optimum level of exploitation. Fishing changes the absolute and the relative abundance of mature fish in a stock, and the effect of this upon the number of recruits in future years has often been considered only in the most general manner. The points of view encountered usually range from an assumption of direct proportion between size of adult stock and number of recruits, to the proposition that number of recruits is, for practical purposes, independent of the size of the adult stock. The possibility of a *decrease* in recruitment at higher stock densities has less often been considered.

The scarcity of information on this subject is quite explicable, since it usually requires many years of continuous observation to establish a relation between size of stock and the number of recruits which it produces. However, it has become an urgent problem to have a scientific description of the regulation of abundance of fish stocks, in order to complete the basis for predicting optimum levels of exploitation. This paper attempts to summarize some of the theoretical and factual information available, both from fish populations and from other animals, and to provide a stimulus to studies which will eventually put the subject on a solid foundation.

THEORY OF POPULATION REGULATION

Basic in any stock-recruitment relationship is the fact that a fish population, even when not fished, is limited in size; that is, it is held at some more or less fluctuating level by natural controls. Ideas concerning the nature of such controls were first clarified and systematized by the Australian entomologist Nicholson

(1933). He showed that, while the level of abundance attained by an animal can be affected by any element of the physical or biological environment, the immediate mechanism of control must always involve competition, using that word in a broad sense to include any factor of mortality whose effectiveness increases with stock density[2]. The term *density-dependent* mortality was used for the same concept by Smith (1935). More strictly, density-dependent causes of mortality should include both those which become more effective as density increases and those which become less so. The former are the ones which provide control of population size; and they have been called *concurrent* (Solomon, 1949), *compensatory* (Neave, 1953) or *negative* (Haldane, 1953). The opposed terms are *inverse, depensatory* and *positive*, all referring to density-dependent factors which become *less* effective as density increases.

No sharp line can be drawn between the kinds of mortality which are compensatory and those which are not, although Nicholson, Smith and others have felt that as a rule biological factors tend to predominate among the former, and physical agents among the latter, for insects at least. Among fishes, extremes of water temperature, drought and floods, are physical agents which may often cause mortality whose effectiveness is independent of stock density; whereas deaths from such biological causes as disease, parasitism, malnutrition and predation will usually become relatively more frequent as stock density increases. Yet exceptions to the rule above are sufficiently numerous to make the rule itself of doubtful applicability to fishes. For example, the biological factor of predation may have a uniform effectiveness over a considerable range of prey abundance, or at times may even become more effective at lower prey densities; and most if not all physical causes of mortality are compensatory when stock becomes dense enough that some of its members are forced to live in exposed or unsuitable environments. In addition, it is of course often difficult to ascribe a death to any single cause.

There is no necessary relation between the relative magnitudes of the causes of mortality existing at a given time, as measured by the fraction of the stock which each kills, and their relative contribution to compensation. An important and deadly agent of mortality may be strongly density-dependent, or weakly so, or not at all; and different agents may have their maximum compensatory effect over quite different ranges of density.

AGE INCIDENCE OF COMPENSATORY MORTALITY

Density-dependent causes of mortality could affect the abundance of either the existing adult stock or the young which it produces. In this paper we will

[2]This almost axiomatic proposition is implied in the writing of various earlier authors back as far as Malthus, but Nicholson was the first to formulate it explicitly and to emphasize its importance: Haldane (1953) calls his inspiration "a blinding glimpse of the obvious". The theorem and its diverse consequences were elaborated mathematically by Nicholson and Bailey (1935). Subsequent writers have developed it with varying emphasis, but nothing very substantial seems to have been added. Solomon (1949) gives a useful review of this literature, and Varley's (1947) quantitative assessment of various agents controlling the abundance of a trypetid fly population is outstanding.

consider mainly the effects of density dependence in the mortality which strikes the younger members of a population—among fishes, the eggs, larvae, fry and fingerlings. That is, the relative abundance of a brood will be considered to be determined by the time the first of its female members begin to mature: subsequent mortality is assumed to be non-compensatory.

This distinction between immature and mature stages of the life history, and the restriction of compensatory mortality to the former, is the principal difference between the thesis of this paper and that of earlier treatments of effects of density upon reproduction (e.g., Hutchinson, 1948; Haldane, 1953; Fujita and Utida, 1953; and many others). These have usually taken the Verhulst logistic equation as a point of departure:

$$\frac{dN_t}{dt} = \frac{bN_t(K - N_t)}{K}$$

(N_t is abundance at time t; K is equilibrium abundance; b is the instantaneous rate of increase of the population at densities approaching zero.) This expression implies a continuous tendency for the population to adjust itself toward the equilibrium size K, whether it is currently less than or greater than K. The adjustment would evidently have to involve compensatory mortality among the adult stock when its density is greater than K. Though such mortality is not impossible, there is in fish populations, at least, little indication of it; and as a matter of fact this aspect of the logistic equation has never been applied to any concrete biological situation.

Our assumption that *no* compensatory mortality occurs among the mature stock is unlikely to be strictly true of any species, and may not be even approximately true of some. Nevertheless it has seemed worth while to follow out the consequences of making this distinction between mature and immature, with fishes particularly in mind. The opportunities for compensatory effects are so much greater during the small, vulnerable early stages of a fish's life, that restriction of compensation to those stages seems likely to have wide applicability as a useful approximation.

To further simplify the initial approach to the problem, fishing is considered to attack only mature individuals, so that the *recruitment* produced by a given density of mature stock means both the number of commercial-sized fish, and the number of maturing fish, which result from its reproductive activity. The term *reproduction* will be used in a similar but slightly more general sense, to mean the number of young surviving to any specified age after compensation is practically complete; it will *not* mean the initial number of eggs or newborn young.

The above conditions and definitions will be assumed to apply unless exception is made specifically.

KINDS OF POPULATION CONTROL MECHANISMS

Compensatory types of mortality can be of various kinds. Some of the more likely possibilities are as follows:

1. Prevention of breeding by some members of large populations because

all breeding sites are occupied. Note that territorial behaviour may restrict the number of sites to a number less than what is physically possible.

2. Limitation of *good* breeding areas, so that with denser populations more eggs and young are exposed to extremes of environmental conditions, or to predators.

3. Competition for living space among larvae or fry, so that some individuals must live in exposed situations. This too is often aggravated by territoriality—that is, the preemption of a certain amount of space by an individual, sometimes more than is needed to supply necessary food.

4. Death from starvation or indirectly from debility due to insufficient food, among the younger stages of large broods, because of severe competition for food.

5. Greater losses from predation among large broods because of slower growth caused by greater competition for food. It can be taken as a general rule that the smaller an animal is, the more vulnerable it is to predators, and hence any slowing up of growth makes for greater predation losses. Since abundant year-classes of fishes have often been found to consist of smaller-than-average individuals (Hile, 1936), this may well be a very common compensatory mechanism among fishes (cf. Ricker and Foerster, 1948; Johnson and Hasler, 1954).

6. Cannibalism: destruction of eggs or young by older individuals of the same species. This can operate in the same manner as predation by other species, but it has the additional feature that when eggs or fry are abundant the adults which produced them tend to be abundant also, so that percentage destruction of the (initially) denser broods of young automatically goes up—provided the predation situation approaches the type in which kills are made at a constant fraction of random encounters (cf. page 609, below).

7. Larger broods may be more affected by macroscopic parasites or micro-organisms, because of more frequent opportunity for the parasites to find hosts and complete their life cycle.

8. In limited aquatic environments there may be a "conditioning" of the medium by accumulation of waste materials that have a depressing effect upon reproduction, increasingly as population size increases.

Not all of the above compensatory effects need exist, or be important, in any given population, or in the same population every year. To a considerable extent they are likely to be complementary, so that if, for example, exceptionally favourable conditions permitted a good hatch of even a large spawning of eggs, a reduced growth rate of the fry would permit increased predation and so reduce survival in that way.

TYPES OF REPRODUCTION CURVES

Whatever the various kinds of compensatory and non-compensatory mortality acting on a brood may be, the average resultant of their action, over the existing range of environmental conditions, is represented by the average size of maturing brood, or recruitment, which each stock density produces. A graph of this relationship between an existing stock, and the future stock which the existing

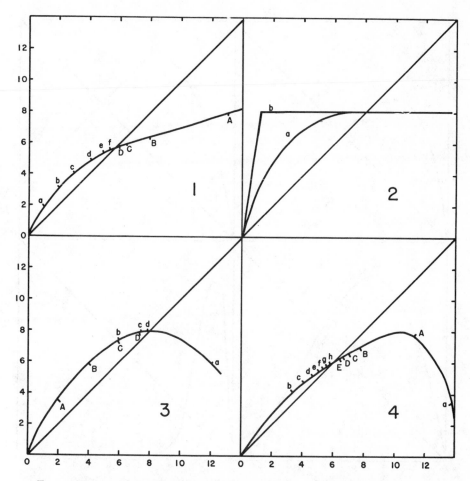

FIGURES 1–4. Stock-reproduction relationships characterized by a stable equilibrium. *Abscissa*— number of eggs produced by parent stock in a given year; *ordinate*—number of eggs produced by the progeny of that year.

stock produces, will be called a "reproduction curve". It is most convenient to label the axes in terms of the *eggs* in present and future generations, respectively[3]. The abscissa represents the mature eggs produced by the current year's stock. The ordinate represents the total of mature eggs produced by the progeny re-

[3]The argument is developed here in terms of populations of oviparous fishes which spawn once a year, but it can readily be modified to apply to other kinds of animals; an example for a viviparous animal is given on page 596. Among mammals, choice of the most suitable census age for plotting on reproduction curves may require care. If newborn young are used, effects of stock density upon frequency of conception and uterine mortality may be overlooked. If number of mature females is used, it should preferably be adjusted to take care of age variation in litter size or frequency, as the age structure of the population changes.

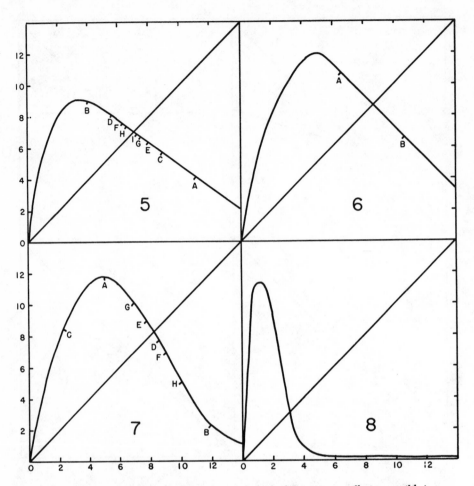

FIGURES 5–8. Stock-reproduction relationships in which there is an oscillating equilibrium, or (in 5) an oscillating approach to stable equilibrium. Axes as in Figures 1–4.

sulting from the current year's reproduction (obtained by summing over such period of time as the current year's hatch is a component of future years' stocks). In a state of nature, or with a stable fishery, the average size of parental and filial egg production, defined as above, tends to be equal over any long period of years, although striking changes may occur between individual years, or generations.

Figures 1–8 show a number of possible types of reproduction curve. In each of them the straight diagonal constitutes a useful boundary of reference which will be called the "45-degree line". Any curve lying wholly above this line describes a stock which is increasing without limit, hence such a curve cannot exist in practice. Similarly a curve below the 45-degree line describes a stock that will decrease to zero in a few generations[4]. The 45-degree line itself would describe a stock in which density dependence is absent, the filial generation tending always to be equal to the parental, except as factors independent of density deflect it. Such a stock would have no mechanism for the regulation of its numbers: if only density-independent causes of mortality exist, the stock can vary without limit, and must eventually by chance decrease to zero. Thus the first qualification of a reproduction curve is that it must cut the 45-degree line at least once—usually only once—and must end below and to the right of it.

Figures 1–8 indicate some of the types of recruitment curve that might exist in actual populations. All are characterized by having a region above the 45-degree line in which reproduction is more than adequate to replace the existing stock, and a region below the 45-degree line in which reproduction is inadequate to replace existing stock.

In the population of Figure 1, rate of reproduction (ratio of filial to parental eggs) decreases continuously as size of stock increases, although the slope of the curve becomes stabilized soon after the 45-degree line is crossed, and the actual number of young produced continues to rise indefinitely. Beyond the 45-degree line, however, this number is inadequate to fully replenish the stock.

In the population of Figure 2a, rate of reproduction also declines continuously; the actual number of recruits produced reaches an asymptotic level and thereafter does not change. Curve 2b differs from 2a in that it rises more steeply and it is initially straight; i.e., at low stock densities rate of reproduction is large, and constant. The range of densities over which there is constant recruitment is quite broad in b, simulating the condition loosely described by the statement "there are always as many recruits as the grounds can support".

Figures 3–8 show reproduction curves in which numbers of recruits begin to decrease after stock reaches some large magnitude. Such curves have an ascending left limb, a dome and a descending right limb. The descending right limb provides of course a more severe control of stock size at the higher densities. The differences in position of the dome and in the slope of the two limbs are of importance in determining changes in stock abundance, as described below.

All these curves are meant to represent the net effect of the sum total of density-dependent mortality factors acting upon the population. The reproduction of any actual year is affected also by density-independent factors, so that the actual number of young produced will deviate from the number indicated by the curve. In a later section the effect of such density-independent factors will be

---

[4]This statement is likely to be literally true only if the reproduction curve is plotted and fitted on logarithmic axes. On arithmetic axes a curve lying wholly below the 45-degree line *could* describe a stable population (having more than one age in the breeding stock) provided there were moderate to large random deviation from the average relationship. This point is discussed further on page 578, where it is evident that it would be advantageous to fit reproduction curves on logarithmic axes so that deviations above and below the curve would be in better balance. However such plotting has the serious disadvantage that statements of parent-progeny relationships in terms of the slopes of the two limbs of the curve become too complex to be practical. The best way out of this difficulty would be to fit the curve on a logarithmic plot, then transform it to arithmetic axes for interpretation.

examined. The first task will be to consider how a population behaves, under the strict control of a reproduction curve, when it is given some initial randomly chosen position.

This subject, and others later, will be considered for two situations separately. The first is where the currently hatched brood will constitute the whole of the breeding stock of· a subsequent generation; that is, there is no mixing of ages in the spawning stock. This situation is fairly common among insects; for example, many mayflies, stoneflies, caddis flies, etc., have a single brood per year, usually with a one-year life cycle. Among vertebrates this condition is exceptional, but it exists in some fishes. There the length of life is commonly more than one year, so that two or more separate populations or "lines" exist concurrently.

The other situation to be considered is of course that where two or more broods contribute to the spawning stock at any given moment. Among vertebrates at least, this is the usual state of affairs.

REPRODUCTION IN THE ABSENCE OF DENSITY–INDEPENDENT VARIABILITY

SINGLE-AGE SPAWNING STOCKS

In all of the relationships shown in Figures 1–8, the stock is in equilibrium at the density at which the reproduction curve cuts the 45-degree line—that is, the stock is then producing enough progeny, and only enough, to replace its current numbers. In Figure 1, an initial deflection of abundance to either side of the equilibrium point is compensated by a gradual, asymptotic return to equilibrium, for example by paths A–D or a–f. In Figure 2a things work the same way to the left of the equilibrium point; to the right of it any deflection, no matter how great, is returned to equilibrium in a single generation. In Figure 2b deviations to the left, as far as 1.5 units of stock, are also returned immediately to the equilibrium level.

In Figure 3 the equilibrium point is at the top of the dome of the reproduction curve. The curve resembles 1 as regards deflections to the left of the equilibrium point; but a displacement to the right is followed by an immediate return across the 45-degree line to the ascending limb, after which it "climbs" this limb to the equilibrium point (a–d). The curve of Figure 4 is similar, but the dome lies to the right of the 45-degree line; A–E and a–h are possible paths to the equilibrium point.

In Figure 5, a stock deflected from equilibrium to any position along the descending limb will oscillate back and forth about the equilibrium point with decreasing amplitude, for example by the route A–I. If the deflection is great enough, the stock may have to climb the left limb before it is swung over to the right limb and begins the oscillating phase.

In Figure 5 the right limb has a downward (negative) slope numerically less than −1. In Figure 6 this slope is exactly −1 after the dome is passed. Here any moderate deflection along the straight part of the right limb results in a swing back across the 45-degree line of exactly the same magnitude, so that the deflection tends to be perpetuated indefinitely; for example, A and B are two such conjugate points. The intersection of the 45-degree line is a point of "indifferent" equilibrium; it is itself stable, but there is no inherent tendency for the stock to return to that level.

Finally, when the slope of the right limb of a reproduction curve lies between -1 and $-\infty$, equilibrium at the 45-degree line is not merely indifferent, it is unstable. That is, any deflection from equilibrium, no matter how small, initiates a series of oscillations along the right limb whose amplitude increases until the dome of the curve is reached or surpassed. The latter event usually sends the stock back to the right limb and the cycle begins again. No matter where they begin, all such cycles eventually reach the dome of the curve, and a stable oscillation series is established for which the dome is a convenient starting point. The cycle in Figure 7 is A, B, C, D, E, F, G, H, A, etc. (cf. Fig. 11D, below); the number of stages in this cycle depends upon the exact shape of the curve, chiefly upon whether or not one stage lands close to the 45-degree line. Figure 8 represents a more extreme situation, in which substantial reproduction is obtained over only a narrow range of stock densities considerably below the equilibrium level, and the stock would be subject to violent oscillations.

MULTIPLE-AGE SPAWNING STOCKS

When a spawning population consists of two or more age-groups, the young produced in a given year contribute to the stock of more than one future year, and the results of a deflection from equilibrium abundance are much modified. A fairly plausible example is where each brood contributes to the spawning stocks of four future years, in the ratio 2:3:3:2, and first spawning occurs 4 years after a brood was produced (Table I). Figure 9 shows the result of an initial deflection of such a stock to an abundance of 12, on some of the reproduction curves of Figures 1–8. The course of events for the most part reflects what was learned in the single-spawning situation. Those based on Figures 1–6 all end up at the stable equilibrium level; this being reached by direct approach for 1, 2 and 3, with one hesitation for 4, and by a series of damped oscillations for 5 and 6. From Figures 7 and 8 series of undamped oscillations are obtained, that is, permanent cycles of abundance.

It may seem surprising that curve 6 too does not generate permanent oscillations, but the mixing of year-classes gradually brings the stock to a steady level, in the absence of any tendency toward divergence.

The permanent oscillations of the type produced by curves 7 and 8 will repay more extended discussion. Figure 10 depicts series of oscillations, based on Figure 7, but with each brood contributing to the spawning stock for only two years. Time of first maturity is successively delayed one, two and more years, so that the average contribution to reproduction is made 1.5, 2.5, 3.5, etc., years after the brood in question existed as mature eggs. In every case stable cyclical fluctuations exist, just as in Figure 9F. Their *period* is always double the mean interval from egg to egg, that is, 3, 5, 7, etc., years.

The *amplitude* of the cycle varies. If the fish spawn first in the year after their appearance, amplitude is very small, but it quickly increases if maturity

TABLE I. An illustration of how to obtain the definitive distribution of population abundance for given conditions and a given reproduction curve, in this case Figure 7. Four years are assumed to elapse between a brood's existence as fertilized eggs and its first year of contribution to reproduction, while its total egg production is divided among the 4 successive years of its mature existence in the ratio 2 : 3 : 3 : 2. An arbitrary stock density (measured in terms of its egg production) is taken as a starting point, in this case 2.0 units (column 5). The reproduction corresponding to this abscissal value is read off from the ordinate of Figure 7, namely 7.9 egg units (column 6). These eggs are divided among the 4 years in which they are actually laid, beginning 4 years after the year in which the fish that carry them were hatched. For example, the 10.2 units of "recruitment" in line 5 are divided among line 9 (2.0 units), line 10 (3.1 units), line 11 (3.1 units), and line 12 (2.0 units). Breeding potential for each year (column 5) is obtained by horizontal addition of the contributions of the 4 age-groups in columns 1–4. In this example the definitive 11-year period of oscillation is achieved in the first peak-to-peak interval, while the average definitive amplitude, approximately 3.6 to 11.3, is apparent in the second peak-to-peak interval. Adjustment to the conditions imposed by the reproduction curve is not always as rapid as this.

| Eggs produced at successive ages | | | | | | Eggs produced at successive ages | | | | | |
IV	V	VI	VII	Stock	Recruit-ment	IV	V	VI	VII	Stock	Recruit-ment
0.4	0.6	0.6	0.4	2.0	7.9	0.6	1.3	2.0	2.0	5.9	11.4
0.4	0.6	0.6	0.4	2.0	7.9	0.6	0.8	1.3	1.4	4.1	11.5
0.4	0.6	0.6	0.4	2.0	7.9	1.0	0.9	0.8	0.8	3.5	10.7
0.4	0.6	0.6	0.4	2.0	7.9	1.7	1.4	0.9	0.6	4.6	11.8
1.6	0.6	0.6	0.4	3.2	10.2	2.3	2.6	1.4	0.6	6.9	10.3
1.6	2.4	0.6	0.4	5.0	11.9	2.3	3.4	2.6	1.0	9.3	6.4
1.6	2.4	2.4	0.4	6.8	10.4	2.1	3.4	3.4	1.7	10.6	4.2
1.6	2.4	2.4	1.6	8.0	8.7	2.4	3.2	3.4	2.3	11.3	3.1
2.0	2.4	2.4	1.6	8.4	8.0	2.1	3.5	3.2	2.3	11.1	3.5
2.4	3.1	2.4	1.6	9.5	6.1	1.3	3.1	3.5	2.1	10.0	5.3
2.1	3.6	3.1	1.6	10.4	4.6	0.8	1.9	3.1	2.4	8.2	8.3
1.7	3.1	3.6	2.0	10.4	4.6	0.6	1.3	1.9	2.1	5.9	11.4
1.6	2.6	3.1	2.4	9.7	5.8	0.7	0.9	1.3	1.3	4.2	11.6
1.2	2.4	2.6	2.1	8.3	8.2	1.1	1.0	0.9	0.8	3.8	11.1
0.9	1.8	2.4	1.7	6.8	10.4	1.7	1.6	1.0	0.6	4.9	11.9
0.9	1.4	1.8	1.6	5.7	11.5	2.1	2.5	1.6	0.7	6.9	10.3
1.2	1.4	1.2	1.2	5.2	11.9	2.3	3.1	2.5	1.1	9.0	7.0
1.6	1.7	1.4	0.9	5.6	11.7	2.2	3.5	3.1	1.7	10.5	4.4
2.1	2.5	1.7	0.9	7.2	9.9	2.4	3.3	3.5	2.1	11.3	3.1
2.3	3.1	2.5	1.2	9.1	6.8	2.1	3.6	3.3	2.3	11.3	3.1
2.4	3.5	3.1	1.6	10.6	4.2	1.4	3.1	3.6	2.2	10.3	4.8
2.3	3.6	3.5	2.1	11.5	2.8	0.9	2.1	3.1	2.4	8.5	7.8
2.0	3.5	3.6	2.3	11.4	3.0	0.6	1.3	2.1	2.1	6.1	11.2
1.4	3.0	3.5	2.4	10.3	4.8	0.6	0.9	1.3	1.4	4.2	11.6
0.8	2.0	3.0	2.3	8.1	8.5	1.0	0.9	0.9	0.9	3.7	11.0

is delayed. More generally, appreciable amplitude in cycles of this type depends upon a preponderance of the reproduction of a brood occurring after one or more spawning periods have elapsed from the time of its birth.

With the longer intervals of lag between hatching and first spawning the cycles of Figure 10 become less regular; minor peaks appear, and the pattern is duplicated more closely at 2-cycle intervals, a feature which can be detected also in Figure 9F. However, these tendencies are much less apparent when more than two ages occur in the spawning stock, and only the dominant peaks would be detectable with any combination of average age of spawners and average age at first maturity that is apt to occur in nature.

Endless examples of reproduction-curve cycles can be constructed. Any desired combination of period and amplitude can be obtained, in more than one way, by selecting appropriate combinations of reproduction curve and age distribution of breeding. The general characteristics of such cycles can be summarized as follows:

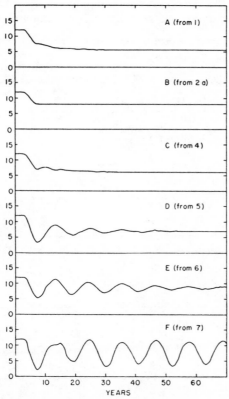

FIGURE 9. Change in abundance of the stocks of Figures 1, 2a and 4–7, following an initial sustained deflection to an abundance of 12 units, when the spawning stock is composed of 4 year-classes and first spawning is in the fourth year after hatching. *Abscissa*—years (generations); *ordinate*—relative abundance (egg production) of the mature stock.

1. Cycles occur when the outer part of the reproduction curve slopes downward, provided this slope begins at some point above the 45-degree line.

2. Cycles are damped and eventually disappear when the slope of the outer limb of the reproduction curve lies between 0 and −1. They are permanent when the slope is numerically somewhat more than −1, the exact critical limit depending upon the amount of mixing of generations in the spawning stock and the interval to first spawning.

3. Period of oscillation is determined by the mean length of time from parental egg to filial eggs, being twice that interval or close to it. It is independent of the exact shape of the reproduction curve, and also independent of the number of generations in the spawning stock, provided there is more than one.

4. Amplitude of oscillation depends partly on the exact shape of the reproduction curve.

5. Amplitude of oscillation tends to decrease with increase in the number of generations comprising the spawning stock.

6. Amplitude of oscillation increases rapidly with increase in number of generations between parental egg and the first production of filial eggs, up to a

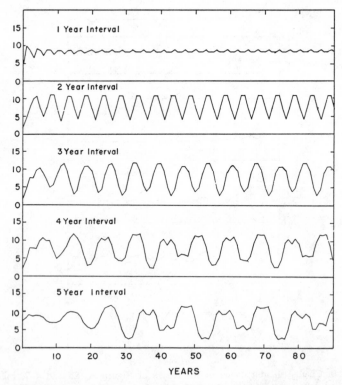

FIGURE 10. Population oscillations determined by the reproduction curve of Figure 7, when there are two ages in the spawning stock and spawning first occurs after 1, 2, 3, 4 and 5 years, respectively, from deposition of the parental eggs. Axes as in Figure 9.

limit imposed by the shape of the reproduction curve. When reproduction by a brood begins strongly in the generation following its birth, the oscillations are so weak that they could not be recognized in practice.

VARIATIONS IN REPRODUCTION PRODUCED BY FACTORS INDEPENDENT OF DENSITY

GENERAL

A comparison of density-dependent and density-independent reproduction is desirable in order to find possible means of distinguishing the two by their effects on population abundance, particularly since it has been suggested that some of the apparently periodic variations in animal numbers may reflect random variability alone (Hutchinson, 1948; Palmgren, 1949; Cole, 1951, 1954).

As an introduction, it is known that quantitative events selected completely at random tend to have an average peak-to-peak interval of exactly 3, provided they are classified finely enough that like values do not occur in adjacent positions. Cole (1951) demonstrates this mathematically and found approximately this period in a selection from Tippett's table of random numbers.

Cole uses a sequence of Tippett's numbers as a model with which to compare cycles of animal abundance in nature. Such a model, however, is not appropriate for our present purpose, because it must be interpreted as a reflection of random variation in the capacity of the environment to sustain the animal in question, rather than random variation in number of mature progeny produced per female. This is true for the following reason. When using Tippett's table or any similar assemblage, the number of numbers available is finite. The smallest number can as easily be followed by the largest as by any other; the largest cannot be followed by anything larger. The biological characteristics of a corresponding population are that it must have a very large potential rate of increase (since the greatest possible abundance can follow directly on the least), and it must have severe compensation at the higher stock densities (since they are close to a ceiling of abundance which cannot be exceeded). Few if any populations could meet both these conditions in the course of a single reproductive period, so that Cole's model, to have verisimilitude, must be applied only to situations where a census is taken at intervals of several generations. Thus it is not appropriate, for example, to fish populations when censused yearly, since these usually spawn only once a year.

In any event, a simple series of random numbers is not a suitable model of random variation *in success of reproduction*. What then *is* a suitable model? Since extremes of environmental conditions are less common than conditions approaching normal, our first postulate will be that a normal frequency distribution provides a fairly realistic picture of the relative frequencies of the resultants of the various independent factors making for success or failure in reproduction; from which resultants a random selection is to be made. Table 8.6 of Snedecor (1946) was used for this purpose, selection of t-values being made corresponding to the figure in the ".00" column closest to each of a sequence of selections from Tippett's random table. The series obtained begins as follows: 1.8, 0.5, 0.2, 0.1, 0.2, 0.0,

165

1.0, 0.1, etc. Tippett's table was again used to divide these into items representing reproduction above and below the replacement level, respectively; this gave the series below, decreases being followed by d.

$$1.8d, 0.5, 0.2, 0.1, 0.2d, 0.0, 1.0d, 0.1, \text{etc.}$$

The next question is to decide how these are to be applied to an existing reproductive potential. An illustration will be of assistance. Suppose that a breeding population matures 10 billion (10^9) eggs; and let us define "average" environmental conditions here as those which will permit the maintenance level of reproduction, that is, 10 billion future mature eggs produced by the progeny of the current year's sexual activity. Variations in environmental conditions add to or delete from this average production. The most severe conditions imaginable can reduce reproduction to zero. The most favourable conditions possible will permit the survival of all the eggs and thus, assuming equality of the sexes, produce future reproductive units to the number of 5 billion times the total expectation of egg production of a female fish just maturing—which expectation might easily be 2,000,000 in the case of the cod, for example. The mean and the extreme limits of reproduction would thus be:

	Mature eggs	No. of fish (at first maturity)
Lower limit:	0	0
Average:	10,000,000,000	10,000
Upper limit:	10,000,000,000,000,000	10,000,000,000

This shows that in a fairly typical instance the stock produced by average environmental conditions is located very asymmetrically with respect to the two extreme limits, on this arithmetic scale. On a logarithmic scale the average production becomes much more nearly central, if a minimum of 1 fish is assumed:

	Log No. of fish
Lower limit:	0
Average:	4
Upper limit:	10

From this and other considerations it appears likely that our symmetrically distributed environmental variations will tend to act in relative rather than absolute fashion; so that, for example, if a given negative deviation produces a decrease in reproduction to half of the average level, the same positive deviation will increase reproduction to twice the average level. In other words, the figures representing environmentally caused deviations from the reproductive norm must be multiplicative rather than additive. To put them into this form, unity is added to each, so that deviations indicating below-average conditions (the d-items above) become divisors, those indicating above-average conditions become multipliers, and zero deviation is multiplication or division by unity. The result is shown in Table II. The absolute magnitude of the items shown is of course arbitrary, and it can be varied at will by multiplying the original random series (before unity was added) by any desired integer or fraction.

TABLE II. Series of numbers selected randomly from a population having the positive half of a normal frequency distribution with a standard deviation of unity, each number augmented by unity, and the series divided randomly into multipliers and divisors (the latter indicated by the suffix "d").

2.8d	1.1d	3.0d	2.2	1.9d	1.3d
1.5	1.1d	1.9	1.1d	1.0	1.6d
1.2	1.9d	1.4	1.0	2.0d	1.2
1.1	2.0d	1.1	1.4	1.1d	2.2d
1.2d	2.4	2.4d	1.7d	2.8	1.3d
1.0	1.5d	1.7d	1.7d	2.7	2.3d
2.0d	2.2	1.3	1.6	2.8d	1.2
1.1	1.3d	2.7d	1.8	3.5d	1.3
2.3d	1.2	1.5d	1.4d	1.2d	3.0
1.8d	1.9	1.9	1.0	1.2d	1.3
1.1d	2.0d	1.9d	1.8	1.7	2.2d
1.7	2.1d	2.1d	1.0	1.9d	3.2d
1.8d	1.5	1.8d	2.2	2.6	1.4d
2.5	1.3d	1.8	1.0	2.4d	1.5
2.0	1.6d	2.6d	1.5d	2.5	1.9d
2.0d	1.5d	2.3	2.2d	2.0d	1.1d
2.0	1.0d	1.7d	1.6d	2.2d	2.6
2.2d	1.9d	1.1	1.9	2.1d	2.5
2.2	1.9d	2.8	1.7	1.9	2.1d
1.3	1.3	1.3	2.6	1.8d	2.6d
1.4	2.1d	1.1d	2.4d	2.8	2.1d
1.2d	2.1	1.6	1.5	1.3	1.8
2.0d	1.2	1.2d	2.0d	1.7d	2.6d
1.7d	1.8	1.0d	1.7d	1.6	1.3
1.1d	2.0	1.7	1.0	1.7d	2.5d
1.5	1.1d	1.4	1.2d	1.8	2.7d
2.2	2.3	1.1	1.5	1.0	1.1
1.6	1.2	2.1d	2.0	2.0d	1.3
1.4d	2.0d	2.1	1.3d	2.0	1.7d
2.0d	1.6d	1.9d	1.8d	1.6	1.1d
1.3d	2.4d	1.2d	2.3	1.0d	2.0
3.4	1.1d	1.1d	1.6	1.3d	1.6
1.3	1.7d	1.6	1.3d	3.7d	3.3d
3.2d	1.9d	1.5d	1.7d	1.3	2.3d
1.3	2.7d	1.3	1.3d	1.0	1.5d
2.2d	1.5	2.2d	1.7	1.3	1.6d

SINGLE-AGE SPAWNERS

Since the figures of Table II constitute a model of variation in success of reproduction under the influence of density-independent factors alone, a picture of corresponding population fluctuations is obtained by multiplying some initial stock density by each item in succession. This is done in the lower line of

Figure 11A; the upper line is from a different random series. The lines fluctuate a good deal and at times diverge considerably from the original abundance: at the end of the 70 generations shown in Figure 11 the Table II line has changed by a factor of 260; while if the same is continued through the whole 216 generations of the Table the net change is 5.5×10^{-7}, representing a relative decrease

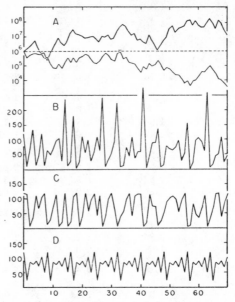

FIGURE 11. Fluctuations in ideal populations of single-age spawners. *Abscissa*—generations; *ordinate*—number of units of stock.

    A.   Two series of changes in a population from an initial density of 1,000,000, when it varies under the influence of density-independent factors only. The lower line corresponds to the first 70 entries of Table II, multiplied in sequence.

    B.   Changes in a population from an initial abundance of 119, as determined by the random factors of Table II and the reproduction curve of Figure 7, when the action of the latter precedes the former.

    C.   As in B, but with the random factors acting before the compensatory mortality shown by the reproduction curve.

    D.   Population cycle generated by Figure 7 alone.

from 55,000,000 to 1. This large change occurs in spite of the fact that Table II is constructed in such a way that, if it were continued over any really long period, the population should be above its initial level about as often and as much as it is below it[5]. The net change results partly from an excess of d-items in Table II, and partly from the fact that the d's happen to be larger numbers, on the whole.

[5]A series of 1,000 "generations" constructed in a different but analogous fashion by Hutchinson and Deevey (1949, fig. 4) shows similar major long-period trends.

By chance the change indicated by Table II is considerably greater than would be expected to occur often in a series of only 216 generations. However it is easy to show that as the number of generations increases, the most probable divergence of population size from the original level also increases continuously. When the number of generations approaches anything appropriate to a geological time scale the likely change is very great indeed. For example, after 10,000 generations the standard deviation of the number of increases (or decreases), from the most probable number 5,000, is $\sqrt{10,000 \times \frac{1}{2} \times \frac{1}{2}} = 50$. This means that there is about 1 chance in 3 that the excess of increases or decreases will be 50 or more at that time, while the chances of the excess having been 50 or more at some time *during* the 10,000 generations are much greater. With an average change factor of 1.67 from one generation to the next (as in Table II), this means that the population is fairly certain to have been increased or decreased by a factor of about 1.67^{50}, or 140,000,000,000, somewhere along the line. Without wasting thought on the impossibility of a population of any sizable organism *increasing* this much, such a *decrease* would obviously bring even the most numerous fish stock, for example, to extinction. Thus no population of single-age spawners can survive if its reproduction is completely at the mercy of density-independent influences in the environment. In the same way a gambler will in the long run lose everything if his opponent has equal skill and unlimited resources. (In practice, of course, the effects of random environmental factors would usually vary as between different parts of the animal's range, and local extinction could be followed by recolonization from adjacent areas.)

The average interval between peaks of Table II should be close to the 3 which is characteristic of a random series; in actuality it was 3.09 when ties were randomly divided into higher and lower values. However this is not what determines the period of a series such as Figure 11A. In the latter a peak occurs whenever a multiplier is followed by a divisor, and the average interval between peaks is nearly 4 years, again adjusting for ties. The theoretical or long-term average period can be determined as follows: since a peak or trough occurs whenever there is a change from multiplier to divisor or divisor to multiplier, the expectation that the first such change will occur between adjacent cells is one-half, the expectation that it will occur between a cell and the next but one is one-fourth, that it will occur at the third cell is one-eighth, and so on. The average of these intervals 1, 2, 3, etc., weighted as to frequency, is exactly 2, which represents the average peak-to-trough interval. Since peaks and troughs are equally common, the average peak-to-peak or trough-to-trough interval is therefore 4.

However it must be emphasized that this 4 represents the average interval between *all* peaks, regardless of size. Casual inspection of Figure 11A might give a different impression. The eye tends to ignore the smaller humps, and to impose a certain regularity among the rest by magnifying those of intermediate size when they happen to fall into a sequence with large ones, and diminishing them when they do not. In this manner the lower line could "suggest" a cycle of 11 or 12 years, with 4 peaks and 4 troughs actually showing.

MULTIPLE-AGE SPAWNERS

To obtain a model of the consequences of random fluctuations upon reproduction of multiple-age stocks, the random values of Table II have been applied to a "population" constructed on the same basis as Figure 9—that in the spawning stock there are 4 ages, that each year-class produces eggs in the ratio 2:3:3:2 in the 4 years of its reproductive activity, and that each fish spawns for the first time 4 years after it was itself a fertilized egg. The resulting curve is shown in Figure 12A (it starts from an age distribution characteristic of the reproduction-curve cycle of 12D).

The most distinctive characteristic of the line of Figure 12A is that it tends to rise. This increase occurs in spite of the fact that divisors happen to be in excess in Table II, and if the series is continued a little farther much higher values are encountered. The fact is that, if continued, the population will increase without limit. The reason is that the contributions of the several year-classes to a given year's spawning must be added arithmetically, while expectation of

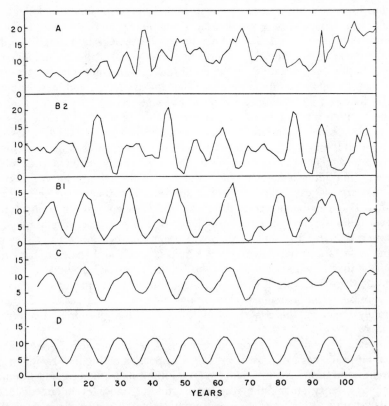

FIGURE 12. Fluctuations in a population of multiple-age spawners, as determined by the random series of Table I (curve A), by the reproduction curve of Figure 7 (curve D), and by combinations of these (curves B and C). Axes as in Figure 9.

survival of an egg is proportional to random environmental effects having a symmetrical distribution so constituted that any increase is greater, absolutely though not relatively, than the corresponding decrease. It would be difficult, of course, to prove that environmental effects upon reproduction are actually distributed in just this manner. On the other hand it was shown above that they are very unlikely to be merely additive; and any intermediate condition will result in a slower but still unlimited upward trend to a graph such as Figure 12A.

It thus appears that any likely random sequence of environmental changes, even one which for single-age spawners causes a catastrophic decline, can be made to produce instead a large increase in abundance of stock by the simple device of dividing the spawning of each year-class among several calendar years. Even though wholly non-compensatory reproduction cannot occur in nature, this circumstance may have considerable advantage for any population in which random variability is large and/or the left limb of the reproduction curve not very steep (e.g., as in Figs. 1, 2a, 3 or 4). Indeed it is easy to show that in the presence of random variability a multiple-age spawning stock can maintain itself even when its reproduction curve lies wholly below the 45-degree line—though of course it cannot be *too far* below. Thus there is a sound ecological basis for the customary occurrence, in nature, of reproductive assemblages consisting of more than one age-group.

As regards period, we find again that the peaks and troughs of Figure 12A have a certain apparent regularity in spite of their random origin. Discounting the first 15 years as being still somewhat under the influence of the reproduction-curve cycle, and counting all peaks no matter how small, there are 17 peak-to-peak intervals in the figure, of which no less than 10 are of 5 years' duration—the average interval is 5.2 and the range 2 to 7. The fact that the average interval is longer than the 3 which is characteristic of a simple random series (and longer than in the series for single-age spawners) is explained by the method used to obtain it. This is quite similar to taking a "running average" of four items at a time, and Cole (1951, fig. 1) has shown that applying such a procedure to a random series increases the mean peak-to-peak interval. "Artificial" trends of really long period are not as apparent in Figure 12A as in 11A, but a longer series might have confirmed them.

In summary, the characteristics of population changes which would result from density-independent factors acting alone (if that were possible) are as follows:

1. A population of single-age spawners would vary widely above and below its initial abundance; eventually, after a few thousand generations at most, it would either become extinct or, more likely, be fragmented temporarily into small independent units.

2. Populations of multiple-age spawners would increase in abundance indefinitely, though not without ups and downs.

3. The average peak-to-peak period for lines of single-age spawners would be 4.

4. The average peak-to-peak period for multiple-age spawners would be

more than 3, and would be the greater, the greater was the number of ages in the spawning stock; in the example used it was about 5 years.

5. For single-age spawners, and probably for multiple-age spawners as well, "cycles" having longer periods than these would tend to be apparent in graphs of abundance, because of conscious or unconscious mental suppression of small peaks and troughs, and regularization of large ones.

COMBINATIONS OF COMPENSATORY AND NON-COMPENSATORY MORTALITY

In natural populations non-compensatory mortality is superimposed upon the reproduction expected from density-dependent factors, and the curve of population fluctuation is the resultant of both. The random effect can be introduced either before or after the amount of reproduction indicated by the curve is written down, according as the random factors are thought to act before, or after, the compensatory ones. The latter procedure gives greater influence to the random element, and may correspond better to events in nature, though the two types of mortality may of course act concurrently, and to a considerable extent probably do.

SINGLE-AGE SPAWNERS

Combinations of Table II or similar series can be made with the various kinds of reproduction curve shown in Figures 1–8. It seems unnecessary to reproduce examples of most of these. With a flat-topped curve such as Figure 2b, and the scale of random variability shown in Table II, the resultant reproduction is practically always equal to the product of the equilibrium abundance and the multiplier or divisor for the year in question. Only an improbably large deviation could shift the population over to the ascending part of the curve for a year. Relatively slight modifications of this situation are obtained when broad-domed curves having a stable equilibrium point are used (e.g., Figs. 3 or 4); the principal difference being that, on the whole, reproduction is somewhat less.

Fluctuations produced by a combination of the reproduction curve of Figure 7 with the random series of Table II have some of the characteristics of either component (Fig. 11). The regular series A–H of Figure 7, repeating at 8-year intervals, is replaced by an irregular one. The average peak-to-peak period is about 3.5—slightly less than that of the random series (which was close to 4) and greater than the 2 of the reproduction-curve sequence. The average amplitude of the "cycles" (ratio of peak to preceding trough) is greater than what either component series exhibits, but there are of course no extreme trends in abundance such as resulted from random causes alone. There is a pronounced tendency for each peak to be followed immediately by a trough, which is a peculiarity also of the random-curve series.

The above description applies whether the random influence operates before the reproduction-curve (Fig. 11C) or after it (Fig. 11B). The two kinds of series are quite similar, but the latter of course has the greater amplitude of changes.

172

MULTIPLE-AGE SPAWNERS

Combination of compensatory and non-compensatory reproduction in multiple-age stocks will be considered only for the case where the former precedes the latter, and will be illustrated by means of a population in which contributions to reproduction are spread over four years in the ratio 2:3:3:2, as in Figure 9.

In general, the kind of fluctuation which results from any combination of compensatory and non-compensatory mortality factors depends upon the relative magnitudes of the two components. The action of the same series of random factors is shown in Figures 12B and 12C at two levels of intensity which are in the ratio of 5:1; Table II is used for B, and values one-fifth as large for C, i.e., 1.36d, 1.10, 1.04, etc. In each case they are combined with the reproduction indicated by Figure 7. The same initial age distribution is used for all series of Figure 12.

By itself, Figure 7 yields the steady oscillation of Figure 12D. At the lower intensity of random effects (line C) the population cycle determined by the reproduction curve is not too seriously altered: there is variation in amplitude, and the peaks move out of phase by as much as a fourth at times, yet the prevailing periodicity can be determined fairly accurately from even a short series. At the higher intensity of random effects, however, much greater disturbance is evident; it is followed through in two lines of Figure 12, beginning with B1 and continuing in B2. As regards *period*, the 11-year cycle is first increased to 15–17 years, then reduced to 7 or 8 (counting major peaks only). Though the average over a long period may thus tend to 11 years, peaks and troughs move out of and back into phase with the reproduction-curve cycle and also with the random "cycle", which are its determiners. More serious is the fact that the smaller peaks and troughs of what appears to be the main series are in some cases impossible to distinguish from the "artifacts" resulting from random fluctuation: hence the average 11-year period above could not be discovered in practice, even with a very long record. As regards *amplitude*, the maximum ratio of peak to adjacent trough is much greater than is found in either the reproduction-curve cycle or the random series.

When the random element is given still greater relative importance, the reproduction-curve element becomes unidentifiable as such. However the latter continues to make an important contribution to the resulting population changes. It makes peaks and troughs much less numerous than they would otherwise be, and it provides a control of limits of abundance—that is, the progressive upward tendency of the random curve is effectively curbed.

The relative importance of random and compensatory factors in determining population abundance also depends somewhat upon the number of ages represented in the spawning stock. When this is large, random factors are less effective in disturbing reproduction-curve cycles.

EFFECTS OF REMOVAL OF MATURE STOCK

SINGLE-AGE SPAWNING STOCKS

Many natural populations have a part of their members removed by man, either because they are useful to him or, in the case of harmful species, because he hopes to reduce their abundance. The case where only *mature* stock is taken by the fishery is considered in examples to follow. Many fisheries take a portion of the immature stock also, but this appears not to add any new principle to what is considered below, provided the region of compensatory mortality is not invaded.

The effect upon the various population parameters of removing a constant percentage of the mature stock, before reproduction, is shown for four types of single-age-group populations in Figures 13–16. In Figure 14, for example, a spawning stock OQ produces filial spawners equal to PQ. However, after the fishery has removed 40 per cent of the mature stock the number remaining is equal to OQ, which again produces PQ, and so on. Thus P is the equilibrium position for 40 per cent

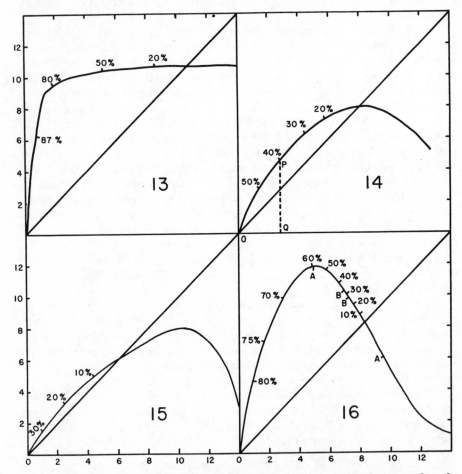

FIGURES 13–16. Equilibrium densities of stock for the various rates of exploitation indicated by percentages on the reproduction curve. Stock density before fishing is the ordinate value corresponding to the point on the curve opposite the percentage exploitation in question. Axes as in Figures 1–4.

exploitation. The fraction of the mature stock which must be removed prior to spawning, in order to maintain equilibrium at any point on the reproduction curve, is equal to the complement of the reciprocal of the slope of a line joining that point to the origin.

The general effect of exploitation is to move the point of equilibrium abundance to the left on the reproduction curve. In Figures 13–15 this means that under exploitation the equilibrium population is always smaller than it is under natural conditions. In Figure 16 (which is the same as 7), removal of part of the spawning stock at first results in a *larger* equilibrium level of stock, and the same is true of Figures 5, 6 and 8. When rate of exploitation becomes large enough, however, the stock is again reduced. Equilibrium points for various rates of exploitation are indicated on Figures 13–16.

Oscillating equilibria are much changed by exploitation. Their amplitude is reduced, and the number of positions through which they swing is decreased, or oscillation may be eliminated entirely. However, stable oscillations persist when exploitation is light to moderate. In Figure 16, for example, with 10 per cent removal the stock traces the cycle 11.9, 4.0, 10.9, 5.5, 11.9, etc.; with 20 per cent removal it alternates between the levels of 6.1 and 11.9 units (A–A); with 30 per cent removal the oscillation is reduced to between 9.9 and 10.3 (B–B), so that it would not be distinguishable in practice; and with 60 per cent removal it disappears completely[6].

The point of maximum sustained yield can easily be computed from graphs like Figures 13–16. For example, in Figure 16 it is at 65 per cent exploitation; in Figure 13 it is close to 80 per cent; in 15 only about 18 per cent. In general, the greater the area between the reproduction curve and the 45-degree line above the latter, the greater is the optimum rate of exploitation.

Beyond the level of maximum yield lies a level of maximum permissible exploitation above which the stock is progressively reduced to zero. This limit is a function of the maximum angle made by any line joining a point on the reproduction curve to the origin; and as noted above, the maximum rate of removal which can be sustained is equal to the complement of the cotangent of that angle.

If a man's interest in a single-age population, of a noxious insect for example, is to reduce it to as low a level as possible by increasingly intensive destruction, he can be sure of making *direct* progress toward the desired goal only if the reproduction curve is one of the types shown in Figures 1 and 2. If it is like Figures 3 or 4, and stock happened to be on the outer limb, some moderate rate of removal would dampen a crash which otherwise was imminent. With curves of types 5–8 a paradoxical situation develops. Moderate destruction of adults will in general tend to stabilize the population at or about some magnitude *greater* than its primitive average abundance. In Figure 16 maximum abundance is con-

---

[6]When equilibrium is unstable the *average* size of the populations of the cycle tends to be less than the equilibrium level. In Figure 16 the average is 7.7 for no exploitation, 7.8 for 10 per cent, and 9.0 for 20 per cent, as compared with the equilibrium values of 8.2, 8.6 and 9.4. For 30 per cent or greater exploitation, equilibrium stock and average stock are the same.

sistently achieved when as much as 60 per cent of the spawning stock is destroyed each year. To reduce the population below the primitive average, more than 73 per cent must be destroyed. Thus although sufficiently intensive effort will be successful, moderate destruction of the mature population is worse than no action at all[7]. Furthermore, if an intensive campaign of continuing control of a variable species is decided upon, it is most efficient to begin it at a time that the population is at a low point of its cycle, i.e., when the pest may be doing no particular damage.

MULTIPLE-AGE SPAWNING STOCKS

The effect of a fishery upon stocks containing more than one age-group of spawners parallels what has just been found for single-age-group stocks. If the stock is one for which a stable equilibrium exists (Figs. 1–5), adding a fishery does not disturb the stability, but the abundance of the population is changed (in 2b the change begins only after the inflection point is reached). With the curves of Figures 1–4 the change is in the direction of a decrease in abundance, but with Figure 5 exploitation at a moderate rate *increases* the adult stock, until the dome of the reproduction curve is reached.

Stocks which perform regular oscillations in the absence of a fishery (Figs. 7 and 8) are changed in three respects when exploitation begins: (1) their average equilibrium abundance is at first increased, but later decreases again if exploitation becomes sufficiently intensive; (2) the amplitude of oscillation decreases; and (3) the period of oscillation tends to decrease slightly.

1. The first-named effect can be estimated rather easily, since the average abundance of multiple-age stocks proves to be practically the same as the equilibrium abundance. The latter has been calculated for Figure 16 using various rates of exploitation, as shown on that Figure and in Table III. The maximum abundance is at the dome of the curve, and maximum catch is obtained slightly to the left of it, at 65 per cent exploitation.

2. The advent of fishing not only affects the total abundance of a stock, but also gives it a younger age composition, because survival rate from year to year is reduced. In the example of Table IV, based upon Figure 16, the fish are assumed to mature first at age III. Fishing takes place just prior to spawning and rate of fishing is the same for all ages. Natural mortality occurs between successive fishing-and-spawning seasons, and is 20 per cent from age III to age IV, 20 per cent from IV to V, 30 per cent from V to VI, and 100 per cent after the spawning at age VI. The average weight of a fish at time of fishing-and-spawning is 2 units at age III, 4.28 units at IV, 7.14 at V and 9.52 at VI; and egg production is proportional to weight. Under these conditions the *equilibrium* distribution of the contributions of the several age-groups to egg production, at different rates of exploitation, is shown in Table IV.

[7]A similar conclusion is reached in Nicholson and Bailey's (1935) detailed analysis of host-parasite interaction in insects. Varley (1947) concluded that a trypetid fly population was maintained at about 10 times the density it would otherwise have had, because of the presence of non-compensatory causes of mortality which killed host and parasite equally.

TABLE III. Average or equilibrium abundance of stock, by weight, in a stock having more than one age at maturity, and the catches at various rates of exploitation; from Figure 16.

Rate of exploitation	Stock before fishing	Catch
%		
0	8.2	0
10	8.8	0.9
20	9.5	1.9
30	10.1	3.0
40	10.8	4.3
50	11.5	5.8
60	11.9	7.1
65	11.4	7.4
70	9.7	6.8
75	7.2	5.4
80	5.0	4.0

TABLE IV. Relative weights and contributions to reproduction of the age-groups in the stock described on page 583.

Rate of exploitation	Fraction of eggs contributed by age			
	III	IV	V	VI
%				
0	0.200	0.300	0.300	0.200
10	0.233	0.314	0.283	0.170
20	0.273	0.327	0.261	0.139
30	0.320	0.336	0.235	0.109
40	0.377	0.339	0.204	0.081
50	0.446	0.333	0.165	0.055
60	0.526	0.315	0.127	0.032
70	0.619	0.278	0.083	0.020
80	0.737	0.221	0.042	0.000

The first line of Table IV was arranged to have the same distribution of contributions to spawning as used for Figure 9F. In Figure 17 are shown cycles obtained from the 10 per cent and 20 per cent lines of Table IV, starting from the equilibrium distribution characteristic of no fishery[8]. At 10 per cent exploitation the oscillation of the population is maintained, though at reduced amplitude and about a higher mean level. At 20 per cent exploitation oscillation gradually decreases to an inconsiderable amplitude, and at higher rates of exploitation it

[8]This procedure is the only practicable one, but it ignores the fact that the age distributions of Table IV would themselves be completely realized only after a period of 4 years from the beginning of fishing. Consequently the equilibrium situations in Figure 17 would actually be approached more gradually than the figure indicates.

177

disappears completely. However even at 20 to 40 per cen⁺ exploitation the oscillation persists through several cycles.

3. The length of a cycle was found earlier to be twice the mean length of a generation (interval from egg to egg). In the first line of Table IV this is 4½ years, and the cycle is 9 years. With the higher exploitations of Table IV a shift of age distribution toward younger fish occurs, and we accordingly expect

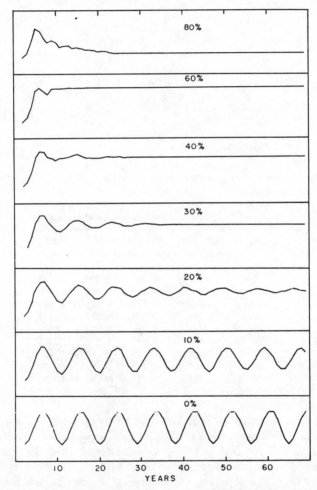

FIGURE 17. Reactions of an hypothetical population to exploitation. The population is one regulated by the reproduction curve of Figure 7, with ages III, IV, V and VI contributing to egg production in the ratio 2:3:3:2 when there is no exploitation. The stable cycle characteristic of no exploitation is shown by the 0% line. The other lines show the effect of annual removal of the indicated percentage of the mature stock from a population which initially had the age distribution characteristic of the start of the ascending phase of the no-exploitation cycle. (See also footnote 8.)

178

a shorter cycle to appear. In this example, however, the cycles disappear before much decrease in period can take place. At 10 per cent exploitation average age of contribution to reproduction is reduced to only 4.39, and at 20 per cent it is 4.27. The corresponding predicted cycle lengths are 8.8 and 8.5 years, and the peaks shown in Figure 17 do in fact become progressively a little closer together. In a somewhat more realistic example there would have been more than four age-groups in the stock under conditions of light exploitation or none, and in that event the reduction in period of oscillation with increased fishing would be greater. However it is doubtful whether in nature the reduction would ever be great enough to be identifiable with certainty, before increasing exploitation removed oscillation of this type entirely.

When the reproduction curve is dome-shaped, *control* of an undesirable population of multiple-age spawners presents some of the same difficulties as described earlier for single-age stocks. However the initial favourable reduction in abundance will tend to last longer—i.e., until the increased broods of young can grow to the size at which they cause damage. Only when this first phase is over, and the fishing effort has to cope with the larger broods which come from the reduced spawning populations, will it be apparent whether or not control can be permanently effective. Few recorded attempts to reduce nuisance fishes have lasted beyond the initial stage of removal of old stock. Foerster and Ricker (1941) have described the rather easy success achieved in this preliminary phase of control of squawfish (*Ptychocheilus*) in a small lake. The experiment was subsequently continued until one or more large broods of young began to grow into the damaging size range, but it did not last long enough to see if these could be kept sufficiently reduced by appropriate effort. In Lesser Slave Lake, Alberta, unlimited fishing for ciscoes began in 1941 with the hope of thus reducing the lake's population of the tapeworm *Triaenophorus*, and as a result the spawning population has been changed gradually from one 5–7 years old to one now (1951) mostly 2 years old. However recruitment and rate of growth have increased sufficiently to maintain a high level of catch and a fairly large population mass, so that control is not yet effective for the purpose intended (Miller, 1950, 1952).

When non-compensatory variability is added to the effects of exploitation and the reproduction curve, the kind of result obtained differs little from what was discovered from Figure 11. The amount of disturbance introduced depends of course upon the relative magnitude of the random factors.

EXAMPLES OF REPRODUCTION CURVES

A wider recognition of the different possible kinds of reproduction curves, and their different properties, will be of value mainly as a guide to current and future research. Such curves should be useful not only as a possible clue to the nature of observed fluctuations, but still more in estimating expected average recruitment when stock is reduced (or increased!) by fishing it, at different intensities. Knowledge of the reproduction curve will thus complement information on rate of growth and natural mortality, and permit more accurate prediction and regulation of fish catches.

Some examples are given below of reproduction curves, for fish and a few other animals, which can be plotted from available data. In several instances information is available concerning only the first part of the life history, but since that is where most compensatory mortality is expected, the data are included. A disadvantage of such curves, however, is that the "45-degree line" can usually be located only approximately, if at all.

PACIFIC HERRING (*Clupea pallasi*)

The herring of the west coast of Vancouver Island are a natural unit which has been given special study. Recently Tester (1948) examined the relation between the intensity of spawning in the brood year, and an estimate of resulting year-class strength based on catch per unit effort in the various years that each brood contributes to the fishery. Tester had two indices of spawning; using "Index A", which was available for five years longer than "Index B", he computed an inverse correlation of −0.30 between spawn deposited and resulting year-class strength for the period 1931–43, though this figure does not differ significantly from zero.

Mr. J. C. Stevenson has very kindly added data for more recent years, and these are plotted in Figure 18. In this figure the more elaborate "Index B" is used, but the five years 1931–34 are included (open circles) using values from the regression of B on A for the years of overlap. The negative correlation −0.16, for the years 1931–49, differs from zero even less than the one computed by Tester, principally because a series of abundant year-classes from 1943 to 1947 has increased the range of variability. However it is still true that the largest

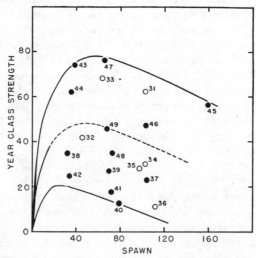

FIGURE 18. Relative strength of year-classes of a Pacific herring stock (ordinate) plotted against relative amount of spawn deposited in the brood year (abscissa). The two solid lines include the observed limits of variation, while the dotted one is an approximate middle value. Data from Tester (1948) and Stevenson (MS).

broods have been produced by smaller-than-average spawnings, whereas the smallest broods have been produced by larger-than-average spawnings.

Dr. Tester and others have also considered the possibility that trends in herring recruitment may exist corresponding to some as yet unidentified rhythm of the ocean environment; and the 1943–47 stanza of good year-classes is in fact fairly well set off from earlier (1934–42) and later (1948–49) stanzas of lower production (Fig. 18). If there is really any systematic effect here, one may notice that *within* each of the two longer stanzas a negative correlation between spawning and recruitment is indicated.

Pink Salmon (*Oncorhynchus gorbuscha*)

Among pink salmon there are two completely separate reproductive lines[9], so that is is very easy to identify the progeny of any year's spawning when it matures two years later. From Pritchard's (1948a, b) summary of the reproduction of the stock which spawns in McClinton Creek, Graham Island, British Columbia,

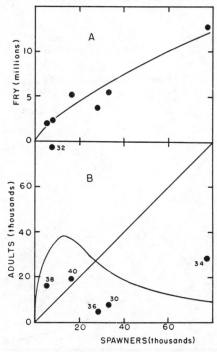

Figure 19. A—Fry produced in millions; and B—returning adults in thousands; plotted against the number of pink salmon spawners in the brood years indicated, at McClinton Creek, B.C. Data from Pritchard (1948a, b).

[9]Separate or nearly separate breeding stocks of salmon are commonly called "cycles", "years" or even "races" in western North America. The term "lines" was proposed by Huntsman (1931), and is used here because it avoids confusion with other kinds of cycles, etc. (cf. Clemens, 1952).

the reproduction curve of Figure 19B is plotted. The progeny shown does not include the commercial catch, which is a considerable but unknown quantity, and hence all points would be moved upward on a graph that represented total adult production of the spawnings indicated. No very clear idea of the shape of the reproduction curve can be obtained, but it seems to slope downward from 15–20 thousand spawners.

Dr. Pritchard also had records of fry production, and these are plotted in Figure 19A. The fact that *percentage* fry production decreases as population increases indicates that compensatory mortality occurs in the freshwater part of the life history. That compensation occurs also during saltwater life is suggested by the change from a generally upward slope in Figure 19A to a generally downward slope in 19B. Obviously too there is much random variation in mortality at both stages, but particularly in the ocean. The ocean phase includes the effect of exploitation by the fishery, which is certainly variable and might tend to be

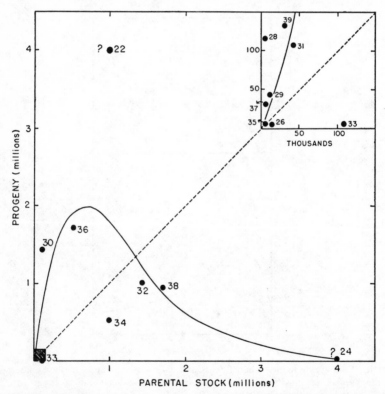

FIGURE 20. Reproduction curve for pink salmon of the Karluk River, Alaska, based upon fence counts except in 1922 and 1924 (see text). The shaded square in lower left is shown on a larger scale in the upper right corner. Commercial catches are not included in the progeny. (From unpublished data of the United States Fish and Wildlife Service, Pacific Salmon Investigations.)

compensatory, because more seiners are attracted to the bay off McClinton Creek when fish are numerous there.

A reproduction curve having a very wide range of parent-stock densities is that for pink salmon of the Karluk River, Alaska. This is shown in Figure 20, plotted from data supplied by Messrs. C. E. Atkinson and C. J. Burner of the U. S. Fish and Wildlife Service. "Progeny" figures do not include the commercial catch; adding the latter could not alter the general form of the curve, but would raise it farther above the 45-degree line. The 1926–40 data are from complete or nearly-complete fence counts. The 1924 escapement was estimated as "over 4,000,000" (Barnaby, 1944, p. 257). Unfortunately there is no numerical estimate of the 1922 run; however, since it is called "large" on the same page as the 1924 run is called "tremendous" (Barnaby, p. 249), it might be assigned a figure of about one million, and on that basis a point for 1922 is plotted in order to better indicate the vertical range of reproduction.

Coho Salmon (*Oncorhynchus kisutch*)

The experiment at Minter Creek, near Tacoma, Washington, has provided information on the relation of spawning stock to smolt output (Smoker, 1954). Production of year-old smolts is related to number of female spawners in Figure

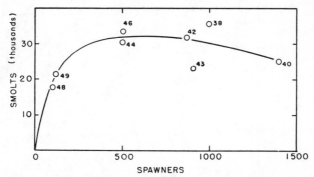

Figure 21. Smolts produced by adult female coho salmon spawners of the brood years indicated, at Minter Creek, Washington. Data from Smoker (1954).

21, based on the same data as the upper half of Smoker's figure 29. The productive capacity of the stream appears to be 25–35 thousand smolts, depending mainly on volume of flow (Smoker's fig. 29, lower half), and if adjustment is made for the latter, the already-flat outer limb of Figure 21 becomes even flatter, with the deviations of the points from the line considerably reduced.

Estimates of total adult production from these smolts are not possible; there is no record of catch, and even the number of returning spawners is available for only four years.

Sockeye Salmon (*Oncorhynchus nerka*)

The sockeye of Karluk Lake, Alaska, mature at 3 to 8 years of age, the mean being a little over 5 years. Barnaby (1944, fig. 3) has plotted a graph of spawning stock against adult progeny produced; the latter was obtained by summing the contributions of the 1921–29 year-classes to subsequent catches and breeding stocks. This graph is practically equivalent to a reproduction curve based upon eggs, the only difference being that possible variation in the number of eggs produced by females of different ages is not taken into account.

Barnaby separates "spring" and "fall" fish in his plotting. Either in this form, or when the data for each year are combined (Fig. 22), there is much variability in production, but a peak apparently occurred when the total spawning stock (spring plus fall) was between 1 and 1½ million fish. Unfortunately, both before and after this period the sockeye stock of the lake decreased, so the 1921–29 results cannot apply either to the earlier period when *catches* of 2–3.5 million fish were customary, or to the modern period when the total stock has fallen below a million. The causes of this decrease are still under study.

An opportunity to document the *ascending* phase of a sockeye reproduction curve is afforded by the "late Shuswap" or Adams River race on the Fraser River. A blockade in 1913 reduced this and other Upper Fraser races of sockeye to a level that was only a small fraction of their natural abundance. These fish mature preponderantly at 4 years of age, so that an adequate reproduction curve

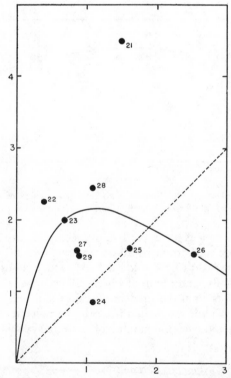

FIGURE 22. Number of mature sockeye (catch plus escapement) produced by the year-classes of 1921–29, at Karluk Lake, Alaska. Data from Barnaby (1944).

can be plotted using estimates of spawners and catches every fourth year (Fig. 23). Sizes of the spawning population are based on estimates of Fishery Inspectors and, in recent years, on figures given in Annual Reports of the International Pacific Salmon Fisheries Commission; the catch is taken as the total Fraser commercial catch each year, less 1,200,000 fish representing the approximate production of other races in this series of years. The curve has a much steeper ascending limb than the Karluk curve; there is a suggestion of a maximum at 1–1.5 million spawners, and possibly a start on the downward path beyond that point.

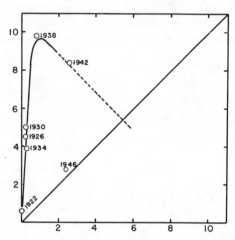

FIGURE 23. Reproduction curve for the late Shuswap race of sockeye salmon, based upon estimated numbers of spawners and an approximate division of the commercial pack. *Abscissa*—estimated number of spawners in the years indicated; *ordinate*—resulting catch plus escapement. Both scales are in millions of fish. The year of spawning is shown beside each point. From data of the B.C. Department of Fisheries and the International Pacific Salmon Fisheries Commission.

The steepness of the left limb of the Shuswap curve is worth noting, being of the same order as that of Figure 8. If the right limb has any considerable slope downward, it would have a tendency to mould chance fluctuations into swings back and forth across the 45-degree line. In this connexion one may note that the original dominant line of sockeye on the Fraser, of which the late Shuswap race was a component, is supposed to have been considerably less abundant in 1909 than in either 1905 or 1913 (Babcock, 1914). This might represent two swings across an equilibrium position, under the then conditions of a fairly low percentage exploitation of this line by the fishery (compare A–A of Fig. 16). Under present conditions the fishery normally takes enough to keep the spawning stock near the dome of the curve, and special measures are taken to ensure this much escapement when below-normal reproduction has occurred, as from the 1946 spawning (Royal, 1953).

HADDOCK (*Melanogrammus aeglifinus*)

Herrington's (1941, etc.) analysis of the George's Bank population is particularly valuable because he was able to obtain from market reports an estimate of the smallest commercial size (scrod—mostly age III) independently of and in addition to the larger sizes. This yields an estimate of recruitment from year to year, which in turn is a reflection of success of reproduction. When size of spawning stock in the years 1912 to 1929 is plotted against scrod produced (solid circles, Fig. 24), a steep outer limb of a recruitment curve is apparent. The 45-degree line has been located very approximately by considering that a pound of scrod during those years had an expectation of egg production, throughout its life, equivalent to the actual egg production in one year of 10 pounds of mature haddock; and by assuming that scrod were half as vulnerable as were adult fish

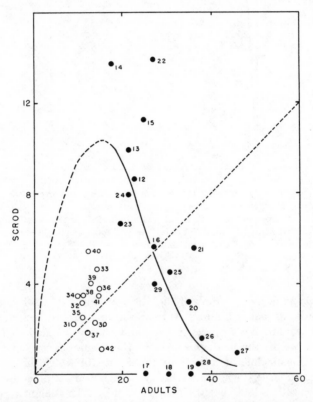

FIGURE 24. Abundance of adult haddock on George's Bank during spawning season (February–April), related to abundance of scrod (age III, mostly) at the same season three years later. Both scales are in terms of thousands of pounds per day caught by trawlers of a certain size class, the figures for adults being smoothed. Black circles represent the years 1912–29, open circles 1930–43. The reproduction curve is drawn for the earlier period only. The diagonal is an *approximate* 45-degree line for that period (see text). Data from Herrington (1948).

to the trawlers whose catches supply these data. Figure 24 is the same as Herrington's figure 11 of 1948, except that we have used catches of the February-April period only, for both scrod and adults. Herrington was of the opinion that the critical time for competition between adults and young, during those years, was when the young were about a year old (his fig. 12); however the two graphs are much the same and do not provide a basis for decision on this point. Either one supports Herrington's interpretation of the fluctuations in adult stock over the period in question, which consisted of two complete oscillations (Fig. 25). The trough-to-trough periods were 9 and 8 years, and trough:peak amplitude was about 1:2. The observed period implies a median age of 4 or 5 years for the production of eggs by the fish of each brood, and this seems consistent with what is known of the size- and probable age-structure of the haddock population in those years.

Thus Herrington has described the only example yet known in which oscillation of a multiple-spawning-age fish population might plausibly be ascribed to the simple effect of a steep reproduction curve. His account of the probable cause of the inverse stock-recruitment relationship emphasizes competition for food between adults and young. Intra-brood competition at an earlier age seems another possible factor, but the feeding habits of haddock are said to preclude significant cannibalism.

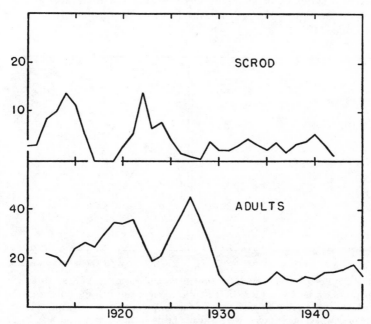

FIGURE 25. Catch per unit effort of adult haddock, related to that of the scrod (age-III fish) produced by them. The years shown on the abscissa are those in which the adult catch was made; the scrod on the same ordinate were taken three years later. From the same data as Figure 24.

187

Points for the years 1930–43 are plotted as open circles in Figure 24. About 1930 the haddock fishery changed radically within a few years; total effort increased, and otter trawls to a large extent superseded baited-hook methods so that smaller fish became relatively more intensively exploited. Although adjustments for these effects were made as well as possible by Herrington, quantitative comparisons between the period before and after 1930 appear uncertain. In particular, the relative level of recruitment indicated after 1930 seems too low to have sustained the continuing fairly high level of catch—that is, the 45-degree line indicated on Figure 24 probably does not apply to these later years.

What is of most interest for us is that the increase in rate of exploitation about 1930 was sufficient to eliminate oscillation from the population (Fig. 25). It was discovered in Figure 17 above that, in a stock subjected to increasingly heavy fishing, reproduction-curve oscillation may disappear while rate of exploitation is still rather low; and further, that the transition from marked oscillation to unrecognizably small oscillation occurs rather abruptly. Possibly accidental, but also in accord with expectation in a fishery of increasing intensity, is the fact that the second cycle in Figure 25 is shorter than the first (*two* years shorter in Herrington's representation).

FRUIT FLY (*Drosophila melanogaster*)

An optimum population density for reproduction was found by R. Pearl and S. Parker in a series of experiments made during the early 1920's. The reproduction curve of Figure 26A is synthesized from graphs summarizing their work, given by Lotka (1925).

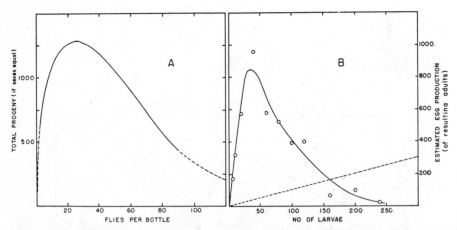

FIGURE 26. Reproduction curves for *Drosophila melanogaster*. A—From data of Pearl and Parker, as presented by Lotka (1925). The ordinate represents half the product of the density in question, times the progeny produced per female per day, times the mean length of life. The range over which observations were actually made is shown as a solid line. B—From data of Chiang and Hodson (1950, table 9 and fig. 13). The abscissa is the number of 24-hour larvae used to start a culture, and the ordinate is the potential egg deposition of the resulting adults. The dotted line is an approximation to the 45-degree line, which ignores possible mortality in the egg to 24-hour larva stages.

Recently Chiang and Hodson (1950) have investigated *Drosophila* reproduction in more detail. Figure 26B shows their graph of "potential fecundity" against initial larval density. Some of the factors involved in decreasing the reproduction with increasing density were greater larval mortality, increased failure to pupate, and smaller size and lesser fecundity of adults produced. The flattening of the curve near the right end was due partly to a secondary *increase* in pupal and adult size at the highest densities.

AZUKI BEAN WEEVIL (*Callosobruchus chinensis*)

Extensive studies of this species have been made by S. Utida, but the only account I have been able to consult is that of Fujita and Utida (1953). There is no overlapping of generations. A reproduction curve is shown in Figure 27, obtained under experimental conditions. The right limb trends downward with a slope numerically less than unity, which would make the population subject to damped oscillations. Such oscillations were actually observed by Utida (Fujita and Utida, p. 494).

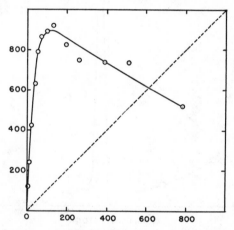

FIGURE 27. Reproduction curve for the azuki bean weevil under experimental conditions. *Abscissa*—size of parental generation; *ordinate*—number of mature progeny. Replotted from figure 4 of Fujita and Utida (1953).

WATER-FLEA (*Daphnia magna*)

Pratt (1943) performed experiments on the effect of stock density upon the production of young and upon longevity, in *Daphnia* grown in 50 cc. of culture medium with excess food. These can be used to construct reproduction curves, shown in Figure 28.

The ordinates in Figure 28 are equivalent to the total number of young produced in an average lifetime at the temperature and density in question (Pratt, p. 135). The identity of these values with those required for a reproduction curve can be shown as follows. Consider the life of a *Daphnia* to be divided into a series of "generations", corresponding to the years of an ordinary fish's life. These could be made of any length, but for convenience they are taken as equal to the approximate time from hatching to first maturity (6 days at 25°C., 14 days

189

at 18°C.). Then consider a 10-*Daphnia* population at 25°. It produces young which live 26 days on the average (Pratt's fig. 6), or 4.3 "generations". During the first of these the *Daphnia* are immature, so that their mature life is 3.3 "generations". The total young produced by an individual in an average lifetime is, however, 23.1 (Pratt, p. 135), or 7.0 per generation. Consequently a population of 10 mature individuals produces $10 \times 7.0 = 70$ young per "generation", and these survive through 3.3 generations of maturity on the average, so that the total reproductive potential produced by the parent generation is $70 \times 3.3 = 231$ units. These "units" are numerically the same as the total number of young which would be produced throughout the life of the 10 parent individuals living at constant density. However the "units" are to be regarded not as young actually produced, but as a measure of the generation's potential contribution to the mature population of several future generations, in exactly the same way as eggs in mature females were used for fishes. In actuality, if stock density were to change by the time these 231 "units" were in the mature population, the actual production of young by the progeny of the original generation would be greater or less than 231.

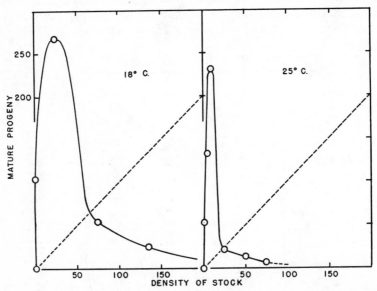

FIGURE 28. Relation between density of adult *Daphnia magna* and their expectation of progeny, at 18° and 25°C. Computed from data of Pratt (1943): 18° curve from his figs. 7 and 8; 25° curve from his figs. 5 and 7.

The reproduction curves of Figure 28 are of a steeper type than any of those discovered among fishes, particularly the 25° curve. Both curves are in fact likely to be even steeper than they are drawn, since the highest point of observed production of young is not necessarily the maximum. This steepness has been believed to be associated with the small volume of habitat and consequent likelihood of waste accumulation—although the medium was changed every second day.

Pratt (1943, figs. 1, 2) also followed the growth of *Daphnia* populations over a period of 6–7 months under the same culture conditions as outlined above. Starting in each case from 2 parthenogenetic females not more than a day old,

TABLE V. Simplified calculation of ideal population changes of *Daphnia magna* at 25°C. under the conditions of Pratt's experiments, by 6-day periods. Average mortality is considered as zero for the juvenile 6 days, one-fourth from the first to the second 6 days of maturity, one-third from the second to third, one-half from the third to fourth, and 100 per cent after the fourth. The number of "reproductive units" produced by a given stock in 6 days is defined as the product of the number of mature animals produced times the average number of 6-day periods each lives. Column 6 shows the total units produced by the stock of column 5, values being taken from Figure 29B. These units are divided among the appropriate periods to give the stock present, as shown in the first four columns. The population starts as two newly mature adults, one of which survives for two 6-day periods, the other for four. Total adult stock thereafter is the horizontal sum of the units in the first four columns. The total stock (column 7) is the adult stock (column 5) plus the number of juveniles, the latter being the same as the initial number of matures, as indicated in column 1 one 6-day period later.

1	2	3	4	5	6	7
\multicolumn{4}{}{Fraction mature during successive 6-day intervals}				Adult stock	Reproductive units produced	Total stock
0.4	0.3	0.2	0.1			
..	..	..	..	2	65	2
..	..	..	..	2	65	28
26	..	..	..	27	20	53
26	20	..	..	47	15	55
8	20	13	..	41	15	47
6	6	13	6	31	20	37
6	4	4	6	20	40	28
8	4	3	2	17	90	33
16	6	3	2	27	20	63
36	12	4	2	54	10	62
8	27	8	2	45	15	49
4	6	18	4	32	15	38
6	3	4	9	22	25	28
6	4	2	2	14	200	24
10	4	3	1	18	65	98
80	8	3	2	93	4	119
26	60	5	2	93	4	95
2	20	40	2	64	6	66
2	1	13	20	36	15	39
3	1	1	6	11	230	17

1	2	3	4	5	6	7
\multicolumn{4}{}{Fraction mature during successive 6-day intervals}				Adult stock	Reproductive units produced	Total stock
0.4	0.3	0.2	0.1			
6	2	1	0	9	230	101
92	4	1	0	97	4	189
92	69	3	0	164	4	166
2	69	46	2	119	4	121
2	1	46	23	72	6	74
2	1	1	23	27	20	30
3	1	1	0	5	160	13
8	2	1	0	11	230	75
64	6	1	0	71	6	163
92	48	4	0	144	4	147
3	69	32	2	106	4	108
2	2	46	16	66	6	68
2	1	1	23	27	20	30
3	1	1	0	5	160	13
8	2	1	0	11	230	75
64	6	1	0	71	6	163
92	48	4	0	144	4	147
3	69	32	2	106	4	108
2	2	46	16	66	6	68
2	1	1	23	27	20	30

the 18° cultures rose to a peak of about 200 animals usually after about 70 days, decreased to about 120 animals, then showed a slight increase. At 25° the cultures grew more rapidly, reached a peak in 20 days, and then as rapidly declined; sometimes they died out, but in two instances the oscillation was repeated four times before the experiment terminated. Figure 29 shows one example at each temperature.

The steep reproduction curves of Figure 28 immediately suggest the possibility of oscillations, and they have been used to obtain the computed population curves (B and D) of Figure 29. These start from 2 newborn animals, like Pratt's figures 1 and 2.

Construction of the curves of Figure 29 required certain assumptions and simplifications. The life of a *Daphnia* was divided into "generations" as before. The total number of "reproductive-units" indicated by the ordinate of Figure 28 was divided among successive generation-long periods of life as follows: for 25°C., in the ratio 0:4:3:2:1; for 18°C., in the ratio 0:22:20:18:14:11:9:5:1. These ratios correspond approximately to the average survival indicated by Pratt's figures 5 and 8. The calculation then proceeded on the same basis as in Table I, the sequence for 25° being shown in Table V. Because Pratt's censuses include immature individuals, the mature stock of column 5 is increased by the number of immatures in the culture (considered equal to the first generation of matures, in column 1 one line below) to give the total stock of column 7.

Obviously this computation differs from actuality in the assumption of an age distribution of mortality that is independent of density (cf. Pratt's figs. 5 and 8). The difference is most serious at 25°, and if a more complex calculation were made to take account of it, the initial effect would be to make the troughs deeper because of the shorter lives of denser stocks.

The predicted course of events in Figure 29 has considerable resemblance to what was actually observed in respect to period, and to a less extent in amplitude. (Pratt's observed curves also vary a lot among themselves, as would be expected when the stock starts out from, and at 25° is periodically reduced to, a very few individuals.) Any judgment concerning the significance of this resemblance would have to be of the subjective type which Smith (1952) rightly deplores, but the comparison has the merit that the factual bases of the two curves of each pair are from different sources—there are no "fitted constants".

At 18° the computed cycles are soon damped to an inconsiderable amplitude, but the suggestion of a second low peak in the observed series (Fig. 29A) would probably have been confirmed if the cultures had been maintained a month or so longer. At 25° the computed oscillation reached a stable equilibrium pattern having a large amplitude by the time of the fifth peak, and the two surviving observed series were still oscillating strongly when they were discontinued. The striking difference between 18° and 25°, in both the calculated and the observed curves, is ascribable partly to the steeper reproduction curve at 25°, partly to the fact that a larger number of "generations" are mixed in the 18° mature population.

However there is another factor which enters into these *Daphnia* fluctuations. At the higher temperature the young produced under crowded conditions are permanently less fertile than those from uncrowded cultures[10]. This phenomenon

[10]This was demonstrated by segregating newly mature specimens from crowded cultures, keeping them in isolation, and comparing their production with that of specimens reared singly, but the data cannot be applied quantitatively to these examples. Pratt did not perform the complementary experiment of crowding mature specimens which had been reared singly. However, the very sharp restriction of reproduction which occurs in such event is sufficiently indicated by the small number of progeny produced by the first young born in a culture. These live their adult (but not their juvenile) days under crowded conditions (cf. Pratt's fig. 3).

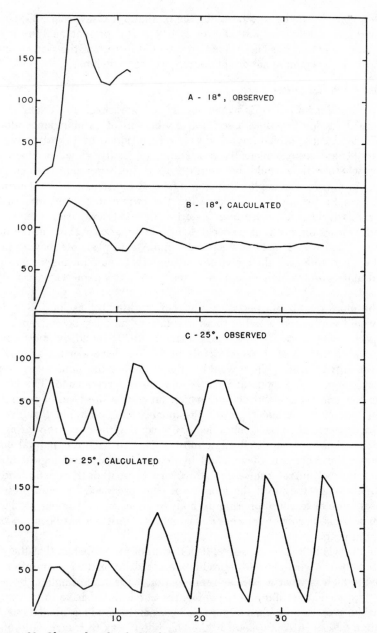

FIGURE 29. Observed and calculated fluctuations in *Daphnia* populations, from data of Pratt (1943). Curves A and C are directly from Pratt's observations, grouped by 14-day intervals at 18°C. and 6-day intervals at 25°C. Curves B and D are calculated from the reproduction curves of Figure 28. *Ordinate*—number of *Daphnia*; *abscissa*—successive 6-day or 14-day intervals.

has been considered the sole basis for the oscillations observed at 25°, both by Pratt and by Hutchinson and Deevey (1949). The present analysis makes it seem more likely that this effect is not an essential feature of the system, and that *actual* density is the more important depressor of reproduction.

STARFISH (*Asterias forbesi*)

No reproduction curve is available for this starfish or any other. However Burkenroad (1946) describes a series of oscillations of its abundance along the New England coast, which have had an average period of 14 years over about 90 years. These seem sufficiently remarkable to justify some speculation concerning whether they could be ascribed to a steep reproduction curve like Figure 7 or 8. If so, the implication would be that the starfish make their median contribution to reproduction 7 years after the zygote stage, and also that they make a relatively small contribution during the first few years after they are hatched. During the peaks of its cycle A. *forbesi* apparently exists in the massive numbers which would make a steep outer limb seem plausible, having in mind that starfish readily eat their own young and that reproduction is partly suppressed among individuals of crowded and poorly fed populations (cf. Vevers, 1949).

The more doubtful aspects are whether starfish live as long as a 14-year cycle would imply, and whether there is sufficient time lag before reproduction. In respect to the latter, Galtsoff and Loosanoff (1939) found the most abundant size group of A. *forbesi* at Wood's Hole in April to have a mode at only 3 cm. (between tips of arms); these would scarcely mature the same year, and even the next group, with a mode at 7 cm., could not be very fecund. The maximum size for this species is about 20 cm., which suggests a life span of several years at least, since rate of linear growth presumably falls off rapidly at the larger sizes, particularly when the stock is dense. Concerning their year-to-year mortality rate little is known except that adult starfish seem to have few natural enemies, so that for the most part they should live out a normal physiological life-span, whose length would of course probably vary a great deal from individual to individual. While they live, the number of eggs produced by each presumably increases more or less in proportion to body volume, so the number produced by a given year-class probably reaches a maximum only after it has been in existence for several years.

On the whole it seems possible, if perhaps not too probable, that the median contribution of a year-class to reproduction could be made 7 years after it was hatched, which is what a 14-year reproduction-curve cycle implies. Because so many ages would contribute to spawning, the curve would have to be very steep —probably even steeper than Figure 8—in order to produce oscillations of the amplitude described by Burkenroad, i.e., having a peak to trough ratio of about 20 to 1. There would likely be minor complications resulting from differences in growth rate between times of abundance and times of scarcity, since growth of starfish is known to vary sharply in response to availability of food. Systematic sampling for size and (if possible) for age, over a period of one or more cycles, should provide the key to the situation. Galtsoff and Loosanoff's samples were taken in 1936–37 at a low point in the cycle, and contained large numbers of young, as expected. At a peak of abundance large individuals should predominate, and young of the year should be very scarce toward the end of the year, or sooner.

Sinuous Reproduction Curves

Dr. Pritchard's data on pink salmon reproduction at one site have already been summarized, as well as the adult counts made on the Karluk River. More recently, Neave's (1952, 1953) review of pink salmon reproduction in a number of localities has led him to suggest a rather complex reproduction situation. The species is peculiar in that there are a few areas where one of the two "lines" is completely absent in a series of streams. In other areas one line exists at a level much below the other, and the two lines will have maintained approximately the same relative position for decades, though not necessarily for the whole period of record. In still other areas there is no consistent difference between the lines in respect to abundance. Furthermore there have been cases where one line has dropped suddenly from a high to a low level of abundance and has remained at that level for a considerable period; in one or two instances such a reduced stock has bounced back up to its former level.

Dr. Neave feels that density-dependent mortality in this species must occur mainly during its freshwater life—from the time the adults appear in spawning streams in August-October to the time the fry leave in April or May. Further, there are factors causing both compensatory and "depensatory" mortality at this stage, the latter being mortality which kills a relatively larger fraction of the fish present when their numbers are small than when they are large. Experiments have shown that the activity of stream predators (trout, etc.) upon pink fry is depensatory —the fry apparently being so vulnerable that the predators eat all they can, then leave the rest alone.

This hypothesis accounting for two fairly discrete and fairly stable levels of population in pink salmon implies the existence of anomalous reproduction curves; two possible types are shown in Figures 30 and 31. (The part to the right of the dome does not enter into the argument.) In Figure 31, at some intermediate level of stock, depensatory mortality is at a maximum relative to compensatory mortality, producing the dip in the left limb of the reproduction curve. For example, a population which adhered closely to the curve of Figure 30 could be in equilibrium with a fishery taking 50 per cent of the adult fish at the two different levels of abundance indicated, but between those levels equilibrium rate of exploitation falls to as low as 20 per cent. Since 50 per cent is the maximum permissible limit of exploitation at the higher level of stock, any slightly greater exploitation will send the stock tumbling to the lower level. There it will stay unless exploitation is sharply restricted or temporarily discontinued—merely reducing it to 50 per cent again will not restore the higher equilibrium position.

More generally, when random variation in reproductive success is taken into account, the existence of a fishery makes it easier for a population to slip down

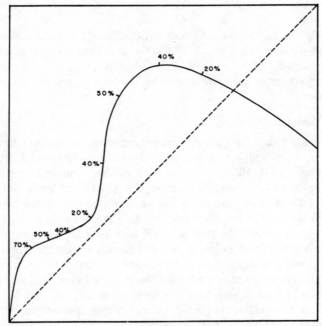

FIGURE 30. Hypothetical reproduction curve, with equilibrium positions for several rates of exploitation. See text.

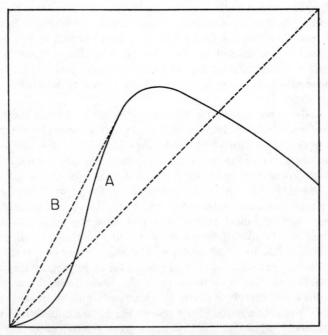

FIGURE 31. Hypothetical reproduction curves. See text.

from the higher to the lower abundance level in a bad year, and harder for it to get back up in a good one. Of course, if the reproduction curve falls off to the right of the 45-degree line, as is actually suggested in Figure 30, the *first* effect of a new fishery would be to increase population abundance.

Figure 31A differs from 30 in that there is no lower level of strong resilience to exploitation. It might represent pink salmon reproduction in areas where one of the two lines ("cycles") does not exist at all. Accidental reduction of the population past a certain low level in that case initiates a progressive decrease and results in eventual extinction. A similar situation is implied in a hypothetical net-change curve of Haldane (1953, fig. 3), and for several animals there is some experimental evidence of reduced effectiveness of reproduction at very low densities (cf. Allee, 1931; Hutchinson and Deevey, 1949).

Dr. Neave's ingenious hypotheses, illustrated by the above curves, are supported by several lines of evidence, but have not as yet been firmly established for any pink salmon stock. McClinton Creek, described earlier, is a stream which has no pinks at all in odd-numbered years, hence might be expected to have a reproduction curve like Figure 31A. The actual curve, however, is of a simple type as far as it is known (Fig. 19). If the left-hand portion of this curve falls below the 45-degree line, it must do so at densities considerably less than 5,000 female spawners—that is, in the extreme lower left corner of Figure 19B. At Karluk there is some indication of an inflexion of the curve at low stock densities (Fig. 20, upper right), but it is not clear that it is sufficient to play the role described above.

A difficulty with the curves of Figures 30 and 31A is that their initial phases are quite different, and hence it seems improbable that both should be applicable to the same species in the same general region. They might be reconciled if the beginning phase of 31A, in varying degrees of development, were commonly found tacked onto the initial part of the Figure 30 curve, producing a doubly sinuous ascending limb for the curve as a whole. If so, hypotheses as to the nature of the compensatory and depensatory mortality would have to be correspondingly more complex. A simpler but less satisfactory compromise type would have merely a straight left limb, as in Figure 31B; this would account for any observed lack of resilience in a population after a certain level of exploitation had been reached, since one density would be as stable as another once the equilibrium point had been shifted to the ascending limb of the curve.

DISCUSSION

Doubtless other investigators will be able to add to the stock of reproduction curves from data now available, and we may expect more in the future. Considering the curves for fishes, cited above, it is unfortunate that in most cases only a section of the curve is available.

A steep ascending limb, as in Figure 2b or 8, is implied whenever a fishery consistently takes some large fraction of a stock before it has a chance to reproduce, as notably among various salmon stocks. Actual points on this ascending limb are available in a few instances where a population has risen from or declined to a very low ebb; the Shuswap sockeye are the best example.

No curve yet discovered suggests that the maximum of recruitment is at or to the right of the 45-degree line (as in Figs. 1, 3, 4), but on several of the curves this line cannot be located.

Concerning the right limb, an indefinite horizontal extension like that of Figure 2 has often been suggested for sea fishes on *a priori* grounds (e.g., Baranov, 1918; Kesteven, 1947). Tester (1948) found this hypothesis consistent with the Vancouver Island west coast herring data, and this is still true (Fig. 18). The same can be said of some of the other curves examined (Figs. 19B, 20, 22). However, without any exception, the lines which best fit these outer limbs slant downward at least slightly, and the weight of their combined evidence

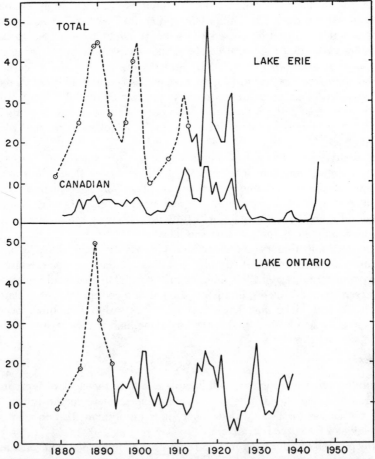

FIGURE 32. Cisco catch from Lakes Erie and Ontario, in millions of pounds. (From Gallagher *et al.*, 1943; and Scott, 1951.)

indicates that negative slopes of at least moderate magnitude are real and are the common situation. On the other hand, an approximation to the flat curve of Figure 2 might well be expected in a species like the coho, which is aggressively territorial and whose stream habitat is limited.

The best-documented example of a really steep descending limb in a natural fish population is provided by the 1912–29 New England haddock (Fig. 24), where the curve is approximately like Figure 7. Even steeper curves characterize the invertebrate populations (Figs. 26–28).

Left to themselves, steep reproduction curves produce such beautifully regular oscillations in population abundance that it is rather disappointing to discover how easily this regularity is disturbed by non-compensatory mortality (Fig. 11B). This means that we should rarely expect to be able to identify the contribution of a reproduction curve to any given series of observations of animal abundance, even though such contribution may be a major factor in many examples of marked fluctuation. And it is a fact that most of the well-known examples of fluctuating abundance cannot be a *simple* resultant of a steep reproduction curve. The 9- or 10-year cycle of grouse and hares, for example, could not be of the simple type shown in Figure 10 because there is not the necessary lag in reproduction—the young are mature in the next reproductive period after they are born; in any event it is doubtful if average longevity in these species is sufficient to put a brood's mean time of contribution to reproduction at 4.5 or 5 years. Certain major changes in fish populations—for example, cod off Greenland, herring off Norway, or mackerel off eastern North America—have occurred on too great a scale and with too long a period to have been governed by any single reproduction curve. Fish population changes having shorter period, like those of the ciscoes of Lake Ontario (Fig. 32), might be strongly influenced by their reproduction curve, but the peaks and troughs are too irregular, and basic population information is insufficient, to permit even a good guess. If the cisco rate of exploitation is moderately great, as seems probable, then direct reproduction-curve periodicity is not likely to be recognizable (cf. Fig. 17).

We are perhaps lucky to have, in Herrington's haddock, even one example of an oscillation in which the effect of the reproduction curve can be identified with fair likelihood.

OTHER REPRODUCTION SITUATIONS

Up to this point all examples and interpretations have been predicated on the postulates that it is only animals of mature size whose abundance affects compensatory mortality; and that where there is human exploitation of the stock, it takes only mature individuals. What is the effect of relaxing each of these restrictions?

Larger Immature Fish Taken by the Fishery

If a fishery takes fish of smaller than mature sizes, it makes little difference to the theory of population control outlined above *as long as the sizes subject to compensatory mortality are not touched.* Details of the equilibrium level of population at different intensities of exploitation are of course affected, but not the general course of events. In actuality it has fairly often been found that spawning populations can be reduced to a very low level while the catch is maintained at a fairly high level by immature fish, as for example the cod in the North Sea. In this and similar instances data for plotting a reproduction curve do not seem to be available, but the general situation strongly suggests that the right limb of the curve must slope downward and that the left limb must be very steep. Until reproduction curves are available and future recruitment can be predicted, proposals for increasing catch by reducing the fishing intensity in the North Sea and elsewhere will lack that final symmetry which would make them wholly convincing.

EXPLOITATION DURING THE COMPENSATORY PHASE

Exploitation that takes fish at an age when natural mortality is still compensatory means, for practical purposes, a fishery for young during the first year or two of their life—the earlier the better. The removal of such young is at least partly balanced by increased survival and/or growth of the remainder; in fact, the effects of removals at this stage are equivalent to reduction of the spawning stock which produced the brood in question. If the reproduction curve for the population is of any of the types 3–8, such reduction will at first *increase* net production of recruits, which will produce more eggs and permit a larger catch of young in future years. This ascending spiral of abundance may continue until the level of stock is reached which produces maximum recruits.

There can of course be no general rule indicating a single optimum time for exploitation for all fisheries. In any particular instance, maximum yield may be obtained by taking young, or adults, or both; and in balancing the alternatives, consideration should be given to the relative value per pound of the two sizes and the ease with which they are caught. But it is clear that any *general* prejudice against exploiting young fish is unsound. Each case should be considered on its merits. The situation most favourable for juvenile exploitation is evidently that where the fish reach a *relatively* large size before compensatory mortality ceases to be important, so that the catch taken during this period can be large in total bulk.

Relatively few fisheries now exist in the temperate parts of the world which attack really young individuals, but this can be ascribed mainly to the (usually) lesser value per pound of small fish as compared with large ones, and the (usually) greater difficulty of catching these young. In Japan a number of species are eaten as larvae—for example the sand lance (*Ammodytes*). There and elsewhere the exploitation of various clupeids in post-larval, but still early, life is fairly common.

One such fishery on this continent is that for "Quoddy" herring in southern New Brunswick (Huntsman, 1952, 1953). A large primitive stock of mature herring in this region apparently fluctuated in abundance during the middle years of the last century, but no details of period or amplitude are available. About 1880 mature herring declined rather abruptly and they have not reappeared in comparable numbers since. About the same time the weir fishery for "sardines" (age 0 to age II herring, mainly) developed enormously and has yielded large and fairly steady catches ever since—catches which bulk two or three times as great as did the early fishery for mature individuals.

If we postulate that the weirs take the young partly during the time that compensatory mortality is operative, and partly after compensation is no longer important, a formal explanation of this situation can be made. The sardines taken young, while the mechanism of compensation is still in operation, represent a diversion to human use of stock which would otherwise be lost to predators or other natural causes. The sardines taken at larger sizes, after compensation is largely finished, plus any larger stock taken, reduce the potential breeding population to a point far below the numbers it would otherwise attain (and did actually attain in years gone by). This reduced spawning stock produces many more young fish than a large one would, the implication being that in decreasing in abundance it has climbed the right limb of a dome-shaped reproduction curve such as Figure 8. In this way the present distribution of sizes and numbers could be maintained.

COMPENSATION BY IMMATURE MEMBERS OF THE POPULATION

In the restricted habitat of ponds and small lakes cannibalism is one of the most likely methods of population regulation for abundant predacious species, e.g., bass (*Micropterus*) or crappies (*Pomoxis*). Indeed, the first stages of all centrarchids are so tiny that even a normally insect-eating species like the bluegill (*Lepomis macrochirus*) may eat eggs or fry of its own species during their very early life. Immature individuals tend to be most active in this, since they normally frequent the shallow water where most spawning occurs, and gather around nests of their own or other species. Adults are less frequently in contact with young, and the males actually guard eggs and young as long as they are in the nest, or even longer; hence adults are less effective in reducing the oncoming generation. As an example, two similar ponds at the Tri-lakes Hatchery, Indiana, were stocked in spring with similar weights (13–15 kg.) of bluegills: in Pond 2 they consisted almost wholly of mature fish, in Pond 4 about half (by weight) were small fish of the previous year's brood. By November the young of the year produced by Pond 2 numbered 19,150 and weighed 36.5 kg., whereas Pond 4 produced only 230 fingerlings weighing 0.18 kg.

To adequately illustrate situations of this sort it would be necessary to construct a reproduction diagram showing the three-way relationship between quantity of mature stock, quantity of immature stock, and resulting recruitment. Present information does not permit this, but the general effect of participation of immatures in compensation must be to spread the reproduction-depressing influence of any successful brood over a longer period of time, and to make it almost impossible to have the "lag" effect which produces population oscillation. In such a situation, however, cycles of fish *size*, and to some extent of poundage, might occur if individuals of a successful brood nearly all died off during the same calendar year (cf. Thompson, 1941, p. 209).

TOWARD A THEORY OF RECRUITMENT

The justification for using any reproduction curve must in the long run come from observation. However it would be most useful, if it were possible, to formulate some general theory of reproduction which might lead to a standard type of reproduction curve applicable in a majority of situations.

THEORY OF PREDATION

An approach to this goal can be made by way of a consideration of predation upon the young of a species, whether by other animals or by older individuals of its own species. The theory of predation was briefly considered by the writer in a recent paper (1952), and the following quotation will provide a basis for the further argument here:

It is convenient to distinguish three types of numerical relationship between predators and a species of prey which they attack.

A. Predators of any given abundance take a fixed number of the prey species during the time they are in contact, enough to satiate them. The surplus prey escapes.

B. Predators at any given abundance take a fixed fraction of prey species present, as though there were captures at random encounters.

C. Predators take all the individuals of the prey species that are present, in excess of a certain minimum number. This minimum may be determined in different ways: (1) There may be only a limited number of secure habitable places in the environment, so that some prey are forced to live in exposed situations where capture is inevitable. The number of such secure niches may be partly governed by territorial behaviour of the prey. (2) The maximum "safe" density of prey may be the one at which predators no longer find it sufficiently rewarding to forage for them, and move to other feeding grounds.

The three situations above tend to intergrade, of course, but it is useful to keep their differences in mind.

SITUATION A

This is likely to occur when a prey species is temporarily massed in unusual numbers, for example, adult herring in spawning schools, or newly-emerged fry of pink and chum salmon going downstream. The main characteristic of such situations is that the number of prey eaten depends on the abundance of predators, but not on the abundance of prey. Hence such situations cannot last long, and the predators cannot make the prey in question their principal yearly food; otherwise they would almost surely increase in abundance and the situation would change to type B or type C. If a type A situation persisted for long, it would come to an abrupt end with the extermination of the prey.

SITUATION B

Here the number of the prey species eaten is proportional to the abundance of predators and to the abundance of prey. Unlike A, this type of predation can easily occur over long portions of the year; and the prey species may comprise the larger portion of the predators' annual ration. This situation was observed at Cultus Lake, British Columbia, for predation of squawfish, char, coho and trout upon fingerling sockeye, over a wide range of abundance of the latter (Foerster and Ricker, 1941).

SITUATION C

The classical example of this situation was described by Errington for bob-white in Iowa, where a given range would winter safely a fixed number of birds, practically independently of the number which were present in autumn, the surplus being taken by predators. Studies conducted from the Atlantic Biological Station of the Fisheries Research Board of Canada,

St. Andrews, N.B., suggest that in some rivers predation upon Atlantic salmon parr tends toward type C, because a marked increase in the number of young fish planted was followed by only a relatively small increase in number of surviving smolts (Elson, 1950). In type C situations the predators must tend to have ample alternative foods, and they are often mobile so that they can leave an area where the food supply has been "cleaned up" for the season.

Consider a prey species which is subject to type B predation over some part of its life history. Assume first that predation causes the whole of the mortality that the prey is subject to during that period. Then the prey species decreases in abundance according to the well-known exponential formula

$$N/N_0 = e^{-it}, \tag{1}$$

where: N_0 is initial abundance, N is abundance at time t, $e = 2.718 \ldots$, and i is a statistic representing the fraction of the prey which would be eaten in a unit of time if its abundance were held constant for that long; i is often called the *instantaneous* mortality rate.

Under the conditions postulated, instantaneous mortality rate is directly proportional to the abundance of predators. This can be illustrated by considering a situation where predators attack a prey population of 1,000,000 individuals under type B conditions, and they inflict losses corresponding to $i = 0.8$, where the unit of time, t, is the whole season that predator and prey are in contact. Equation (1) indicates that in, for example, 1/1000 of the season, the predators will eat $i/1000 = 0.0008$ of the prey present, or 800 fish. During the next thousandth of the season, the predators eat $i/1000$ of the surviving prey, or $999,200 \times i/1000 = 799$ fish. The following interval they eat $998,401 \times i/1000 = 799$; then $997,602 \times i/1000 = 798$. This continues until all the thousand time-intervals have elapsed, at the end of which there are:

$$1,000,000(1 - 0.0008)^{1000} = 472,400$$

survivors, or 47.2 per cent. What happens if the number of predators, is doubled? In that event, during the first thousandth of the season twice as many fish will be eaten, i.e., 1,600, leaving 998,400. In the next thousandth the fraction eaten is likewise double, namely 0.0016; multiplied by the number of survivors this gives 1,597; and so on. At the end of the season $1,000,000(1 - 0.0016)^{1000}$ survive, which is 201,900, or 20.2 per cent.

Thus doubling the number of predators doubles the instantaneous mortality rate (which is true generally), but it increases actual mortality by only $27.0/52.8 = 51$ per cent (which is true of only this particular example).

The general relation between predator abundance and actual mortality, or survival, is most conveniently expressed as:

$$\frac{p_2}{p_1} = \frac{\log s_2}{\log s_1}, \tag{2}$$

where p_2 and p_1 represent two levels of predator abundance, and s_2 and s_1 are the corresponding survival rates for the prey. This expression can easily be derived from (1), since p is proportional to i, and $s = N/N_0$ when $t = 1$.

Cannibalism

Of all the methods of population regulation listed earlier, cannibalism is the one in which the abundance of the control agent is most closely and inseparably allied to that of the population controlled. That is, an increase in mature stock not only increases the number of eggs laid or young born in a given reproductive season, but it also decreases the rate of survival of those young. What is the combined effect of these two opposed influences?

Let the size of a stock be measured by the number of eggs it lays, and let this size be proportional to its *instantaneous* efficiency in cannibalism (the fraction of young eaten by the adults in a short interval of time). Consider two situations characterized by different sizes of stock and let the ratio of the second to the first be w, corresponding to the p_2/p_1 of equation (2) of the quotation above. Finally, let all other sources of mortality, whether they occur before, after or during the period of the cannibalism, be density-independent, their total effect being to reduce survival rate of eggs and young to the fraction k of what it would otherwise be. The number of eggs laid, in situation 1, is E_1; in situation 2 it is therefore $E_2 = wE_1$. The reproduction (absolute number of recruits produced) in situation 1 is:

$$R_1 = E_1\, ks_1; \tag{3}$$

and in situation 2 it is:

$$R_2 = E_2 ks_2 = wE_1 ks_2. \tag{4}$$

From equation (2) of the quotation above,

$$w = \frac{\log s_2}{\log s_1}\ ; \quad \text{hence } s_2 = s_1^{\,w}. \tag{5}$$

From (4) and (5), the reproduction in situation (2) becomes:

$$R_2 = wE_1 ks_1^{\,w}. \tag{6}$$

Expressed as a fraction of the reproduction in situation 1, this is:

$$ws_1^{\,w-1}. \tag{7}$$

Plotting numerical values of (7) against w, for various values of s_1, yields a family of curves each of which has an origin at zero, a dome, and an extended right limb which approaches the abscissa asymptotically. No matter what survival rate is chosen for s_1, each curve describes the whole range of possible survival values as w is varied; hence only one curve of the family is needed. The convenient one to choose is that for which (7) is a maximum when $w = 1$. To locate this maximum, (7) is differentiated with respect to w and equated to zero:

$$ws_1^{\,w-1} \log_e s_1 + s_1^{\,w-1} = 0; \tag{8}$$

whence,

$$-\log_e s_1 = 1/w; \quad \text{or, } s_1 = e^{-1/w} \tag{9}$$

The value of s which makes (7) a maximum is $e^{-1/w}$, and if this is taken as the initial value of s (that for which $w = 1$), s becomes equal to $1/e$ or 0.3679. Substituting $1/e$ for s_1 in (7), we thus finally obtain the expression

$$we^{1-w}. \tag{10}$$

This shows the actual level of reproduction as a fraction of the maximum, when w represents the ratio of the actual density of mature stock to the density which gives maximum reproduction. Values of (10) are plotted as curve B of Figure 33.

From the equations and the curve the following conclusions are evident:

1. Since the curve approximates to the abscissa as w is increased, then *if the breeding stock is made sufficiently large, cannibalism reduces reproduction practically to zero*, in spite of the greatly augmented egg deposition. In theory this is true no matter how small the original rate of cannibalism; in practice, a type B predation situation between adults and young could not be maintained over *too* wide a range of stock densities.

2. Since, as w approaches 0, e^{1-w} approaches e, it follows that at minimal densities the survival rate from cannibalism is e times the survival characteristic

of maximum reproduction. In other words, the instantaneous mortality rate at maximum reproduction is greater by unity than it is at vanishingly small stock densities (since $- \log_e (1/e) = 1$; cf. expression (1) of the quotation above).

Curve B of Figure 33 can be changed into a reproduction curve by changing the scale of each axis to represent actual numbers of parents and progeny measured in comparable units—for example the egg production unit described in an earlier section. Equation (10) gives no information concerning the steepness of this actual reproduction curve, which will depend chiefly on the magnitude of k, the survival rate from factors other than cannibalism. On Figure 33 a number of possible curves have been drawn, all based on equation (10) but using different ratios of abscissal to ordinate scales. Referring each to the 45-degree line indicated, shapes reminiscent of several of the arbitrarily drawn types of Figures 3–8 can be identified, as well as most of the curves observed in actual populations. Curve E, of course, could not describe a stable population because it lies well below the 45-degree line, while even D would be rather precarious.

In addition to cannibalism, the argument of this section can be extended to the analogous situation where strife between mature and young animals occurs and results in frequent killing of the latter, but without necessarily any eating of one by the other; this has been described among muskrats, for example (Errington, 1954). In so far as these contacts are of type B—with the older animals killing a fixed fraction of the young they encounter—they would have the same population status as described above.

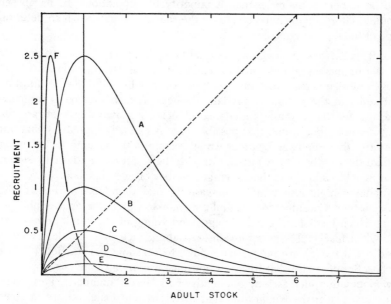

FIGURE 33. Graphs of we^{1-w} ("recruitment") plotted against w ("adult stock"). Curve B corresponds to the axes as labelled; the other curves are obtained by varying the ordinate and (for F) abscissal scales.

205

It is not easy to assess the probable importance, in nature, of cannibalism and allied effects. In most of the fish populations whose reproduction curves were examined earlier there is little or no evidence of cannibalism or of other direct injury to young by their parents, and among *Oncorhynchus* this is impossible because the parents die. However, in assessing the possible role of cannibalism (or any other agent of predation) in population regulation, it is not possible to weight it directly on the basis of the number of units which it kills. The latter will depend mainly upon the stage at which it is active. For example, suppose that cannibalism by mature trout fell upon their fingerling progeny following an average density-independent egg mortality of 50 per cent and fry mortality of 90 per cent (95 per cent in all); and that the adult trout ate 63 per cent of the 5 per cent remaining, or 3 per cent of the original brood. In that event cannibalism would seem to be quite insignificant either from the point of view of the percentage of the total young which it killed, or from the point of view of the frequency of occurrence of small trout in stomachs of larger trout. In spite of this, under the conditions just postulated cannibalism would be the sole mechanism regulating the abundance of the population.

PREDATION BY OTHER ORGANISMS

Even granting that lethal contact between adults and young is the most direct and least fallible population-regulating mechanism, and that it is quite apt to exist undetected, the writer's present opinion is that it will probably not prove to be important in more than a minority of populations. In its absence, other compensatory agents must take over the role of regulators which determine the shape of the reproduction curve.

TYPE B SITUATIONS. A predator can qualify as an agent of population control only if it can increase in density or effectiveness as the abundance of prey increases, and vice versa. One way this can occur is by migration of additional predator units to the region of predation when prey abundance is large. For example, greater-than-average abundance of herring spawn might attract birds from a wider area than usual. It is possible (not necessarily probable) that, after gathering together, such aggregations would remain in the vicinity for a few days longer than the supply of eggs justified it, with the result that an initially superior spawning would yield a less-than-average number of eggs hatched.

A predator might also actually increase its total abundance with that of the prey in question. Consider, for example, a plankton predator or assemblage of predators which feeds on the pelagic eggs and young of mackerel. Suppose that in years when mackerel eggs approach zero abundance these predators (which have additional foods) are numerous enough to consume the fraction $(1 - s_1)$ of the eggs. Suppose also that for each unit increase in egg numbers the average predator population increases by one-tenth of its "basic" number. (This average would be taken over the whole time that predators and mackerel are in contact because, just before the eggs are available, initial predator abundance is assumed to be the same in all years.) During the later stages of the predator-prey contact the abundance of predators would have "overshot" that of the prey, with the result again that mackerel broods initially larger than average would end up smaller than average. From expression (2) or (7) it is easy to calculate that relative reproduction is equal to

$$Zs_1{}^{1+Z/10},\qquad(11)$$

where Z is the density of mackerel eggs immediately after spawning in a given year, in the unit mentioned above, and the predator density corresponding to no eggs is taken as unity. Values of (11) are calculated in Table VI, which shows the relative number of mackerel surviving the pelagic stage, for a series of initial prey densities and for two values of s_1. Sette (1943) estimated total survival of

TABLE VI. Survival rate and absolute number of survivors when average predator abundance varies with initial prey density in the manner indicated by columns 1 and 2. Pairs of survival rates and survivor numbers are shown for initial survival rates of 0.1 and 0.0001. The last two rows show the computation of the maximum number of survivors for each situation, which occurs when survival is 0.3679 $(1/e)$ of its initial value.

1	2	3	4	5	6
Prey abundance	Predator abundance	Survival rate of prey	Relative number of survivors	Survival rate of prey	Relative number of survivors
			$(1) \times (3) \times 10^3$		$(1) \times (5) \times 10^7$
0	1.00	0.1000	0	0.0001000	0
0.2	1.02	0.0955	19	0.0000832	166
0.4	1.04	0.0912	37	0.0000692	277
0.6	1.06	0.0871	52	0.0000575	345
0.8	1.08	0.0832	67	0.0000479	383
1.0	1.10	0.0794	79	0.0000398	398
1.2	1.12	0.0759	91	0.0000331	397
1.4	1.14	0.0724	101	0.0000275	385
1.6	1.16	0.0692	111	0.0000229	367
1.8	1.18	0.0661	119	0.0000191	343
2.0	1.20	0.0631	126	0.0000158	317
2.5	1.25	0.0562	141	0.00001000	250
3.0	1.30	0.0501	150	0.00000631	189
4	1.4	0.0398	159	0.00000251	100
5	1.5	0.0316	158	0.000001000	50
6	1.6	0.0251	151	0.000000398	24
8	1.8	0.0158	127	0.000000063	5
10	2.0	0.0100	100	0.000000010	1
15	2.5	0.0032	47	...	...
20	3.0	0.0010	20	...	...
30	4.0	0.0001	3	...	...
1.086	1.1086	...	...	0.00003679	399
4.343	1.4343	0.03679	160	...	...

Atlantic mackerel eggs and pelagic larvae, in 1932, to lie between 0.000001 and 0.00001, so that the $s_1 = 0.0001$ column of Table VI would not be unrealistic.

The distributions of columns 4 and 6 of Table VI are of course the same as those of Figure 33. Expression (11), like (7), is a maximum when the rate of survival from the predator in question is $1/e$ of what obtains when egg abundance is close to zero (cf. Table VI). Thus we again conclude that *the instantaneous mortality rate at maximum reproduction is greater by unity than the rate characteristic of a very small population density.* For example, if $s_1 = 0.01$, the instantaneous mortality rate at minimum stock density, from the action of the controlling predator, is equal to $-\log_e 0.01 = 4.61$. At the density of maximum reproduction, mortality from this predation becomes 5.61. Mortality from other factors $(= -\log_e k)$ must be added to this 5.61 to give the total average egg-to-egg instantaneous mortality rate at maximum reproduction. Note however that at the *replacement* density of stock the compensatory mortality is greater than 1, its exact value depending on the steepness of the reproduction curve (relative to the 45-degree line).

The form of expression (11) indicates that the magnitude of the arbitrary factor 10, relating initial egg density to mean predator density, does not affect the general shape of the recruitment curve, though it of course affects its steepness: the larger this factor, the broader is the range of initial egg densities which afford substantial reproduction. Furthermore the rule italicized above holds even if the relation between prey and predator is quite irregular.

Similarly, the magnitude of s_1 affects the shape of the reproduction curve only in that the smaller s_1 is, the greater is the percentage change in reproduction produced by a given change in predator abundance (Table VI).

SITUATIONS OF TYPES A AND C. So far the theory presupposes type B predation situations. Type A situations need not be discussed in detail: for reasons given earlier, they tend to be restricted in space or time, and they lack the qualifications of a population control mechanism. Their effect on reproduction curves would be to introduce irregularities such as are shown in Figures 30 and 31.

Type C situations however *can* regulate a population, and in a pure form they produce reproduction curves like Figure 2b. The horizontal part of the curve corresponds to the limit of surviving young which the habitat will sustain. Type C situations may grade into type B, in which case it would in practice be difficult to distinguish a broad-domed type B curve from the mixed type.

It is also possible for type B and C situations to follow each other, and this is examined in Figure 34. If the type B situation comes first, its typical reproduction curve is truncated by the limit of reproduction imposed later by Type C (curve A of Fig. 34). If the type C situation comes first, it will at a certain point limit the brood to a density which may be less than, equal to, or more than what gives maximum reproduction in the subsequent type B situation; these three possibilities are illustrated by curves B–D of Figure 34. Note that while the type C situation can *reduce* the difference in instantaneous mortality rate between minimal and maximal reproduction, it cannot *increase* this difference.

The various curves of Figure 34 should be looked for in nature when the biology of the animal in question points in their direction. Among the examples given earlier, a limit of environmental capacity is strongly suspected for coho salmon; actually Figure 21 could readily be fitted with a truncated curve like 34A, but there is only one point for the outer decreasing phase.

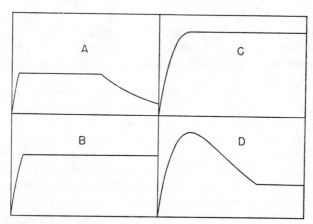

FIGURE 34. Reproduction curves when predation situations of types B and C occur in succession. *Curve A*—when type B precedes type C; *curves B–D*—when type C precedes type B. The type B reproduction curve is the same for all four situations.

SIZE OF PREDATOR AND SIZE OF PREY. It has sometimes been said that large predators (e.g., foxes) cannot control or regulate the abundance of a smaller prey species (e.g., mice), because the rate of reproduction of the prey is so much greater than that of the predator. This argument would be valid only if the predation situation were of type A, with the foxes always able to kill as many mice as interested them—a situation which could scarcely last indefinitely. However if the fox-mouse predation situation were of type B, calculations from (7) show that any increase at all in the number of foxes reduces the survival rate of mice more than proportionately, provided the foxes initially are the cause of an instantaneous mortality rate of 1 or more in the mouse broods; and the greater the increase of the foxes, the less this initial mortality need be to accomplish the result stated. To put it another way, the effectiveness of foxes in reducing the survival of mice will increase more rapidly than does the actual number of foxes, whenever they are already reducing mouse survival to 37 per cent or less of what it would be in the absence of foxes. This effect gives the foxes an advantage which tends to counterbalance the greater rate of reproduction of the mice.

More generally, there is evidently no reason why a large predator, or assemblage of predators, might not regulate the abundance of a smaller and more prolific prey species, provided their abundance can change with that of the prey to *some* extent. Indeed, since a given change in predator abundance has a greater effect upon survival, the smaller the average survival rate of the prey (and hence the greater its fecundity), it could be argued that great fecundity should make a species more—not less—easily controlled by small changes in predator abundance. Against this must be laid the fact that great fecundity makes the prey more likely to occasionally "slip out from under" by becoming so numerous that the type B predation situation cannot be maintained.

OTHER COMPENSATORY AGENTS

We have seen that both cannibalism and "ordinary" predation lead to the Figure 33 type of reproduction curve under type B conditions. For other compensatory agents only a qualitative examination can be attempted here, but they must be mentioned briefly.

The special type of predation situation in which insect "parasites" eat other insects regularly leads to low host reproduction at high host densities, as shown in Nicholson and Bailey's (1935) detailed analysis.

"True" parasites (those which do not regularly and consistently destroy their hosts) and disease organisms can also exhibit non-linear effectiveness which can seriously reduce reproduction at high stock densities—though their activity in this respect must usually be very irregular, if the human situation is a guide.

In the *Daphnia* cultures described earlier no predation or cannibalism existed, nor apparently did disease. The most plausible suggestion which has been made is that some metabolic product acted as a depressor of reproduction. If, for example, a metabolic waste killed a fixed fraction of *Daphnia* embryos per unit concentration and per unit time throughout prenatal life, it would have exactly the same effect upon survival as random cannibalism, and lead to the reproduction curve of Figure 33. The points of Figure 28 can be fitted by expression (10) with good apparent agreement.

Again, consider the compensatory situation where a brood, initially more numerous than average, grows more slowly than average because of competition for limited food; and because of this smaller individual size it is more vulnerable to predation than a faster growing brood at corresponding ages. It is fairly clear that here too there *could* be a lag effect whereby the brood would lose by increased predation more than the numerical advantage which it originally had; though it would be difficult to show that this should (or should not) occur as a general rule.

Finally, consider what happens when spawning facilities are limiting, as for some salmon or trout populations, for example. Natural selection will probably see to it that the population uses the best spawning facilities first. At increasing densities there is both a spread to less suitable gravel in other parts of the stream and a crowding and superimposition of eggs on the favourable grounds. With extremely heavy seeding, fungus may cover and kill almost the whole of the eggs in even the best redds; something of the sort probably decimated the Karluk pinks in the disastrous year 1924 (Barnaby, 1944). The sum of the above effects would usually add up to a domed reproduction curve with an extended right limb. Though an *exact* fit to expression (10) would not be likely, the latter might serve as a picture of the normal expectation.

FITTING THE CURVE $z = we^{1-w}$

A first approximation to a fit of expression (10) to any body of data can be obtained very quickly, because the distribution is completely determined by the position of the maximum and the latter can be selected by eye. The work consists merely of dividing each observed estimate of reproduction (Z) by the estimated maximum Z, and each observed parental stock density W by the density which produced the estimated maximum W; this gives w and z, respectively. From (10),

$$\log_e \hat{z} = \log_e w + (1 - w). \tag{12}$$

A sample calculation is shown in the last line of Table VII.

To find the *best*-fitting curve of this type it would be necessary to calculate the residuals (differences between observed and calculated values) for a series of trial positions of the maximum point. The latter can be varied in two dimensions, so the process might be fairly protracted. The accepted criterion of best fit would be that for which the sum of the squares of the residuals is least, but an easier

TABLE VII. Computation of expected reproduction ($\hat{Z}$) and residuals (R) for a set of stock-reproduction (W–Z) observations, using $z = we^{1-w}$. W = 13, $Z_{max.}$ = 38, is the trial position of the dome of the curve. The last line is a computation of $\hat{Z}$ for an arbitrary W. (See text for further details.)

1 W	2 Z	3 z	4 w	5 $\log_e z$	6 $\log_e w$	7 Unity	8 R	9 R^2	10 $\log_e \hat{z}$	11 $\hat{z}$	12 $\hat{Z}$
5	18	0.47	0.38	−0.75	−0.95	1.00	−0.42	0.18	−0.33	0.72	27
8	78	2.05	0.61	+0.72	−0.49	1.00	+0.82	0.67	−0.10	0.90	34
16	20	0.53	1.23	−0.64	+0.21	1.00	−0.62	0.38	−0.02	0.98	37
28	5	0.13	2.15	−2.02	+0.77	1.00	−1.64	2.69	−0.38	0.68	26
33	8	0.21	2.54	−1.56	+0.93	1.00	−0.95	0.90	−0.61	0.54	21
78	28	0.74	6.00	−0.30	+1.79	1 00	+2.91	8.47	−3.21	0.04	1.5
Sum	..	..	..	..	..	..	+0.10	13.29	..	..	..
50	..	..	3.85	..	+1.35	1.00	..	..	−1.50	0.22	8

criterion—that the sum of the residuals should approximate to zero—would be useful in the earlier stages. Table VII shows a computation of residuals (R) for the pink salmon data of Figure 19. The dome of the curve is estimated to be at a parental abundance of W = 13 and filial abundance of $Z_{max.}$ = 38. Dividing the observed values of columns 1 and 2 by these values gives z and w in columns 3 and 4. The residual (R) is the observed value $\log_e z$ less the estimated $\log_e \hat{z}$ from (12), which is equivalent to

$$R = \log_e z + w - \log_e w - 1. \tag{13}$$

The four right-hand terms of (13) are in the order of columns 4–7 of Table VII; adding columns 4 and 5 and subtracting 6 and 7 gives R in column 8, and R^2 is shown in 9. The sum of the R's is close to zero, showing that the trial position of the dome of the curve cannot be too far from the best one; however since more than half of the R^2 total is contributed by the last observation, a somewhat better fit could probably be obtained by moving the dome a little to the right or upward.

Expected reproduction values, $\hat{Z}$, are calculated in the last three columns of Table VII. Column 10 is equal to 8 less 5; or it can be obtained directly by adding columns 6 and 7 and subtracting 4 (cf. formula 12). The latter method is used to calculate $\hat{Z}$ for arbitrary values of W, as shown in the last line of the Table.

Base-10 logarithms can be used in Table VII if columns 4 and 7 are multiplied by 0.4343, but the procedure shown is more convenient when a comprehensive table of natural logarithms is at hand.

Fitted curves have not been used on the observed reproduction figures above (Figs. 18–28). Apart from the work involved, the theory of expression (10) needs further examination, and uncritical use of it now might conceal variant or alternative situations, such as that suggested for Figure 21. All the curves of Figures 18–28 were drawn freehand before expression (10) had been developed, and it is interesting that most of these observations suggested a concave and tapering right limb. Even the original hypothetical curves of Figures 7 and 8 were drawn this way, because of a feeling that reproduction couldn't really decline right to zero as spawners became increasingly numerous.

Returning to the question introduced at the start of this division of the paper, no universally applicable theory of reproduction has been discovered, or is likely to be. However several possible reasons are apparent why recruitment can decline, and usually will decline, at higher stock densities. Also, from reasonable assumptions a simple mathematical expression has been developed which could be used to represent most of the observed reproduction data over the range of densities which they cover.

SUMMARY

1. The general theory of reproduction indicates that density-dependent causes of mortality set a limit to the size which a population achieves. The "reproduction curve" for a fish species is defined as a graph of the average number of eggs produced by a filial generation against the number produced by its parental spawning assemblage, under the existing frequency distribution of environmental conditions for survival (Figs. 1–8).

2. The level of adult stock at which a *maximum number* of recruits (on the average for existing variations in environmental conditions) is obtained is not necessarily the level at which the *replacement number* is obtained (on the average). Maximum recruitment may occur either at the replacement level of adult stock (Figs. 2, 3), or at a higher level (Figs. 1, 4), or at a lower one (Figs. 5–8).

3. When, with increasing stock density, the maximum of recruits exceeds and precedes the replacement number, a population tends to oscillate in abundance. These oscillations are stable if the reproduction curve crosses the 45-degree line (representing the replacement level of reproduction) with a slope between -1 and $-\infty$ (Figs. 7, 8); they are damped if the curve crosses at any numerically lesser slope (Fig. 5).

4. Under the conditions of Figures 7 and 8 a population of single-age spawners has an irregular but permanent cycle of abundance, such as shown in Figure 11D for example, if enivronmental conditions are stable.

5. Under the same conditions a population of multiple-age spawners tends to have a more regular cycle, whose peak-to-peak period is close to twice the median length of time from oviposition by the parent generation to oviposition by all its progeny (Figs. 9F, 10).

6. For such populations to have cycles of appreciable amplitude, it is necessary that reproduction be absent or light during the first year or two of the animal's life (assuming that reproduction occurs once a year).

7. Random fluctuation in reproductive success, *by itself*, leads to very large changes in abundance and eventual extinction or fragmentation, for populations of single-age spawners (Fig. 11A), and to unlimited increase for multiple-age spawners (Fig. 12A).

8. Combinations of random fluctuation in reproduction with reproduction curves produce populations which neither increase indefinitely nor decline to extinction. The cyclical changes produced by steep reproduction curves are maintained when random fluctuation is small or moderate, but they become variable in period and eventually unrecognizable as random effects become more important.

9. Under the combined influence of a steep reproduction curve and variable non-compensatory mortality, populations fluctuate more widely than when controlled by either of these factors alone.

10. When the dome of a population's reproduction curve lies above the 45-degree line, the result of light or moderate exploitation is to *increase* the abundance of the stock in subsequent generations; more intensive exploitation will decrease it (Fig. 16).

11. Another result of exploitation is to reduce the amplitude and complexity of any reproduction-curve oscillations that may be in progress; sufficiently intensive exploitation eliminates such oscillation entirely (Fig. 17).

12. Among multiple-age spawners exploitation tends also to reduce the *period* of oscillation somewhat, because the spawning stock gradually becomes younger.

13. Reproduction curves, or approximations to them, are plotted for fish populations and four invertebrate populations in Figures 18–28. No example was discovered where the maximum of recruitment is at or to the right of the 45-degree line (as in Figs. 1, 3, 4). The left limb was usually steep. The most probable position for the right limb was always sloping downward, sometimes only slightly, sometimes quite steeply; the most typical situation has a slope about as in Figure 5.

14. Cyclic population changes that are apparently the direct result of a steep reproduction curve are those of Herrington's haddock (1912–29) and the *Daphnia* cultures of Pratt. Most other well-known cycles seem not to be of a *simple* reproduction-curve type, but the shape of this curve must profoundly influence the course of population abundance in any animal.

15. More complex reproduction curves have been suggested by indirect evidence (Figs. 30, 31), and a depression of percentage reproduction at extremely low densities of stock may be fairly common.

16. An asymmetrical dome-shaped reproduction curve can be developed from simple assumptions involving random cannibalism or compensatory preda-

tion: if w represents adult stock density as a fraction or multiple of the density which provides maximum reproduction, we^{1-w} represents the actual reproduction at density w, as a fraction of the maximum reproduction. In this event the survival rate at maximum reproduction is always $1/e$ or 37 per cent of what it is at densities near zero; or in other words, at maximum reproduction the instantaneous mortality rate from compensatory activity is 1.

17. Curves computed from the above formula, shown in Figure 33, include most of the types actually observed. However if "environmental capacity" sets an upper limit to number of survivors at some stage of the pre-adult life history, this "typical" curve could either be levelled off at any point of its course, or else truncated (Fig. 34).

ACKNOWLEDGMENTS

A number of individuals have read the manuscript, and numerous changes, deletions and additions have resulted from their suggestions. Those who co-operated in this manner include Y. M. M. Bishop, L. M. Dickie, J. R. Dymond, R. E. Foerster, S. D. Gerking, J. L. Hart, D. J. Milne, F. Neave, A. L. Pritchard, W. B. Scott, W. M. Sprules, J. C. Stevenson, F. H. C. Taylor, F. C. Withler and D. E. Wohlschlag. I am indebted to Messrs. C. E. Atkinson, C. J. Burner and J. C. Stevenson for permission to peruse, and in some cases use, unpublished or incompletely published data. Mrs. D. Gailus has had the difficult task of maintaining the legibility and accuracy of the manuscript through a long series of revisions.

REFERENCES

ALLEE, W. C. 1931. Animal aggregations. Univ. Chicago Press, 431 pp.

BABCOCK, J. P. 1914. Annual Report of the British Columbia Commissioner of Fisheries for 1913, Victoria, B.C.

BARANOV, F. I. 1918. [On the question of the biological basis of fisheries.] N.-i. Ikhtiologicheskii Inst., Izvestiia, 1(1): 81–288.

BARNABY, J. T. 1944. Fluctuations in abundance of red salmon, Oncorhynchus nerka (Walbaum), of the Karluk River, Alaska. U. S. Fish and Wildlife Serv., Fish Bull., 50: 237–295.

BURKENROAD, M. D. 1946. Fluctuations in abundance of marine animals. Science, 103(2684): 684–686.

CHIANG, H. C., AND A. C. HODSON. 1950. An analytical study of population growth in Drosophila melanogaster. Ecological Monogr., 20: 173–206.

CLEMENS, W. A. 1952. On the cyclic abundance of animal populations. Canadian Field-Naturalist, 66: 121–123.

COLE, LaMONT C. 1951. Population cycles and random oscillations. J. Wildlife Management, 15: 233–252.

1954. Some features of random population cycles. Ibid., 18:2–24.

ELSON, P. F. 1950. Increasing salmon stocks by control of mergansers and kingfishers. Fish. Res. Bd. Canada, Atlantic Prog. Repts., No. 51, pp. 12–15.

ERRINGTON, P. L. 1954. On the hazards of overemphasizing numerical fluctuations in studies of "cyclic" phenomena in muskrat populations. J. Wildlife Management, 18: 66–90.

FOERSTER, R. E., AND W. E. RICKER. 1941. The effect of reduction of predaceous fish on survival of young sockeye salmon at Cultus Lake. J. Fish. Res. Bd. Canada, 5: 315–336.

FUJITA, H., AND S. UTIDA. 1953. The effect of population density on the growth of an animal population. *Ecology*, **34**: 488–498.

GALLAGHER, H. R., A. G. HUNTSMAN, J. VAN OOSTEN AND D. J. TAYLOR. 1944. Report of the International Board of Enquiry for the Great Lakes Fisheries, pp. 1–24, Govt. Printing Office, Washington, D.C.

GALTSOFF, P. S., AND V. L. LOOSANOFF. 1939. Natural history and method of controlling the starfish (*Asterias forbesi* Desor). *Bull. U. S. Bur. Fish.*, **49**(31): 75–132.

HALDANE, J. B. S. 1953. Animal populations and their regulation. *New Biology*, **15**: 9–24. Penguin Books, London.

HERRINGTON, W. C. 1941. A crisis in the haddock fishery. *U. S. Fish and Wildlife Serv.*, *Fish. Circ.*, No. 4, 14 pp.

1944. Factors controlling population size. *Trans. 9th North Am. Wildlife Conf.*, pp. 250–263.

1947. The role of intraspecific competition and other factors in determining the population level of a marine species. *Ecol. Monogr.*, **17**: 317–323.

1948. Limiting factors for fish populations. Some theories and an example. *Bull. Bingham Oceanogr. Coll.*, **9**(4): 229–279.

HILE, R. 1936. Age and growth of the cisco, *Leucichthys artedi* LeSueur, in the lakes of the northeastern highlands, Wisconsin. *Bull. U. S. Bur. Fish.*, **48**: 211–317.

HUNTSMAN, A. G. 1952. How Passamaquoddy produces sardines. *Fundy Fisherman*, **24**(24): 5. St. John, N.B.

1953. Movements and decline of large Quoddy herring. *J. Fish. Res. Bd. Canada*, **10**: 1–50.

HUTCHINSON, G. E. 1948. Circular causal systems in ecology. *Annals New York Acad. Sci.*, **50**: 221–246.

HUTCHINSON, G. E., AND E. S. DEEVEY, JR. 1949. Ecological studies on populations. *Survey of Biol. Progress*, **1**:325–359. Academic Press, New York.

JOHNSON, W. E., AND A. D. HASLER. 1954. Rainbow trout production in dystrophic lakes. *J. Wildlife Management*, **18**: 113–134.

KESTEVEN, G. L. 1947. Population studies in fisheries biology. *Nature*, **159**: 10–13.

LOTKA, A. J. 1925. Elements of physical biology. Williams and Wilkins, Baltimore, 460 pp.

MILLER, R. B. 1950. Observations on mortality rates in fished and unfished cisco populations. *Trans. Am. Fish. Soc. for 1949*, **79**: 180–186.

1952. The role of research in fisheries management in the Prairie Provinces. *Canadian Fish Culturist*, No. 12, pp. 13–19.

NEAVE, F. 1952. "Even-year" and "odd-year" pink salmon populations. *Trans. Royal Soc. Canada* (V), Ser. 3, **46**: 55–70.

1953. Principles affecting the size of pink and chum salmon populations in British Columbia. *J. Fish. Res. Bd. Canada*, **9**: 450–491.

NICHOLSON, A. J. 1933. The balance of animal populations. *J. Animal Ecol.*, **2**: 132–178.

NICHOLSON, A. J., AND V. A. BAILEY. 1935. The balance of animal populations. Part I. *Proc. Zool. Soc. London* for 1935, pp. 551–598.

PALMGREN, P. 1949. Some remarks on the short-term fluctuations in the numbers of northern birds and mammals. *Oikos*, **1**: 114–121.

PRATT, D. M. 1943. Analysis of population development in *Daphnia* at different temperatures. *Biol. Bull.*, **85**: 116–140.

PRITCHARD, A. L. 1948a. Efficiency of natural propagation of the pink salmon (*Oncorhynchus gorbuscha*) in McClinton Creek, Masset Inlet, B.C. *J. Fish. Res. Bd. Canada*, **7**: 224–236.

1948b. A discussion of the mortality in pink salmon (*Oncorhynchus gorbuscha*) during their period of marine life. *Trans. Royal Soc. Canada* (V), Ser. 3, **42**: 125–133.

RICKER, W. E. 1952. Numerical relations between abundance of predators and survival of prey. *Canadian Fish Culturist*, No. 13, pp. 5–9.

1954. Effects of compensatory mortality upon population abundance. *J. Wildlife Management*, **18**: 45–51.

RICKER, W. E., AND R. E. FOERSTER. 1948. Computation of fish production. *Bull. Bingham Oceanogr. Coll.*, **11**(4): 173–211.

ROYAL, L. A. 1953. The effects of regulatory selectivity on the productivity of Fraser River sockeye. *Canadian Fish Culturist*, No. 14, pp. 1–12.

SCOTT, W. B. 1951. Fluctuations in abundance of the Lake Erie cisco (*Leucichthys artedii*) population. *Contr. Royal Ontario Mus. Zool.*, No. 32, 41 pp.

SMITH, F. E. 1952. Experimental method in population dynamics: a critique. *Ecology*, **33**: 441–450.

SMITH, H. S. 1935. The role of biotic factors in the determination of population densities. *J. Econ. Entomol.*, **28**: 873–898.

SMOKER, W. A. 1954. A preliminary review of salmon fishing trends on Inner Puget Sound. *Washington Dept. Fish., Res. Bull.*, No. 2, 55 pp.

SNEDECOR, G. W. 1946. Statistical methods. 4th ed., Iowa State College Press, Ames, Iowa.

SOLOMON, M. E. 1949. The natural control of animal populations. *J. Animal Ecol.*, **18**: 1–35.

TESTER, A. L. 1948. The efficacy of catch limitations regulating the British Columbia herring fishery. *Trans. Royal Soc. Canada* (V), Ser. 3, **42**: 135–163.

THOMPSON, D. H. 1941. The fish production of inland streams and lakes. *In* "A Symposium on Hydrobiology", pp. 206–217. Univ. Wisconsin Press, Madison.

TIPPETT, L. H. C. 1927. Random sampling numbers. *Tracts for Computers*, No. 15.

VARLEY, G. C. 1947. The natural control of population balance in the knapweed gall-fly (*Urophora jaceana*). *J. Animal Ecol.*, **16**: 139–187.

VEVERS, H. G. 1949. The biology of *Asterias rubens* L.: Growth and reproduction. *J. Marine Biol. Assn. U. K.*, **28**: 165–187.

Handbook of computations for biological statistics of fish populations.

Appendix I. Development of model reproduction curves on the basis of a theory of predation at random encounters

W.E.Ricker

Bulletin of the Fisheries Research Board of Canada, **119**, 300pp., 1958. Pages 263-270 are reproduced here.

APPENDIX I. DEVELOPMENT OF MODEL REPRODUCTION CURVES ON THE BASIS OF A THEORY OF PREDATION AT RANDOM ENCOUNTERS[1]

Using formula (11.6) of Chapter 11 can be justified only by its applicability to observed information on reproduction. However it is instructive to consider what kind of theoretical situation can produce such a relationship. One such approach[2] is from consideration of predation upon the eggs or young of the species, when this occurs (on the average, if not necessarily on each occasion) as though at random encounters between predator and prey. That is, the relative (or absolute) abundance of predator and prey does not affect the expectation of capture of an individual of the prey species by an individual predator.

The consequences of random searching for insect prey by hymenopterous parasites (really predators) has been examined by Nicholson (1933). Much the same concept is involved here.

Consider a predator-prey system in which the prey density is governed by a predator species or assemblage of species early in its life, and the abundance or effectiveness of these predators is affected by or is related to the abundance of the prey in some direct manner. The predators are not wholly or even mainly dependent for food on this prey. The prey is not necessarily killed by these predators alone, but these include all the predators which are responsible for governing prey abundance: any other causes of death of prey are non-compensatory.

Specifically, think in terms of a fish having pelagic eggs, larvae and juveniles, during which stages all compensation is assumed to occur. This seems fairly realistic because relative year-class strength usually varies little after a fish becomes of commercial size.

For predation to be compensatory, it is necessary that the abundance or effectiveness of the predators in question increase with increase in the abundance of the prey. For the mathematical development below, we assume that the average abundance of the predators (during the time of their contact with the prey) varies, from year to year, as some constant fraction of the *initial* abundance of prey (fish eggs or larvae, etc., at whatever stage compensation begins), but that the predators have a minimum abundance which is sustained by their other foods. This relationship is most easily pictured when the parents of the eggs or larvae in question are themselves the controlling predators, but it seems fairly appropriate for other possible compensatory agents.

It may be objected that predators ordinarily don't take all the food which they encounter, but feed to satiation and then stop. This is perhaps typical of warm-blooded predators, but for fishes a much greater range of daily rations is consistent with maintaining life, and the evidence of stomach analysis seems to be that fishes almost always eat considerably less than they *could*. The same probably applies to other poikilothermic organisms. In any event, if a predator becomes satiated at some density of prey and ceases to function as a control, that merely shifts a part of the compensatory mechanism to some other predator or other agent which can function at the higher density.

An important feature of the conditions just outlined is that there is a certain *time lag* in the adjustment of abundance (or effectiveness) of the controlling predators to the abundance of the prey. For simplicity we have assumed that the predators are able to adjust themselves in response to the *initial* prey abundance, but obviously their numbers might become more nearly proportional to some slightly later stage of the predator-prey interaction. For example, if a

---

[1] Material in this appendix was presented as part of a series of Lectures on Population Dynamics at the Scripps Institution of Oceanography, September 6-8, 1955.

[2] A different approach which, making appropriate assumptions as to details, also leads to the relationship (A13), is developed by Beverton and Holt (1957, p. 55). The situation which they describe is one where competition for food reduces the growth rate of the denser broods during their first year of life, and the rate of predation on a brood at any given interval after hatching is inversely related to the average size of its members at that time.

herring population were regulated by way of the consumption of its larvae by comb-jellies, the abundance[3] of the ctenophores might be able to adjust itself in a given year so that its average level corresponded best not to the initial larval abundance, but to the abundance of larvae a week after the hatch (say). If however (as seems unlikely on biological grounds) the ctenophores were able to adjust their abundance continuously, day by day, to the changes in herring larval abundance, then the resulting reproduction curves would be quite different from the types obtained below[4].

The hypothesis to be examined requires that *with sufficiently high initial prey densities* the predators must be or must become so numerous that in the later part of the predation period they consume enough prey to actually decrease the absolute number of prey survivors (by comparison with survivors of some smaller initial prey density).

The following symbols are used:

M mean abundance of predators which are involved in compensatory predation, during the time they affect the prey. (These will be called "controlling predators".)

M_0 abundance of controlling predators when the prey species is at zero abundance and hence no prey eggs are produced

P abundance of adults of the prey species at spawning

E number of eggs produced per adult (both sexes)

k_1, k_2, K parameters; $k_1 k_2 = K$

F number of adults of a filial generation (of the prey) produced by parental abundance P

F_m maximum reproduction (filial abundance) of prey

P_m parental abundance of prey which results in maximum reproduction

F_r replacement reproduction of prey

P_r parental abundance of prey which results in replacement reproduction ($F_r = P_r$ when, as here, parents and progeny are in equivalent units)

Z F/F_r

Z_m F_m/F_r

W P/P_r

a P_r/P_m

i_0 instantaneous mortality rate of prey from causes other than the controlling predators

s_c survival rate of prey from compensatory predation

s_n survival rate of prey from all non-compensatory causes (including controlling predators up to density M_0)

E used as a subscript, denotes an equilibrium level of stock, exploitation or yield

The average abundance of controlling predators at prey egg density PE and hence adult prey density (at spawning) P is:

$$M = M_0 + k_1 P \qquad (A1)$$

The total instantaneous mortality rate of the prey is:

$$i_0 + k_2(M_0 + k_1 P) = i_0 + k_2 M_0 + k_1 k_2 P \qquad (A2)$$

Hence:

$$s_n = e^{-(i_0 + k_2 M_0)} \qquad (A3)$$

$$s_c = e^{-k_1 k_2 P} = e^{-KP} \qquad (A4)$$

The abundance of the filial generation of prey, F, produced by a parental generation, P, is:

$$F = PE s_n s_c = PE s_n e^{-KP} \qquad (A5)$$

---

[3] Or better, the "predation potential" of the ctenophores, which would involve their size, for example, as well as their abundance.

[4] They are, in fact, of the type described in the Addendum to this Appendix, page 268.

In order to evaluate K, consider the condition for maximum reproduction, found by differentiating (A5) and equating to zero:

$$-PKe^{-PK} + e^{-PK} = 0$$

Since e^{-PK} cannot $= 0$, $1 - PK = 0$ when P produces maximum F; hence:

$$K = 1/P_m \tag{A6}$$

Substituting this in (A5) gives:

$$F = PEs_n e^{-P/P_m} \tag{A7}$$

Maximum reproduction, when $P = P_m$, then becomes:

$$F_m = P_m Es_n e^{-1} \tag{A8}$$

Parenthetically, this indicates that survival rate at maximum reproduction ($= F_m/P_mE$) is e^{-1} or 37% of s_n, that is, 37% of what it is when stock density approaches zero.

Dividing each side of (A7) by the corresponding side of (A8):

$$\frac{F}{F_m} = \frac{P}{P_m} e^{1-P/P_m} \tag{A9}$$

This is the equation (10) of Ricker (1954b), showing reproduction in terms of its maximum. ❙

However it is usually more convenient to put (A9) into terms of replacement reproduction, F_r, because at replacement $P_r = F_r$ and the two axes of the relationship have the same scale. From (A9), at the replacement level of stock:

$$\frac{F_r}{F_m} = \frac{P_r}{P_m} e^{1-P_r/P_m} \tag{A10}$$

Dividing (A10) into (A9) on both sides:

$$\frac{F}{F_r} = \frac{P}{P_r} e^{P_r/P_m-P/P_m} \tag{A11}$$

Since $F_r = P_r$:

$$F = Pe^{P_r/P_m-P/P_m} \tag{A12}$$

(A11) can also be shortened to:

$$Z = We^{a(1-w)} \tag{A13}$$

RELATION OF MAXIMUM REPRODUCTION TO REPLACEMENT REPRODUCTION. Given (A13), maximum reproduction can be found by differentiating and equating to zero:

$$-aWe^{a(1-w)} + e^{a(1-w)} = 0 \tag{A14}$$

Since $e^{a(1-w)}$ cannot equal 0, at maximum reproduction $aW = 1$ and $W = 1/a$. Substituting these in (A13), the ratio of maximum reproduction to replacement reproduction is:

$$Z_m = \frac{e^{a-1}}{a} \tag{A15}$$

This expression can also be used in reverse to compute a from a given value of Z_m. The solution has to be by interpolation in a table of exponentials or logarithms; and for each value of Z_m there are two possible values of a—one greater than unity and one less. Suppose, for example, that we want a curve in which the maximum recruitment is twice the replacement recruitment, i.e., $Z_m = 2$. The maximum can come either when the spawning stock is greater than the replacement level or when it is less. Solving (A15), $a = 2.678$ or 0.232; substituting these in (A13), the two curves can be calculated.

EQUILIBRIUM EXPLOITATION. Consider any reproduction curve, for example that of Figure A1. For any position of a stock *to the left of the 45-degree line*, there is a rate of exploitation which will maintain the stock in equilibrium at that position. Let A be any point on the curve and AC a perpendicular cutting the 45-degree line at B. At equilibrium the portion BC of the

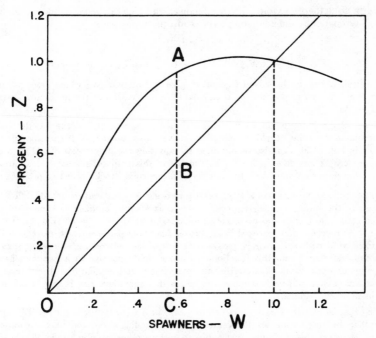

FIGURE A1. An example of a reproduction curve of the type $Z = We^{a(1-W)}$, with $a = 1.119$. The point A is any point on the curve, the distance AB representing the surplus reproduction which must be removed by fishing if the stock is to remain in equilibrium at this level. The distance AB becomes a maximum a little farther to the left on the curve.

recruitment must be used for spawning (since BC = OC), while AB is taken by the fishery. Representing such equilibrium values by the subscript E, the equilibrium rate of exploitation is:

$$u_E = \frac{AB}{AC} = \frac{Z_E - W_E}{Z_E} = \frac{W_E e^{a(1-W_E)} - W_E}{W_E e^{a(1-W_E)}} = 1 - e^{-a(1-W_E)} \qquad (A16)$$

This can be reorganized to indicate directly the number of spawners, W_E, needed to sustain a rate of exploitation u_E:

$$W_E = 1 + \frac{\log_e(1 - u_E)}{a} \qquad (A17)$$

EQUILIBRIUM EXPLOITATION IN TERMS OF SLOPE. The slope, S', of a line joining point A to the origin is AC/OC, or Z/W. From (A16) we may write:

$$u_E = 1 - \frac{1}{Z_E/W_E} = 1 - \frac{1}{S'_E} \qquad (A18)$$

It follows that the locus of any given u_E is a straight line from the origin lying within the top left half of the recruitment graph: a number of these are drawn on Figure 11.2. These loci are applicable, of course, regardless of what may be the shapes of the reproduction curves under consideration.

CATCH AT EQUILIBRIUM. Knowing W_E from (A17), the equilibrium catch, C_E, corresponding to any equilibrium rate of exploitation, u_E, is:

$$C_E = u_E Z_E = Z_E - W_E = W_E e^{a(1-W_E)} - W_E \qquad (A19)$$

MAXIMUM EQUILIBRIUM CATCH. The maximum catch which can be taken under equilibrium conditions is estimated by differentiating (A19) and equating to zero:

$$- aW_E e^{a(1-W_E)} + e^{a(1-W_E)} - 1 = 0$$

$$(1 - aW_E)e^{a(1-W_E)} = 1 \quad \text{(when } C_E \text{ is a maximum)} \tag{A20}$$

For any given a, (A20) can be solved by trial to give the value of W_E which gives maximum equilibrium yield, the corresponding Z_E is calculable from (A13), and the maximum sustained yield is $Z_E - W_E$.

The locus of maximum equilibrium yields is drawn on Figure 11.2 as the fine dotted line. It of course marks the point at which each of the curves has a tangent of 1, corresponding to a slope of 45°; in fact (A20) can also be obtained by equating the differential of (A13) to unity (instead of to zero as was done in A14). To the left of this point, the absolute magnitude of the reproduction decreases more rapidly than does the spawning stock needed to produce it.

SPAWNERS NEEDED FOR MAXIMUM EQUILIBRIUM CATCH. As $a \rightarrow 0$ and the reproduction curve approximates to the 45-degree line, the locus of stock sizes which produce maximum sustained yields, calculated from (A20), approaches a terminal value at $W = Z = 0.5$ (Fig. 11.2). This means that, with reproduction curves of this type, *the size of the spawning stock which gives maximum sustained yield will in no case be greater than half of the replacement number of spawners.* That is, a fully-developed smoothly-functioning fishery should operate with less than half the average spawning stock which characterized the pre-exploitation population. From tests with other reproduction curves of various types, it appears that the above rule is true of practically all of the likely-looking kinds.

On the other hand, with curves described by (A13) best yield is obtained when W is no *less* than about 0.25, because for any smaller values the corresponding curve would become unreasonably steep. However this limit does *not* apply to certain other possible kinds of curves, for example Curve 2a of Ricker, 1954b, p. 564.

REPRODUCTION CURVE CORRESPONDING TO A GIVEN RATE OF EXPLOITATION AT MAXIMUM EQUILIBRIUM CATCH. The rate of exploitation, u, is equal to $(Z - W)/Z$, hence $Z/W = 1/(1 - u)$. From (A13) we may write:

$$Z/W = 1/(1 - u) = e^{a(1-W)} \tag{A21}$$

Substituting this in the condition for maximum equilibrium catch (A20):

$$1 - aW_E = 1 - u_E$$

Hence, at MEC:

$$u_E = aW_E \tag{A22}$$

Taking logarithms of (A21):

$$a - aW = -\log_e(1-u) \tag{A23}$$

Substituting (A22) in (A23), at MEC:

$$a = u_E - \log_e(1-u_E) \tag{A24}$$

Expression (A24) indicates the value of a from which can be computed a reproduction curve whose maximum equilibrium catch will be at a given rate of exploitation u_E.

LIMITING EQUILIBRIUM RATE OF EXPLOITATION. As the equilibrium rate of exploitation, u_E, increases, the corresponding equilibrium abundance of spawners, W_E, decreases. Eventually W_E becomes zero if u_E is sufficiently increased; and this happens at a point where u_E still has some positive value less than unity. This value is the limiting, or maximum, equilibrium rate of exploitation—that which cannot be reached without eventually exterminating the population. Substituting $W_E = 0$ in (A16), the limiting equilibrium rate of exploitation becomes:

$$u_E = \frac{e^a - 1}{e^a} = 1 - e^{-a} \quad \text{(when } W \rightarrow 0) \tag{A25}$$

The limiting equilibrium instantaneous rate of fishing is therefore (in the absence of natural mortality during fishing):

$$p_E = a \quad \text{(when } W \to 0) \tag{A26}$$

It is also easy to show that the slope of the reproduction curve at the origin $(W = 0)$ is equal to:

$$e^a \tag{A27}$$

Some characteristic statistics of each of the 5 curves of Figure 11.2 are shown in Table A1. 1. When $a < 1$, as in Curve D, the dome of the reproduction curve is to the right of the "45-degree line"—that is, maximum reproduction occurs when the spawners are more numerous than their replacement level. Starting a fishery immediately reduces the size of stock, but the stock comes

TABLE A1. Characteristics of 5 reproduction curves of the type $Z = We^{a(1-W)}$, when the replacement level of stock is taken as 1000. The curves are plotted in Figure 11.2, page 239.

	Reproduction curve				
	D	A	E	B	C
Value of a ($= P_r/P_m$)	$\frac{2}{3}$	1	1.25	2	2.678
Maximum stock (replacement stock = 1000)	1072	1000	1027	1359	2000
Spawners needed for max. stock ($1000/a$)	1500	1000	800	500	373
Maximum equilibrium catch (MEC)	198	330	447	935	1656
Spawners needed for MEC	456	433	415	361	314
Stock density (before fishing) at MEC	654	763	862	1296	1970
Rate of exploitation at MEC	0.304	0.433	0.519	0.722	0.841
Limiting equilibrium rate of exploitation	0.486	0.632	0.714	0.865	0.932

to equilibrium at the reduced density defined by (A17), provided the rate of exploitation does not exceed $1 - e^{-a}$ (expression A25). The best rate of exploitation is that which brings the spawning population to the density defined by (A20), and thus gives maximum equilibrium catch. 2. The above is true also when $a = 1$ (Curve A) and the replacement density of stock is also the maximum density. 3. When $a > 1$ (Curves B, C, E), starting a fishery *increases* absolute reproduction for some time, but if exploitation is increased far enough reproduction eventually decreases again. The formulae for maximum catch and limiting rate of exploitation still apply.

ADDENDUM. Beverton and Holt (1957, pp. 49-55) develop a theoretical relationship between egg production, E, and resulting recruits, R, based on the postulate that rate of decrease from egg to recruit is, at successive intervals, proportional to the sum of (1) a density-independent mortality rate and (2) a compensatory mortality rate which is proportional to the number of larvae surviving at successive intervals. These assumptions lead to a relationship with two parameters:

$$R = \frac{1}{\alpha + \beta/E} \tag{A28}$$

With this expression the largest recruitments are at the highest values of E, and in fact recruitment asymptotically approaches the value $1/\alpha$. Inverted, (A28) becomes:

$$\frac{1}{R} = \alpha + \frac{\beta}{E} \tag{A29}$$

One way to fit the expression to data is therefore to plot values of $1/R$ against $1/E$; the slope of the line is an estimate of β and the intercept is an estimate of α. (The fact that $1/R$ instead of R is used to fit the line is an advantage, because $1/R$ is likely to be more symmetrically distributed than is R itself. However the computed R values are reciprocals of the mean of $1/R$ for the given E, rather than a mean of R-values themselves.) Another method of evaluating β (Beverton and Holt, 1957, figures 15.14, 15.18) gives lower expected values of R for high E.

To put (A28) into a form comparable to (A13), values of R can be multiplied by a factor, x, which adjusts them to the same unit as E. Calling the new values R′, we have:

$$\frac{1}{R'} = \frac{\alpha}{x} + \frac{\beta/x}{E} \tag{A30}$$

At the replacement level of reproduction R′ = E; let this particular value of E be represented by E_r. If observed values of R′ and of E are divided by E_r, they then represent reproduction and spawning stock, respectively, in terms of replacement stock. For these quantities we have been using the symbols Z and W; hence:

$$\frac{1}{Z} = \frac{\alpha E_r}{x} + \frac{\beta/x}{W} \tag{A31}$$

Now at replacement Z = W = 1, so that $\beta/x = 1 - \alpha E_r/x$. Letting $A = \alpha E_r/x$ for convenience, (A31) becomes:

$$\frac{1}{Z} = A + \frac{1 - A}{W} \tag{A32}$$

which is an expression with one parameter, comparable to (A13). In inverted form it is:

$$Z = \frac{W}{1 + A(W - 1)} \tag{A33}$$

This expression illustrates the relation of the asymptotic reproduction $1/A$ to the replacement reproduction (which is unity). Its practical use would hinge on obtaining an estimate of this

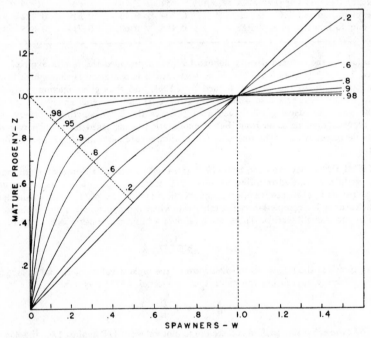

FIGURE A2. Reproduction curves corresponding to the relationship $Z = \dfrac{W}{1 + A(W - 1)}$, for the values of A indicated on the corresponding curves. Both the progeny, Z, and the spawning stock, W, are expressed in terms of the replacement density of stock, which is 1. The diagonal from the ordinate value Z = 1.0 is the locus of maximum equilibrium yields.

relationship, but it is instructive to compare populations described by different values of A but having the same equilibrium density in the absence of exploitation, in a manner similar to Figures 11.2 and 11.4.

Some curves conforming to (A33) are shown in Figure A2. To the left of the replacement density, all the curves are symmetrical about a diagonal running from (0, 1) to (0.5, 0.5). The limiting forms of the curve are the 45° diagonal when $A = 0$, and a right-angled "curve" when A has its maximum value, 1. With values of A greater than about 0.9, expression (A33) describes reproduction curves quite unlike any described by (A13): that is, situations in which reproduction rises rapidly to a point close to *but not exceeding* the asymptotic level and changes little thereafter·

Statistics can readily be derived from (A33) which are similar to those calculated above for (A13). We will note only that the slope of (A33) is given by its differential:

$$\frac{1 - A}{[1 + A(W - 1)]^2} \tag{A34}$$

The position of 45° slope, which is the position of maximum equilibrium yield, is obtained by equating this to 1 (tan 45°). Using the subscript E to represent equilibrium situations, the positive root of the quadratic above is:

$$W_E = \frac{A - 1 + \sqrt{1 - A}}{A} \quad \text{(when } Z_E - W_E \text{ is a maximum)} \tag{A35}$$

The value of (A35), the W_E corresponding to maximum sustained yield, has an upper limit of 0.5 as $A \to 0$. Hence the maximum equilibrium yield is obtained with a spawning stock no greater than half of its primitive replacement level, for all curves of this family—the same conclusion as discovered for the (A13) family. The locus of maximum equilibrium catches is in fact the NW-SE diagonal (Fig. A2).

In contrast to (A13), plausible *simple* biological situations which would result in (A33) are apparently not easy to visualize; but in biology complex situations seem to be at least as frequent as simple ones (Section 11D). On the observational side, the available information for North Sea plaice evidently suggests that recruitment has no detectable trend over a wide range of stock densities (Beverton and Holt, 1957, p. 270), a condition which could be described well by (A33) when A equals 0.95 or more. Because of their large year-to-year variability in reproduction, some existing series of reproduction observations could be described almost equally well by (A33) and by (A13).

Introduction to the later papers in the development of fisheries science

D.H.Cushing

Of the later developments, the first group, Gulland, Garrod and Pope describe the use and limitations of cohort analysis, a method which is widely used today in fisheries science. Gulland's paper gives the original description of virtual population analysis, Garrod gives an example of its application and Pope describes the errors in the system.

Gulland (1965)

The catch equation in numbers was originally applied to the whole stock but it can equally well be applied to single age groups within a year class (see, for example, Ricker, 1948). Jones (1964) and Murphy (1965) had tried to analyze a population with a string of catch equations, but the addition of an age group added another unknown to the system. Gulland solved the problem by summing the catches in the year class to a given age, which is the virtual population to that age and by making educated guesses of fishing and natural mortality in the oldest age group of the yearclass. The calculations proceed back from the oldest age group to the recruiting year class; the system depends on the fact that from the catch at the end of one year the stock at the beginning of the next can be estimated. Gulland named his method *virtual population analysis*, but by usage it now tends to be included in a more general term *cohort analysis*. The great virtue of cohort analysis is that a matrix of fishing mortality emerges by age and year, which was a considerable advance on earlier estimates.

Garrod (1967)

This paper displays in detail the use of cohort analysis on a well studied stock of fish, the Arcto Norwegian cod. The array of data appears formidable but the method can be managed with a good programmable calculator. It is used in many of the stock assessments employed in national and international fora, details of which are rarely published.

Pope (1972)

In this paper Pope showed that the variance of the terminal fishing mortality, the estimate used on the oldest age group of the cohort decreases as the calculation proceeds back up the yearclass, or cohort. The consequence is that the matrices of fishing mortality, or stock, are quite reliable three years back from the present. An appendix to this paper describes Pope's simplification of virtual population analysis which he called cohort analysis, which was needed to calculate variances.

Today cohort analysis is often used in the procedures for establishing quotas in the North Atlantic. The data are complete up to the end of last year and the quota must be effective next year, yet the estimates were most reliable four years earlier. The gap between reliable estimates and the quota is bridged with a relationship between fishing effort observed and estimated fishing mortality. As the matrices lengthen with the years the estimates of terminal fishing mortality improve. Because the system is operated remotely from real estimates of stock, independent measures of stock are needed from time to time with groundfish surveys, acoustic surveys or egg surveys.

The second group of papers comprises two developments of the Schaefer model. Pella and Tomlinson generalize the Schaefer model to some degree and Brown *et al.*, apply it to the multispecies catches in a trawl fishery.

Pella and Tomlinson

Schaefer made the simplest assumption from the logistic equation that the self regulating function of the population is $(P_m - P)$, where P is stock and P_m is maximal stock. Then catch per unit of effort is an inverse linear function of effort, from which it follows that the function of yield on fishing effort is a parabola. However, the early relationships of catch per unit of effort on fishing effort are distinctly curvilinear (Garstang, 1903-05; Thompson, 1951), slightly hyperbolic. Further, Beverton and Holt's (1957) self regenerating yield curve (that is, with their stock and recruitment curve incorporated) is skewed, with the maximum at a slightly lower fishing mortality than would be expected from Schaefer's parabola.

Pella and Tomlinson generalize the Schaefer model by leaving the power function of yield on fishing effort (m) free, rather than restraining it to $m = 2$. They display yield curves and curves of catch per unit of effort on effort for different values of m. They also include a program for estimating constants and the maximum catch which is included here. The importance of the method is that the best estimates may be obtained with simple statistics and that the science of Schaefer is accommodated to that of Beverton and Holt.

Brown et al. (1976)

During the late sixties and early seventies, fisheries biologists in the north eastern United States and eastern Canada became concerned that much mortality was generated in the bycatches of foreign trawlers. Further, such bycatches were not sampled particularly well, but there were records of total catches.

The total biomass is based on catches made by demersal trawls working between Cape Hatteras and the Canadian border. The error about the yield curve is relatively low as compared with that of the usual single species plot. In other words the total catch is less variable than catches of single species, which implies some biological interaction between species. The interactions are necessarily not specified, but one might imagine that differences in recruitment from year to year might be generated by annual difference in predation.

The third group of papers tackles the economic problems of the fisheries and it includes the important paper by Gulland on the marginal yield. Its application is taken from a small section of a paper by Gulland and Boerema (1973).

Gulland (1968)

Economists have formulated the exploitation of fish stocks in value as a function of costs. Superficially their models resemble the Schaefer model and because fishermen work for money, economic models might be preferred to biological ones. However, fisheries are international and quantities such as costs and values are hard to compare between economies, east and west, and almost impossible across the exchanges. During the sixties, there was much discussion on 'bioeconomics' and the most important conclusion was that the maximum economic yield was less than the physical yield, the MSY.

Gulland defined the marginal yield as the increment of yield for an increment of fishing intensity, which must be zero at MSY. Gulland's paper puts the argument formally. I append a small part of Gulland and Boerema (1973) which gives an arbitrary method of defining the point in fishing effort at which the marginal yield is greatest. It provides an estimate of maximum economic yield on the assumption that catch varies as value and fishing mortality as cost. The method is now called $F_{0.1}$ because it is the fishing mortality at which catch per unit of effort is one tenth of the value in the virgin stock at which $F = 0$. The figure from Gulland and Boerema illustrates the rationale which has been used in the North Atlantic since 1973.

Cushing (1981)

The last paper in the selection summarizes present knowledge on the dependence of recruitment upon parent stock which remains the central unresolved problem. In a preliminary sense it was solved by Ricker and by Beverton and Holt but the variability of recruitment is high and fisheries biologists were unable to reach conclusions. When the spawning stock is low steps are taken to increase it through the management procedure. But a yield curve is needed to replace the yield per recruit. Managers continue to conserve stocks and to lose them, and they need secure advice on how recruitment declines under the pressure of fishing.

Because recruitment is determined at an early stage in life history as Ricker assumed nearly thirty years ago, collections are needed on the growth and mortality of larval fish and those just after metamorphosis. They are expensive in research vessel time and cruises are often curtailed for operational reasons. Perhaps this chapter might stimulate a young scientist to step confidently towards the truth.

References

Jones,R.: 1964, 'Estimating population size from commercial statistics when fishing mortality varies with age,' *Rapp. Procès-Verb. Cons. Int. Explor. Mer.,* **155**, 210-214.

Murphy,G.I.: 1965, 'A solution of the catch equation.' *J. Fish. Res. Bd. Can.,* **22**, 191-202.

Ricker,W.E.: 1948, 'Methods of estimating vital statistics of fish populations,' *Indiana Univ. Publ. Sci. Ser.* **15**, 101 p.

Estimation of mortality rates

J.A.Gulland

Annex to the *Northeast Arctic Working Group Report,* 1965.

Introduction

In the previous analysis of the Arctic cod, as presented in the second progress report of the Working Group at ICES in 1959, mortality rates had been estimated in the usual manner as the ratio of the catches per unit effort of the same year-class (or year-classes) in successive years.

As shown in the Figures 15 and 16 in the report, this method gave some extremely variable estimates, though an attempt was made to reduce the variance by omitting certain years where the estimated mortality appeared to be too high. More seriously the method, at least in the simple form, depends on fishing mortality being constant with age. This is clearly not true for the trawl fisheries; thus the 1959 report estimated the fishing mortality (for all years combined) to be about the same, or even higher, for the immature fish as for the mature fish. As the majority of the mature fish are caught outside the feeding areas, mainly by gears other than trawl, the fishing mortality on mature fish caused by trawlers in the feeding area must be quite small, and certainly much smaller than the corresponding mortality on young fish. Such a change in fishing mortality with age will bias, possibly quite seriously, the estimates of mortality rate.

The present Working Group therefore considered that other methods of estimating mortality should be considered. The method of virtual populations (Fry, 1949; Ricker, 1958) was used. This appears to reduce fluctuations due to changes in availability, and the known catches in the mature fisheries provide useful upper estimates to the fishing mortality in the immature fisheries. Also, using methods analogous to those of Jones (1964) preliminary estimates of natural mortality, and of total mortality among the oldest fish, were used to obtain unbiased estimates of the true mortality among the younger fish.

Methods

The following notation will be used:-

$_x C_n$ = catch in numbers, during year n, of the year-class born in year x;

$_x V_n$ = virtual population in year n of the x year-class;

i.e., $_x V_n$ = the total number of fish of the x-year-class which will be caught in the year n or later;

$_x N_n$ = total number of fish of the x-year-class alive at the beginning of year n;

then $_x V_n = {_x E_n} \cdot {_x N_n}$,where

$_x E_n$ = 'exploitation ratio', i.e., the proportion of the fish of the x-year-class alive at the beginning of year n which will, at some time, be caught.

(In the simple constant parameter case $_x E_n$ = constant = $E = F/(F+M)$) In these definitions suffixes have been used to denote different years, and prefixes denote different year-classes. In the following symbols it is more convenient to use prefixes for different age-groups, though retaining suffixes for years;

$_t F_n$ = fishing mortality coefficient on fish at age t in year n;

f_n = fishing effort in year n;

$_t q_n$ = catchability coefficient for fish of age t in year n;

M = natural mortality coefficient (assumed constant).

A first estimate of the survival during year n is given by the ratio of the virtual populations of a year-class at the beginning and end of the year, i.e.,

$$_x S_n = {_x V_{n+1}} / {_x V_n}$$

which if all the mortalities are constant reduces to

$$_x S_n = {_x E N_{n+1}} / {_x E N_n} = e^{-(F+M)}$$

The virtual population also provides, in all situations, an upper limit to the rate of exploitation ('u' in Ricker's notation; $\dfrac{F}{F+M} [1 - e^{-(F+M)}]$, as the rate of exploitation =

$$_x C_n / {_x N_n} < {_x C_n} / {_x V_n},$$ and this upper limit may not infrequently be useful.

More precisely, the catch during any year can be expressed as a function of the fishing and natural mortality rates during the year, and of the population at the end of the year. Thus, in a manner similar to that of Jones (1964), if it is assumed that natural mortality is constant, and some value of fishing mortality among the very old fish is assumed, it is possible for each year-class to proceed year by year backwards from old to young fish estimating the fishing mortality in each year.

Assuming that year-class x is t years old in year n,

let $_x r_n = {^x N_{n+1}} / {_x C_n}$

i.e. r is the population at the end of the year, expressed as a porportion of the catch during the year (thus r can be greater or less than unity)

then $$_x r_n = \frac{^x N_{n+1}}{_x C_n} = \frac{^x N_n\, e^{-(F+M)}}{_x N_n\, \dfrac{F}{F+M}\, (1 - e^{-(F+M)})}$$

where for convenience F has been written for $_tF_n$.

Thus $_x r_n$ is a simple function of $_tF_n$ and M, and if given M, the function

$\dfrac{(F + M)e^{-(F + M)}}{F(1 - e^{-(F + M)})}$ is tabulated for a range of values of F, then once $_xr_n$ is

determined, $_tF_n$ can be at once read off from this table.

Now $_xr_n = {_x N_{n+1}} / {_x C_n} = \dfrac{_x V_{n+1}}{_x E_{n+1} \cdot {_x C_n}}$

$= \dfrac{1}{_x E_{n+1}} \left[\dfrac{_x V_{n+1}}{_x V_n - {_x V_{n+1}}} \right] = \dfrac{1}{_x E_{n+1}} \left[\dfrac{_x S_n}{1 - {_x S_n}} \right]$

i.e. $_xr_n$ is a simple fraction of the apparent survival during year n (as estimated from virtual populations) and the exploitation ratio $_x E_{n+1}$, applicable to the fish of the x-year-class alive at the end of year n.

The exploitation ratio, $_x E_n$, applicable to the fish at the beginning of year n will be the sum of the proportions of fish alive at the beginning of the year caught during the year, and caught later, i.e.

$$_x E_n = \dfrac{_tF_n}{_tF_n + M} \left[1 - e^{-({_tF_n} + M)} \right] + e^{-({_tF_n} + M)} {_x E_{n+1}}$$

Thus, if values of M and $_x E_{n+1}$ are assumed, estimates can be observed in succession of $_xr_n$, $_tF_n$, $_x E_n$, $_x r_{n-1}$, $_{t-1}F_{n-1}$ .... etc. The actual steps in the calculation of mortality rates for the 1948 year-class are set out in Table 1 (values of $M = 0.20$, and E at the 15th birthday of 0.8 were taken).

Results

Table 2 shows Z', the mortality estimated as the ratio of virtual populations at the beginning and end of the year for fish between 4 and 14 years old for the years 1946 – 1962. (The figures are based on preliminary data on the total catches of each age-group, which have since been revised. It is believed that their revision will not alter the estimates of mortality appreciably). Compared with the estimates obtained from catch per unit data these are much less variable; from the method used no negative values can occur, and for fish less than 10 years old the greatest value is only 1.13[1]. Examination of the table suggests, as does the catch per unit effort data, that the fish are not fully recruited until they are six years old; from eight years old there is some recruitment to the mature fisheries, so that an increasing part of the total fishing mortality occurs outside the feeding areas. Accordingly a first estimate of the division between fishing and natural mortality was obtained by relating the apparent mortality Z' among 6 and 7 years old fish

[1] Footnote: in the Working Group Report, higher values are recorded with catch per unit material.

233

Table 1. Calculation of true mortality rates for the 1948 year-class.

Age	S	Z'	$\dfrac{S}{1-S}$	E	r	F	Z	$\dfrac{F}{Z}(1-e^{-Z})$	Ee^{-Z}
14	0.37	0.99	0.588	0.8	0.735	0.79	0.99	0.501	0.298
13	0.39	0.94	0.640	0.799	0.801	0.75	0.95	0.484	0.309
12	0.45	0.80	0.818	0.793	1.03	0.62	0.82	0.423	0.349
11	0.48	0.74	0.923	0.772	1.20	0.56	0.76	0.392	0.361
10	0.40	0.92	0.667	0.753	0.886	0.70	0.90	0.462	0.306
9	0.49	0.70	0.961	0.768	1.25	0.54	0.74	0.382	0.366
8	0.42	0.87	0.724	0.748	0.968	0.65	0.85	0.438	0.319
7	0.51	0.67	1.04	0.757	1.38	0.50	0.70	0.360	0.376
6	0.68	0.39	2.13	0.736	2.89	0.27	0.47	0.215	0.460
5	0.70	0.36	2.33	0.675	3.45	0.23	0.43	0.187	0.439
4	0.78	0.24	3.55	0.626	5.67	0.15	0.35	0.127	0.441
3				0.568					

The right-hand columns are determined quickly, using tabulations, for a range of F, of r, $\dfrac{F}{Z}(1-e^{-Z})$, and e^{-Z}. Included in the table are the values of Z', $(= -\log_e S)$, the first estimate of the total mortality coefficient. In fact, for much of the table Z' is close to the corrected value, Z, though fluctuating rather more widely, and being a distinct under-estimate of Z for the youngest fish.

to the total effort in the feeding area (Regions I and IIb). There is no direct estimate of the combined effort in the two regions. The estimate used was the sum of the total international effort in each area, expressed in English units (millions of ton-hours). Alternatively, because the catch per ton-hour is higher in Region IIb than Region I, by an average factor of 1.5, a better estimate might be

Effort = (Effort in I) + (Effort in IIb) x 1.5.

However, as the trends in effort in the two regions have been similar it is probably that the results would be the same.

Figure 1 shows the plot of apparent mortality of 6 (below) and 7 (above) year old fish against effort; the correlation is very good. As the method tends to under-estimate the mortality when this is low, the total mortality at low levels of effort, and hence the intercept on the y-axis (the estimate of natural mortality) will tend to be low. In Figure 1 the intercepts on the y-axis are 0.05 (for 6 year olds) and 0.20 (for 7 year olds): as these are under-estimates a first estimate of 0.2 for natural mortality was used to calculate by the methods outlined above, better estimates of Z. These are given in Table 3.

Using these estimates, further plots of total mortality against effort are shown (Figure 2). For both ages the correlation is slightly improved: intercepts (i.e. the estimate of M) are 0.21 and 0.40 for 6 and 7 year olds, respectively. The confidence limits of the two estimates of M are 0.15 − 0.27 and 0.30 − 0.50, sug-

Table 2. Total mortality coefficients 'Z' as estimated from the ratio of initial populations at the beginning and end of each year.

Age	1946	1947	1948	1949	1950	1951	1952	1953	1954	1955	1956	1957	1958	1959	1960	1961	1962
14			0.304	0.575	0.861	1.185	1.086	0.708	1.184	0.886	1.020	0.815	0.174	0.409	0.787	1.514	0.992
13	0.536	0.638	0.525	0.835	0.866	0.945	1.073	0.824	0.863	0.836	1.140	1.135	0.491	0.476	0.510	0.943	0.762
12		0.769	0.705	0.743	1.134	0.869	1.659	0.824	0.884	0.818	1.194	0.928	1.070	0.823	0.800	1.108	1.342
11	0.504	1.006	0.720	0.802	1.142	0.883	1.207	0.892	0.924	1.110	1.077	0.968	0.921	0.742	0.935	1.128	1.191
10	0.500	0.643	0.550	0.625	0.582	0.710	0.952	0.671	0.839	0.954	0.882	0.735	0.924	0.940	0.898	0.979	0.889
9	0.425	0.616	0.615	0.506	0.578	0.701	0.750	0.542	0.518	0.736	0.885	0.705	0.565	0.796	0.565	0.817	1.130
8	0.234	0.239	0.435	0.424	0.523	0.564	0.581	0.467	0.524	0.727	0.874	0.839	0.572	0.597	0.580	0.793	0.919
7	0.221	0.329	0.651	0.487	0.500	0.651	0.724	0.425	0.487	0.669	0.798	0.719	0.691	0.601	0.591	0.664	0.790
6	0.156	0.195	0.274	0.462	0.348	0.452	0.691	0.450	0.391	0.716	0.906	0.666	0.742	0.602	0.717	0.638	0.881
5	0.067	0.182	0.107	0.276	0.124	0.355	0.499	0.362	0.370	0.408	0.621	0.288	0.472	0.568	0.533	0.586	0.666
4	0.333	0.072	0.018	0.101	0.031	0.219	0.241	0.210	0.206	0.138	0.283	0.149	0.350	0.294	0.305	0.359	0.366
3	0.013	0.009	0.005	0.022	0.002	0.055	0.034	0.063	0.034	0.029	0.053	0.035	0.109	0.060	0.065	0.133	?

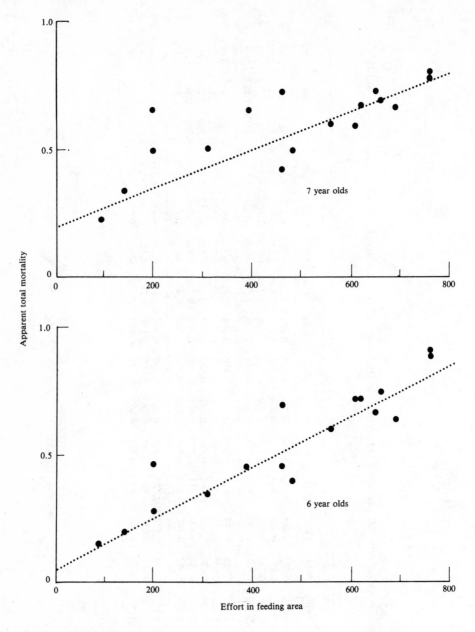

Figure 1. The relation between apparent total mortality and effort in the same year.

Table 3. Corrected estimates of total mortality coefficient, assuming $M = 0.2$.

Age	1946	1947	1948	1949	1950	1951	1952	1953	1954	1955	1956	1957	1958	1959	1960	1961	1962	1963
14									1.15	0.91	1.02	0.85	0.32	0.51	0.81	1.45	0.99	
13								0.86	0.89	0.87	1.11	1.02	0.55	0.55	0.61	0.95	0.79	
12							1.62	0.84	0.90	0.85	1.14	0.85	1.01	0.83	0.82	1.07	1.28	
11						0.93	1.15	0.90	0.92	1.10	1.02	0.95	0.90	0.76	0.95	1.14	1.19	
10					0.65	0.76	0.94	0.72	0.87	0.94	0.90	0.74	0.90	0.94	0.92	0.99	0.93	
9			0.49	0.57	0.65	0.73	0.77	0.61	0.58	0.76	0.88	0.74	0.63	0.83	0.64	0.85	1.12	
8			0.66	0.66	0.59	0.61	0.62	0.53	0.59	0.75	0.85	0.83	0.64	0.64	0.64	0.86		
7		0.41		0.54	0.54	0.66	0.71	0.50	0.56	0.70	0.80	0.72	0.70	0.63	0.66			
6	0.29	0.33	0.38	0.51	0.43	0.50	0.67	0.50	0.47	0.73	0.88	0.68	0.74	0.64				
5	0.24	0.31	0.26	0.37	0.28	0.44	0.52	0.43	0.46	0.49	0.63	0.40	0.53					
4			0.23	0.26		0.34	0.35	0.33	0.34	0.29	0.38	0.29						

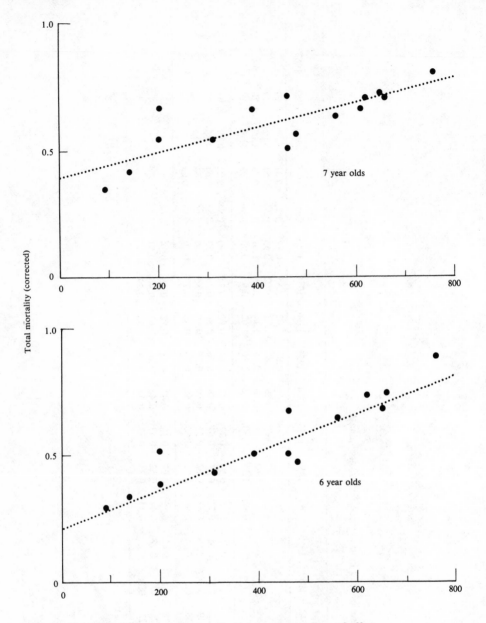

Figure 2. The relation between the corrected estimates of mortality and effort.

gesting that there are some real differences in the mortality/effort relation for the two ages. Though the agreement between the estimates from the two ages is not too good, they suggest that M is between 0.2 and 0.4.

A nearly independent estimate can be obtained from the ratio of the catches per unit effort of certain year-classes in the Barents Sea and (four years later) on the Norway coast; the calculations of this ratio were made in the 1959 report (Table 18 and Figures 18a and b). The value of this ratio depends on the effort units used in the two areas as well as on the mortality between the times when the catches per unit effort are measured (4 – 7 years old in Region I, and 8 – 11 years old in Region IIa). However, if the effort units remain the same then changes in the ratio will be related directly to changes in the mortality. Figure 18b[2] suggests that at the present high levels of effort the logarithm of the ratio is about 2.0 greater than when fishing was zero, i.e.,

$$4\,\overline{F} = 2.0;\ \overline{F} = 0.5$$

where $\overline{F}$ is the average fishing mortality between 4 and 11 years old. The average total mortality over the main ages (4 – 10) and years concerned (1953 – 57) was 0.71; subtracting an F of 0.5 gives an estimate of natural mortality of 0.21.

Comparison with previous results

The total mortality rates obtained here are, for the immature fish (under say 10 years old) considerably smaller than those given in the previous report. This is due to the real decrease with age in the fishing mortality in the trawl fisheries. The decrease can be estimated by dividing the total fishing mortality on each age into that occurring in the spawning area (Region IIa) and in the feeding areas (Regions I and IIb) in the ratio of the catches in the two areas. That is F^1, the fishing mortality in the trawl fisheries is given by

$$F^1 = F \times \frac{\text{Catch in I and IIb}}{\text{Total catch}}$$

The relevant calculations for the 1948 year-class between 4 and 11 years old are given in Table 4. F^1 increases between 4 and 7 years old, and then decreases. These estimates of F^1 cover the years from 1952 to 1959, during which the effort has changed, and the more important measure is the changes with age of q^1,

where $$F^1 = q^1 f, \text{ or } q^1 = F^1/_f$$

The mortality estimated from catch per unit effort data may differ from the true mortality during the year for two reasons: the decrease in catchability, q, with age, and any change in true mortality rates. The magnitude of these effects can be determined from the equation

[2] Footnote: in the Working Group Report not quoted here.

Table 4. Estimation of fishing mortality in feeding areas.

Age	Year	Total F	Numbers caught x 10^{-6}			Trawl F $= F^l$	Effort	^tq
			I + IIb	Total	% Trawl			
14	1962	0.79	57	315	18.1	0.143	758	0.189
13	1961	0.75	285	542	52.6	0.394	691	0.571
12	1960	0.62	363	1,859	19.5	0.121	609	0.199
11	1959	0.56	1,307	3,715	35.2	0.197	556	0.354
10	1958	0.70	3,785	10,772	35.1	0.246	657	0.374
9	1957	0.54	10,297	18,619	55.3	0.299	649	0.461
8	1956	0.65	34,532	50,473	68.4	0.445	764	0.582
7	1955	0.50	78,636	82,534	95.3	0.476	616	0.773
6	1954	0.27	80,382	80,811	99.5	0.269	479	0.562
5	1953	0.23	109,197	109,197	100	0.230	455	0.505
4	1952	0.15	98,068	98,068	100	0.150	456	0.329

Table 5. Estimation of the apparent total mortality rate in the feeding areas, and comparison with observed values in I and IIb.

Ages	$^tq/_{t+1}q$	B	Z_t	A	Apparent Z	Observed Z	
						I	IIb
8 – 9	1.263	0.23	0.85	− 0.05	1.03	1.42	1.46
7 – 8	1.328	0.28	0.70	+ 0.06	1.04	1.12	0.82
6 – 7	0.727	− 0.32	0.47	+ 0.10	0.23	0.44	0.73
5 – 6	0.899	− 0.11	0.43	+ 0.02	0.34	− 0.12	0.33
4 – 5	0.651	− 0.43	0.35	+ 0.03	− 0.05	− 0.33	0.05

Note: A = correction for change in mortality = $\dfrac{Z_{t+1}(1 - e^{-Z_t})}{Z_t(1 - e^{-Z_{t+1}})}$

B = correction for change in catchability = $\log_e {}^tq/_{t+1}q$

$$\log_e {}^nt/_{n_{t+1}} = Z_t + \log_e \frac{Z_{t+1}(1-e^{-Z_t})}{Z_t(1-e^{-Z_{t+1}})} + \log_e {}^tq/_{t+1}q$$

where n_t, n_{t+1} are the catches per unit effort of a given year-class in successive years, and $_tq$, $_{t+1}q$ are the values of q for that year-class in the two years

$$\text{or } \log_e {}^nt/_{n_{t+1}} = Z_t + A + B$$

where A = correction for changes in mortality

B = correction for changes in q.

Table 5 shows these corrections, the resulting expected value of the apparent total mortality based on catch per unit effort data, and also the observed apparent mortalities in Regions I and IIb. These last are each the average of the estimates based on English and on U.S.S.R. data. Though the agreement between expected and observed apparent mortalities is not complete, it is reasonably good.

Population dynamics of the Arcto-Norwegian cod

D.J.Garrod

Journal of the Fisheries Research Board of Canada, **24,** 145-190, 1967.

Population Dynamics of the Arcto-Norwegian Cod[1]

By D. J. Garrod

*Ministry of Agriculture, Fisheries and Food
Fisheries Laboratory, Lowestoft*

ABSTRACT

By reason of its geographical distribution, the Arcto-Norwegian cod (*Gadus morhua*) supports three distinct fisheries, two feeding fisheries in the Barents Sea and at Bear Island–Spitsbergen, and a spawning fishery off the Norway coast. In the past this diversity of fishing on the one stock has made it difficult to unify all the data to give an overall description of post-war changes in the stock. In this contribution three modifications of conventional procedures are introduced which enable this to be done. These are: (i) a system of weighting the catch per unit effort data from each fishery to a level of comparability; (ii) a more rigorous definition of the effective fishing effort on each age-group; (iii) a method of estimation of the effective fishing effort on partially recruited age-groups.

Using these methods the analysis presents the effects of fishing on each fishery in the context of its effect on the *total* stock, and at the same time it indicates ways in which factors other than fishing may have influenced the apparent abundance of the stock. The treatment of the data is also used to derive estimates of spawning stock and recruitment of 3-year-old cod for subsequent analysis of stock–recruitment relationships.

INTRODUCTION

DURING recent years the upward trend in fish production from the North Atlantic has passed a total of 10 million tons of all marine species. Of this, $4\frac{1}{2}$ million tons are gadoid species, and in 1962–63 almost 1 million tons (8% of the grand total) were taken from the single stock of Arcto-Norwegian cod (*Gadus morhua*).

This stock supports three major fisheries, two in the feeding areas of the Barents Sea, and on the Spitsbergen Shelf northward from Bear Island, and a spawning fishery in the vicinity of the Lofoten Islands. In addition to this important geographical distinction the feeding fisheries are exploited principally by trawlers whereas the spawning fishery is more heterogeneous, including trawls, gill nets, purse seine, handlines, and longlines. The Lofoten "skrei" fishery for spawning cod has been carried on for centuries by Norwegian fishermen, but it was not until the late 1920s that other European countries began regularly to exploit the same stock on its feeding grounds to the northeast. Since World War II this has continued and intensified with the expansion of the fishing fleet of the USSR, so that throughout recent history the Arcto-Norwegian cod has been one of the most important fishery resources of the North Atlantic. As such it has been the subject of a considerable amount of ecological research by fishery biologists in the USSR (see Maslov, 1960a, and

---

[1]Received for publication August 22, 1966.

for a bibliography of studies carried out in the USSR) and in Western Europe (Rollefsen, 1953; Trout, 1957); indeed, it was from his observations of the Lofoten fishery that Hjort (1914) first appreciated the significance of fluctuations in year-class strength to the yield of a fishery.

The recognition that the cod in these fisheries of the northeast arctic constitute a single unit stock has been the culmination of a long series of studies of the spawning biology and migration patterns of the fish in each of the three fisheries. The results of this work, which was begun by Sars in 1862, are summarized by Cushing (1966). Briefly cod eggs spawned off the Norwegian coast drift northeastward and become distributed between the Barents Sea and Spitsbergen Shelf by the North Cape and West Spitsbergen currents, respectively. The young cod adopt the demersal habit at the end of their first year and spend the early years of their life in much the same area, carrying out limited annual northeast–southwest migrations according to hydrological and feeding conditions in the areas. These fish mature at between 7 and 13 years of age and then migrate annually to the Norway coast to spawn.

In recent years the Norwegian Government has expressed concern at the downward trend in the catches of the Lofoten spawning cod fishery, and studies of the dynamics of the Arcto-Norwegian cod have been intensified with special reference to the interactions between the various fisheries. In particular all the relevant international fishery statistics and biological data have been collected and analysed by the Arctic Fisheries Working Group established by ICES (Beverton, 1960). This report outlined the magnitude and causes of mortality in the Arcto-Norwegian cod, and estimated the effects that would follow specified changes from the 110-mm minimum size for codend meshes of trawls which had been in force since 1954. As a result, the Permanent Commission recommended that, as from January 1, 1963, the minimum mesh size for codend meshes of trawls fished in this area should be raised from 110 to 120 mm (International Fisheries Convention, 1946). However, catches of Arcto-Norwegian cod continued to decline after this measure came into force, as might have been expected so shortly after the change, but nevertheless renewed concern at the uninterrupted trend of decline of the Norway coast fishery led to a reappraisal of the position by the Arctic Fisheries Working Group in 1965 (International Council for the Exploration of the Sea (ICES), 1965). The findings of this group are given in more detail in the Discussion.

The conclusions reached by the first working group depended upon conventional analyses of catch and effort statistics, in which mortality was estimated as the ratio of catch per unit effort in successive years (Ricker, 1940, 1944, 1958; Beverton and Holt, 1957) but in the 1965 reappraisal more use was made of virtual population analysis (Fry, 1949; 'Bishop, 1959; Jones, 1964; ICES, 1965). The initial catch and effort analyses encountered some difficulty in consolidating the statistics from the different fisheries. In this contribution these catch and effort data have been reanalysed with a different weighting system, to demonstrate their close similarity with the results subsequently obtained by the "virtual population" analysis and to clarify important uncertainties regarding fluctuations in year-class strength. A tentative link between the catchability (q) of the stock and environmental changes is also suggested.

METHODS

It has been pointed out that the Arcto-Norwegian cod supports fisheries in three distinct-geographical areas, each of which contains only a part of the total stock. As a result of this, measures of the total abundance of the stock, and hence its mortality, cannot be derived from any one fishery. Thus, whilst it has been possible to estimate rates of decrease (i.e. mortality + net migration) in each part of the stock separately, it has not previously been possible to form a satisfactory synthesis of all the data, particularly when including that of the Norway coast fishery, where changes in the exclusive territorial limits have influenced the estimates of abundance of the spawning stock based on the trawl data.

In this paper modified methods of analysis are developed which take account of the different age compositions of the catches in the different fisheries and permit the total stock to be estimated in terms of its abundance in the feeding fisheries. The results are then used to interpret the data from the Norway coast fishery, and subsequently the data from all three fisheries have been combined to analyse mortality in terms of the fluctuating abundance of the total stock when it is distributed in the feeding fisheries.

TOTAL CATCHES, CATCH PER UNIT EFFORT, AND THE ESTIMATION OF EFFECTIVE FISHING EFFORT IN THE FEEDING FISHERIES OF THE BARENTS SEA, AND AT BEAR ISLAND–SPITSBERGEN

Precise analyses of mortality are based upon fluctuations in the numerical strength of the stock, whichever method of analysis is adopted. The total number caught in each age-group is therefore a fundamental statistic which, in this catch/effort analysis, must be associated with estimates of fishing effort. In international fisheries such as this the efficiency and objectives of the different fleets are so varied that it is not possible to obtain an adequate measure of effort by the simple addition of the effort statistics of each fleet. In the standard notation, this limitation recognizes that

$$qf(\text{Fleet A}) \neq qf(\text{Fleet B}).$$

In order to overcome this inequality it is customary to assess the total effective fishing effort, $\bar{f}$, in terms of the fishing effort of one fleet thus:

$$\text{Total effective effort} = \frac{\text{Total international catch} \times \text{fishing effort (Fleet A)}}{\text{National catch (Fleet A)}};$$

or, alternatively,

$$\text{Total effective effort} = \frac{\text{Total international catch}}{\text{Catch per unit effort (Fleet A)}}.$$

This technique is usually applied to the total catch of all age-groups in individual calendar years, but this contains the implicit assumption that the relative catchability of different age-groups is constant between fleets, i.e.

$$\frac{qf(\text{age}_x)}{qf(\text{age}_{x+1})} \text{ Fleet A} = \frac{qf(\text{age}_x)}{qf(\text{age}_{x+1})} \text{ Fleet B},$$

and, very strictly speaking, it assumes that within each fleet

$$qf(\text{age}_x) = qf(\text{age}_{x+1}),$$

even though it takes account of the overall inequality of the two fleets.

Neither of these assumptions is accurate, because different fleets have different objectives and may concentrate on a different size-group of fish; furthermore, the catchability of two successive age-groups may differ within one fleet, and the ratio between these age-groups may be different again in another fleet. This complex can be simplified by assuming only that $qf(\text{age}_x)/qf(\text{age}_{x+1})$ is constant from year to year in one fleet, and estimating the effective effort on each age-group as

$$\text{Total effective effort (age}_x) = \frac{\text{International catch (age}_x)}{\text{National catch per unit effort (age}_x)}.$$

In this way differing objectives between fleets will be accounted for by fluctuations of the effective effort on age-groups within 1 year, and trends in q with age-groups within the selected fleet will be reflected as a trend in natural loss from the fishery with age of the fish.

In this analysis the catches per unit effort of English, German, and USSR trawlers were available, but only the English and USSR data give full coverage in all post-war years, and only the English data contain a correction for secular increases in fishing power of the trawler fleet (Gulland, 1956). The catch per 100 ton-hours of the English trawler fleet has therefore been used to derive estimates of the effective fishing effort on each age-group. The discrepancies between the English and USSR data are discussed in Appendix A.

The Estimation of Comparable Indices of Catch Per Unit Effort and Effective Effort on the "Part Stocks" in the Feeding Fisheries

Considering the effect of fishing over very short time periods the basic theory of population dynamics (Gulland, 1964a) gives the relation

$$N = \frac{A}{q} \times \frac{C}{f}$$

where N is the number of fish, A is the area occupied by these fish, C is the catch in numbers, q is their catchability, and f is the fishing effort. This may be written

$$C = \frac{f}{A} q N$$

where f/A represents the fishing intensity per unit area. It is usually taken that the annual mean distribution area of a stock is uniform, so that

$$C = q f N.$$

Hence fishing mortality is usually taken as proportional to fishing effort, rather than to fishing intensity as is required by the more rigid definition.

However, in the northeast arctic, after the spawning season, the Arcto-Norwegian stock is distributed over the Barents Sea (Region I) and on the Bear Island–Spitsbergen (Svalbard) Shelf (Region IIB), and the areas of these two regions are different, so that when fishing effort is equal in both, the fishing intensity per unit area, and hence fishing mortality, is also different. Consequently neither the catches per unit effort, nor the fishing efforts, in the two regions are directly comparable in terms representing the total stock. In order to combine the data for these two regions it is therefore necessary to weight the statistics for the difference in area of the two feeding fisheries. The second step in this analysis has therefore been to deduce values of q/A for both regions by plotting the catch per unit effort in the English trawl fisheries against the sums of effort in the preceding 3 years (Gulland, 1961). When the catch per unit effort is expressed in numbers of fish this parameter should decline exponentially with increasing effort, since $N_{x+1} = N_x e^{-z}$, provided that all year-classes are of equal strength. This last condition is not met by the data, and of course the discard rates on English trawlers would influence the plot of data grouped by years. To offset this the logarithm of the catch weight per unit effort has been plotted against 3-year sums of effort in Fig. 1. The differing slopes of the fitted regressions reflect the different fishing intensities required to generate the same proportional decline in stock in the two areas. Assuming that the other factors affecting q are constant between the two areas (e.g. the behaviour of cod with respect to the trawl) then, per 10^8 ton-hours, the areas of the two regions are in the ratio 0.2826 of Region IIB = 0.0798 of Region I, or Region I = 3.54 times Region IIB. In Fig. 2 the shaded portions in Regions I and IIB cover the fishing grounds in this ratio. The areas have been extended across Goose Bank to Novaya Zemlya in the east, and to Hope Island Deep in the northwest because, although fishable concentrations of cod may not have appeared in these localities during recent years, they still represent part of the potential range of the stock during the later part of each year, as indicated by Maslov (1960b).

The catch per unit effort and fishing effort data for the two regions can now be combined by supposing that the Bear Island–Spitsbergen "stock"

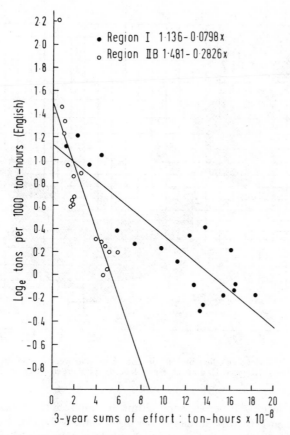

FIG. 1. The relation between stock abundance of Arcto-Norwegian cod and the amount of fishing in the Barents Sea (Region I) and Bear Island–Spitsbergen (Region IIB) fisheries.

was distributed over an area equivalent to that of the Barents Sea "stock." At Bear Island–Spitsbergen the weighted index of abundance would then become c.p.e./3.54, and fishing intensity would be fishing effort times 3.54.

THE ESTIMATION OF TOTAL ABUNDANCE OF EACH AGE-GROUP AND THE EFFECTIVE FISHING EFFORT UPON IT, INCLUDING EFFORT ON THE NORWAY COAST FISHERY

From these basic statistics of catch, fishing intensity, and index of abundance, life-history tables of individual year-classes have been reconstructed, as exemplified by that for the 1948 year-class in Table I. Columns 1–4 give the total numerical catch and catch per unit effort (English) which refer to this year-class. Column 5 is the weighted index of abundance of the total stock of

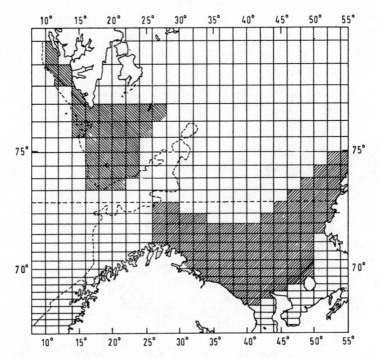

Fig. 2. The feeding distribution of the Arcto-Norwegian cod in the Barents Sea (Region I) and at Bear Island–Spitsbergen (Region IIB).

each age in the feeding fisheries, and column 6 expresses the proportion which is found in Region IIB (it is unreliable for age-groups 3–5 owing to variation in discard rates between the two areas on English trawlers). Columns 7 and 8 give the fishing effort data relevant to each age-group in each fishery, and column 9 the weighted mean fishing intensity on both feeding fisheries. Columns 1–9 thus sum up the statistics from the feeding fisheries; the contribution of the Norway Coast fishery must be added to this.

From the relationships discussed earlier, the effective fishing effort on each age-group at the Norway Coast (Region IIA) can be computed by assuming this catch as part of the total international catch of that age group. In practice it has been estimated as

$$\text{Effective effort IIA (age-group}_x) = \frac{\text{International catch IIA (age-group}_x)}{\begin{array}{c}\text{Best index of abundance (I + IIB)}\\\text{during the spawning fishery}\end{array}} .$$

Any estimate of effective effort at the Norway coast which is based upon the abundance of the age-group in the feeding fisheries in the same calender year, but subsequent to the Norway coast fishery, must under-estimate the abundance in this latter fishery and hence overestimate the true effective effort.

250

TABLE I. Exploited life history of the 1948 year-class of Arcto-Norwegian cod (total catch in numbers × 10⁻³; fishing effort in ton-hours × 10⁻⁸ and catch per effort as numbers/100 ton-hours).

Age (years)	Year	Total catch Reg. I	Total catch Reg. IIB	Catch/effort (= index of abundance) Reg. I	Catch/effort Reg. IIB	Catch/effort Total	Reg. IIB stock as per cent of total	Fishing/effort Reg. I	Fishing/effort Reg. IIB	Mean fishing intensity regions I and IIB	Total catch Reg. IIA	Total Fishing effort Reg. IIA	Fishing intensity (adjusted) Reg. IIA	Fishing intensity (adjusted) Reg. I+IIB	Catch/effort (adjusted)	Total effort I+IIB+IIA	Cumulated effort	Loge catch/effort
		1	2	3	4	5	6	7	8	9	10	11	12	13	14	15	16	17
2	1950		52															
3	1051	24 543	1 379	0.04	0.39	0.43	0.91	3.400	3.536	3.524				3.524	0.43		1.762	−0.844
4	1952	78 245	19 822	4.75	10.17	14.92	0.68	4.100	1.950	2.634				2.634	14.92	3.079	4.841	2.701
5	1953	87 383	21 814	14.46	10.94	25.40	0.43	2.600	1.994	2.339				2.339	25.40	2.486	7.327	3.235
6	1954	74 167	16 215	22.12	9.96	32.08	0.31	3.352	1.628	2.816	429	0.017	0.017	2.816	32.08	2.594	9.921	3.469
7	1955	69 163	9 473	16.97	4.02	20.99	0.19	4.074	2.357	3.745	3 898	0.086	0.086	3.745	20.99	3.366	13.287	3.045
8	1956	26 077	8 510	5.71	1.76	7.47	0.24	4.566	4.835	4.630	15 941	0.759	0.759	4.630	7.47	4.946	18.233	2.011
9	1957	6 949	3 007	1.29	0.47	1.76	0.27	5.387	6.399	5.657	8 322	1.114	1.114	2.150	4.63	4.504	22.737	1.533
10	1958	3 549	253	0.90	0.06	0.96	0.06	3.943	4.218	3.961	6 987	3.970	1.508	3.961	0.96	4.563	27.300	−0.041
11	1959	1 096	232	0.29	0.04	0.33	0.12	3.779	5.800	4.024	2 478	2.582	2.582	4.024	0.33	6.574	33.874	−1.109
12	1960	346	116	0.11		0.11		3.145		3.145	1 535	4.651	4.651	3.145	0.11	8.235	42.109	−2.207
13	1961	244		0.14		0.14		1.636		1.636	542	4.927	4.927	1.636		7.317	49.426	

In this analysis the index used is the catch per unit effort I + IIB (year$_{x-1}$), though this will tend to underestimate the effort. The best estimate would be obtained using the mean abundance of the feeding fisheries before and after each Norway coast fishery. The influence of this upon the final interpretation of the distribution of mortality between the fisheries is discussed more fully further on in this paper.

The estimates of effective effort at the Norway coast are given in Table II. These can be checked from the known maturation rate of Arcto-Norwegian cod. Within reasonable limits, determined by the selectivity of different fishing gears, all age-groups which are fully mature will be exposed to the same effective fishing effort at the Norway coast. Those age-groups which are not fully mature will be exposed to an effective effort at the Norway coast which is proportional to the degree of maturity of that age-group, and in this instance, where the units of measurement of stock abundance and fishing effort are based upon data from the feeding fishery, i.e. they measure the *total* abundance of each group, then the trends in effective effort with age at the Norway coast should correspond to the maturity ogive of Arcto-Norwegian cod.

This comparison is appended to Table II. The mean effective effort on each age-group is given for the entire sampling period, showing the increase in effort with age which reflects the degree of maturity of each age-group. In determining this mean effective effort the four estimates underlined in Table II have been excluded from the average. This can be justified by reading horizontally the effective effort in the same age-group in successive years: the values underlined show particularly high discrepancies from all other readings from that age-group, arising from poor estimates of their abundance in the feeding fisheries. The mean effort per age-group is then referred to the effort on fully matured age-groups in the spawning fishery (ca. 4.000) and compared to a percentage maturity deduced from Rollefsen's data for the years 1941–53 (Rollefsen, 1953). The similarity is illustrated in Fig. 3. There are, of course, variations in the rate of maturation of different year-classes, and in more recent years there appears to be a decrease in the mean age of maturity, but it is not clear in the English data to what extent this reflects fishing near the overwintering grounds of immature cod which are within Region IIA but not on the spawning grounds. Gulland (1964b) estimated 40% maturity at 9 years old, but he discusses various sources of bias that could also account for the discrepancy between his figure and this present estimate of 25% maturity at 9 years. For example, any estimate of maturation rate based upon data collected from the Norway coast spawning fishery may be biassed downward by the effect of fishing in the feeding areas; this mortality will lead to progressive underestimates of the true numbers of potentially late-maturing fish when they do mature and reach the spawning fishery.

The estimates of effective effort at the Norway coast have been carried over into the year-class history tables (see Table I, columns 10 and 11), with a number of minor adjustments where evidently poor estimates of abundance in the feeding fisheries (such as those referred to above) have led to dispro-

TABLE II. A comparison of effective effort on each age-group in the Norway coast spawning fishery, with the maturation rate of Arcto-Norwegian Cod.[a]

Age (years)	1947	1948	1949	1950	1951	1952	1953	1954	1955	1956	1957	1958	1959	1960	1961	1962	1963
4															0.005		
5					0.004			0.002							0.004	0.006	
6		0.004	0.002		0.020	0.004	0.019	0.017	0.007	0.046	0.022	0.022	0.026	0.041	0.031	0.033	0.021
7	0.016	0.011	0.040	0.015	0.056	0.045	0.052	0.093	0.086	0.094	0.099	0.059	0.056	0.071	0.135	0.196	0.146
8	0.088	0.058	0.081	0.177	0.344	0.221	0.163	0.315	0.506	0.759	0.221	0.510	0.484	0.247	0.663	0.697	0.665
9	0.465	0.295	0.217	0.562	1.022	1.209	0.851	0.521	1.154	1.337	1.114	0.674	1.962	1.041	1.825	1.403	1.674
10	0.758	0.538	0.382	0.474	1.484	2.194	1.916	1.595	2.625	1.997	2.222	3.970	3.543	3.774	6.505	2.229	1.575
11	0.503	1.343	1.395	0.714	1.541	3.086	2.556	4.195	1.978	1.780	2.381	3.247	2.582	3.040	15.326	4.841	1.575
12	0.739	0.352	1.795	1.679	0.976	0.969	1.546	9.085	2.810	1.921	2.793	2.392	2.324	4.651		18.450	2.859
13	2.611	0.572	1.143	4.962	5.339	19.298	0.401	3.052	7.264	7.010	6.480	2.922	1.736	4.377			2.523

Age (years)	Mean f/age-group	Percentage max. of 4.000	(Rollefsen, 1953)
4			
5			
6	0.019	+	
7	0.074	0.09	0.03
8	0.365	0.25	0.10
9	1.019	0.49	0.24
10	1.935	0.61	0.42
11	2.438	0.58	0.61
12	2.306	0.93	0.79
13	3.715	1.00	0.92
14+			0.99

[a] Excluding figures underlined in the section above: these are obviously anomalous.

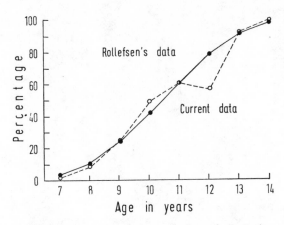

Fig. 3. A comparaison between the rate of maturation of Arcto-Norwegian cod, deduced from fishing effort data, and estimates from otolith studies described by Rollefsen (1953).

portionately high levels of effective effort at the Norway coast. For example, in column 11 of Table I the intensity of fishing on 10-year-olds is extremely high (bearing in mind the mean rate of maturation of a year-class). Reference to column 5 shows that this is generated by an apparently low estimate of abundance in the feeding fishery in 1957. An adjustment to the fishing intensity on 10-year-olds in the Norway coast fishery, in column 12, to bring it into conformity with the effort on adjacent age-groups in the same year of sampling is then carried back to indicate a more probable level of abundance of 9-year-olds in the feeding fishery (column 14). Ten of a total of 102 estimates in the age range 7–12 years have been adjusted in this way.

THE ESTIMATION OF MORTALITY RATES IN FULLY RECRUITED AGE-GROUPS (6 YEARS AND OLDER)

The estimates of total effort, and estimates of total mortality, $Z = (\text{Log}_e \text{ c.p.e.}_x - \text{Log}_e \text{ c.p.e.}_{x+1})$ derived from the data described above are given in Table III. However, for the conventional determination of the regression of Z against f only data from 1950–63 and for 6- to 10-year-old fish have been used. Data for the years 1946–49 are given in Table III but have been excluded from the calculations, owing to some doubt of the accuracy of the English catch per effort in the immediate post-war years. In the absence of a direct sampling program for length at English ports before 1950, and because of greater concentration on haddock in the Barents Sea in the immediate post-war years, the estimate of abundance of cod in the 1946–50 period given by the Arctic Working Group involves two corrections, one to adjust for changes in geographical distribution of English fishing, and a second to obtain an estimate of length composition of English catches from records of fishing by USSR

TABLE III. The correlation between total mortality and the effective fishing effort for the Arcto-Norwegian cod.

Calendar years

Age-groups	46/47	47/48	48/49	49/50	50/51	51/52	52/53	53/54	54/55	55/56	56/57	57/58	58/59	59/60	60/61	61/62	62/63
A	*(Between pairs of age-groups)*																
6 — 7	-0.056	-0.416	0.130	0.885	-0.013	0.152	0.706	0.605	0.424	0.782	0.936	0.574	0.662	1.100	0.792	0.630	1.217
7 — 8	0.823	-0.615	1.382	0.697	0.817	1.055	0.813	0.715	0.623	1.034	0.761	0.794	0.741	1.392	0.351	0.939	1.261
8 — 9	0.247	-0.195	0.363	0.556	1.026	0.875	0.805	0.939	0.729	1.501	0.478	0.494	0.936	0.926	0.344	1.158	1.687
9 — 10	1.076	1.479	0.768	1.134	1.108	1.009	0.615	0.778	0.557	1.184	0.806	1.574	1.429	0.693	1.054	1.728	1.802
10 — 11	0.630	0.888	0.305	0.888	1.252	0.720	1.236	1.085	0.538	1.689	0.542	0.405	1.068		1.387	0.792	0.693
11 — 12	0.983	1.788	0.580	1.609	2.040	1.387	1.408	1.061	1.085	2.110	1.170	1.265	1.541	1.098		0.931	2.485
12 — 13	0.272	1.207	0.761	1.166		1.163		1.058	0.944	1.140							1.179
13 — 14		0.520	0.048	1.425													
Mean:	0.734				0.734	0.773	0.735	0.759	0.583	1.125	0.745	0.859	0.942	1.028	0.635	1.114	1.492
B	*Effective fishing effort on the total stock. Ton-hours* $\times\ 10^{-8}$																
6 — 7	0.872	1.252	1.682	1.861	2.315	2.765	3.006	2.823	3.366	4.733	5.159	4.793	3.709	4.363	4.249	4.601	4.883
7 — 8	0.975	1.363	1.763	2.181	2.827	2.885	2.547	3.135	3.919	4.946	4.937	5.190	4.488	4.744	5.378	5.472	5.755
8 — 9	1.423	1.662	1.761	2.603	3.220	4.008	3.404	2.801	4.653	5.477	4.504	3.764	5.970	3.985	6.913	6.046	7.516
9 — 10	1.723	1.922	1.865	2.319	3.473	4.548	3.947	4.549	4.572	6.069	6.788	4.563	3.678	6.666	8.851	7.255	7.802
10 — 11	1.413	2.690	3.211	2.574	4.738	5.172	4.543	4.002	5.209	5.301	6.595	7.198	6.574		9.418	10.572	10.778
11 — 12	2.071	1.689	3.039	4.115	2.674	9.648	3.863	4.557	4.796	4.129	7.217	5.946	4.873	8.235		10.919	8.348
12 — 13	3.557	2.715	2.832	2.880		4.653		5.308	5.738	5.551							5.634
13 — 14				3.234													
Mean:					2.959	3.551	2.551	3.327	4.127	5.306	5.347	4.577	4.461	4.939	6.348	5.843	6.489

Regression formula 50/51 – 62/63. Age-group 6/7 – 9/10 $Z = 0.33 + 0.119\ f$.

trawlers. Data for older age-groups have also been excluded, owing to suggestions that over the years there may have been secular changes in the accuracy
of sampling of these groups.

The relationship between total mortality, Z, and effective fishing effort, f,
has the regression

$$Z = 0.33 + 0.12f$$

and is illustrated in Fig. 4, showing only the mean values for each year of sampling. The value of the intercept, $0.33 = M$, compares favourably with the
values of 0.21 (6-year-olds) and 0.40 (7-year-olds) estimated by" virtual
population" analysis by the Arctic Working Group (ICES, 1965).

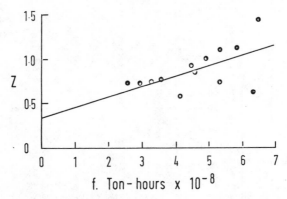

FIG. 4. The relation between fishing effort (f) and
total mortality (Z) of Arcto-Norwegian cod.

The Estimation of the Abundance and Mortality of Partially Re-cruited Age-Groups

The proportion of the total stock which is sampled by a commercial trawler
is determined by the distribution of the different age-groups with respect to
the trawl, and by the selectivity of the codends of the trawl. Excluding special
circumstances, e.g. fisheries based upon spawning fish, it is generally assumed
that during the year the stock sampled in a locality will represent all age-groups
equally well. There is some doubt of this concerning the oldest age-groups of
Arcto-Norwegian cod, and it is certainly untrue of the smallest cod. This,
coupled with gear selection, means that younger age-groups are only partially
recruited to the trawl catches, and in some fleets discarding at sea reduces still
further the proportion recruited to the landings. Inspection of the English catch
per unit effort data indicates that Arcto-Norwegian cod are fully recruited to the
feeding fisheries at 6 years of age, i.e. $q^3 < q^4 < q^5 < q^6 = q^7 = q^8 \ldots$
However, the effect of fishing upon the partially recruited age-groups is a
critically important aspect of mortality analyses which is frequently neglected

through lack of data. For example, estimates of year-class strength based upon fully recruited age-groups (6+ years) may be biassed by undetected changes in the effect of fishing upon younger age-groups (see Discussion).

Some estimate of mortality in these age-groups can be deduced from analysis of the history of individual year-classes (Table I and Appendix Table A).

From the relationship

$$N_{x+1} = N_x e^{-(qf+M)} \qquad \ldots\ldots 1$$

the usual derivation follows

$$\text{Log}_e\, N_{x+1} = \text{Log}_e\, N_x - (qf+M),$$

and hence

$$\text{Log}_e\, N_{x+1} = \text{Log}_e\, N_x - f(q+\frac{M}{f}). \qquad \ldots\ldots 2$$

This implies that a plot of the abundance of a year-class against the accumulated fishing effort to which it is exposed will be linear, with a slope $(q+\frac{M}{f})$ and an intercept $\text{Log}_e\, N_0$, which is an estimate of abundance when the year-class is first recruited to the fishery. This is essentially a method developed by DeLury and quoted by Ricker (1958). Taking the usual assumption that q and M will be constant with age, the form of expression 2 means that the plot will only be linear where f, the fishing effort, is constant from year to year. This limitation also applies to other methods of estimating mortalities based on annual means of pairs of years; any variation of effort between years will introduce a small error (Beverton and Holt, 1957; Paloheimo, 1961).

The further computations used in applying this technique for each year-class are added to Table I (columns 16 and 17) for the 1948 year-class. The plots for each year-class are illustrated in Fig. 5. At low levels of cumulative effort, i.e. in the younger age-groups, in each year-class the estimated abundance lies well below the regression fitted to data for fully recruited age-groups (e.g. year-classes 1950 and 1951). The estimation of effective effort on these age-groups of each year-class is not accurate, but for the present purpose of determining the relation between abundance and cumulative effort in the fully recruited age-groups this does not matter (further corrections for the partially recruited ages are introduced below). The calculated regressions for each year-class are listed in Table IV, showing the decreasing slope with increasing fishing effort per year. Mathematically some degree of curvature must be included in each regression, since there have been systematic trends in fishing effort and hence in the ratio $F/F+M$ during the lifespan of individual year-classes, but this is barely recognizable in Fig. 5. In all year-classes the regressions are good, having approximately the same slope, as would be expected from the theoretical considerations above. It is worth pointing out here that the regressions plotted for

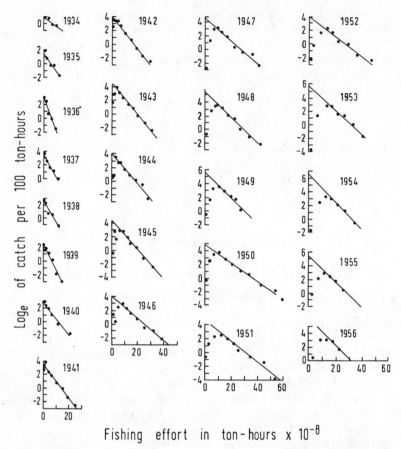

Fig. 5. The relation between catch per unit effort of Arcto-Norwegian cod and the cumulative fishing effort to which individual year-classes are exposed during their exploited life history.

data from each feeding fishery separately will not satisfy this requirement. That the weighted combined data does give a satisfactory interpretation supports the weighting factors used, and the method of treating the data.

In Table I, Appendix Table A, and Fig. 5, it has been assumed that the partially recruited age-groups are subject to the same effective effort as the fully recruited age-groups in the same year. It has already been pointed out that this must overestimate the true level of their exploitation, but the regressions in Table IV provide a method of extrapolating the abundance of younger age-groups if the effective effort to which they are exposed can be determined.

An earlier equation gave the relationship

$$\mathrm{Log_e\ N}_{x+1} = \mathrm{Log_e\ N}_x - f(q + \frac{M}{f}), \qquad\qquad \dots\dots 3$$

TABLE IV. Calculation of the regression of the life history of each year-class of Arcto-Norwegian cod (see Fig. 5).

$$\text{Log}_e \text{ c.p.e.} = \text{Log}_e R_3 - f\left(q + \frac{m}{f}\right)$$

$$x = a - f \ (b)$$

1934	$x = 0.911 - 0.139 f$		1945	$x = 4.386 - 0.223 f$
5	$x = 1.550 - 0.284 f$		6	$x = 4.045 - 0.167 f$
6	$x = 2.915 - 0.500 f$		7	$x = 4.488 - 0.165 f$
7	$x = 3.462 - 0.332 f$		8	$x = 5.397 - 0.186 f$
8	$x = 2.575 - 0.336 f$		9	$x = 5.630 - 0.189 f$
9	$x = 2.673 - 0.389 f$		50	$x = 4.804 - 0.133 f$
40	$x = 2.851 - 0.242 f$		1	$x = 4.543 - 0.149 f$
1	$x = 3.460 - 0.245 f$		2	$x = 3.990 - 0.147 f$
2	$x = 4.061 - 0.243 f$		3	$x = 5.624 - 0.178 f$
3	$x = 4.445 - 0.231 f$		4	$x = 6.355 - 0.197 f$
4	$x = 3.968 - 0.221 f$		5	$x = 5.428 - 0.184 f$

where N_x represents the index of abundance. In back-extrapolation

$$\text{Log}_e N_x = \text{Log}_e N_{x+1} + f\left(q + \frac{M}{f}\right), \qquad \dots \dots 4$$

or, more strictly, where the index of abundance is related to the mid-part of a sampling year, as in Table I,

$$\text{Log}_e N_x = \text{Log}_e N_{x+1} + \tfrac{1}{2}f_{x+1}\left(q + \frac{M}{f}\right) + \tfrac{1}{2}f_x\left(q + \frac{M}{f}\right). \qquad \dots \dots 5$$

The abundance of an age group at the beginning of the preceding year (N_x^1) would be

$$\text{Log}_e N_x^1 = \text{Log}_e N_{x+1} + \tfrac{1}{2}f_{x+1}\left(q + \frac{M}{f}\right) + f_x\left(q + \frac{M}{f}\right). \qquad \dots \dots 6$$

Now $N_x = Y_{N_x}/f_x$, so that

$$\text{Log}_e N_x = \text{Log}_e Y_{N_x} - \text{Log}_e f_x, \qquad \dots \dots 7$$

and, substituting for $\text{Log}_e N_x$ in equation 5

$$\text{Log}_e Y_{N_x} = \text{Log}_e N_{x+1} + \tfrac{1}{2}f_{x+1}\left(q + \frac{M}{f}\right) + \tfrac{1}{2}f_x\left(q + \frac{M}{f}\right) + \text{Log}_e f_x.$$

Rearranging equation 7

$$\frac{Y_{N_x}}{N_{x+1}\, e^{\frac{1}{2}f_{x+1}\left(q + \frac{M}{f}\right)}} = f_x e^{\frac{1}{2}f_x\left(q + \frac{M}{f}\right)}. \qquad \dots \dots 8$$

Given estimates of Y_{N_x}, N_{x+1}, f_{x+1}, and $(q+\dfrac{M}{f})$, expression 8 can be solved for f_x by a graphical evaluation of the right-hand side for known values of $(q+\dfrac{M}{f})$. The value of f_x can then be substituted in expression 5 or 6 to obtain estimates of abundance either at the beginning or midpoint of year x. However-er, rapid changes in the ratio $\dfrac{F}{F+M}$ between years introduce nonlinearity into the regression, and in partially recruited age-groups such changes may be quite large because natural mortality will not be fully accounted for within the regression. As a result the direct substitution of f_x into expression 5 will lead to underestimates of N_x or N_x^1. But, to a close approximation, the regression method gives the value of natural mortality (distributed over the mean fishing effort per year) on the fully recruited age-groups of each year-class ($\bar{f}$), so that where there are large changes in f_x a more accurate value of $Log_e\ N_x^1$ can be calculated by raising the first estimate obtained from expression 5 by a pro-portion $M(1-f_x/\bar{f})$.

The basic data required in order to apply this technique are estimates of $(q+\dfrac{M}{f})$, M, N_{x+1}, and Y_{N_x}. In the present analysis all of these are available except the last, but this, the numerical yield of partially recruited age-groups, is influenced by discard rates. These are imperfectly known, but some progress can be made from an assumption that the discard rate was nil after the decline in stock abundance in the northeast arctic in 1957–58. Then, for the 1955, 1956, and 1957 year-classes the calculations described above can be carried out using the value $(q+\dfrac{M}{f}) = 0.184$ (see Table IV).[2] This calculation is set out in Table VA.

Column O of Table VA compares the true effective effort (g) with the effective effort on the fully recruited age-groups of the same year of sampling. This in effect is a measure of the relative availability of the age-group, and it can be seen that these year-classes give closely comparable results, with a mean 5% of 3-year-olds, 32% of 4-year-olds, and 62% of 5-year-olds.

The calculations detailed above estimate the basic availability of the partially recruited age-groups 3–5. Assuming this to have remained constant over the years it can then be used to revise the provisional effective effort attributed to these age-groups in Appendix Table A. This computation is carried out in Table VB(i). The corrections to the catch per unit effort in each age-group are set out in Table VB(ii), calculated as in Table VA, columns

[2] $(q+\dfrac{M}{f})$ has not been determined for the 1956 and 1957 classes, owing to lack of data from the fully recruited age-groups. Preliminary data for 1956 (see Fig. 5) indicate that the regression will be similar to that of the 1955 class, and it is reasonable to assume that the value for the 1957 class will not be very different.

TABLE Va. Calculation of the availability of partially recruited age-groups of Arcto-Norwegian cod.

Year-class	Age (years)	A $YN \times 10^{-6}$	B $\frac{YN}{f}$	C f	D $\tfrac{1}{2}f(q+M)/f$	E e^D	F(i) 3) Antilog L. 6) ExB.	F(ii) $\frac{Ax}{F(i)x+1}\,f_x^D$	G f_x	H $G\times0.184$	I $\frac{1-fx}{\tilde f}$, $\tilde f=4.797$	J $I\times M$, $M=0.33$	K $\mathrm{Log.}F(i)x+1$ or $Lx+1$	L $H+J+K$	M $K+\frac{L-K}{2}$	N Antilog. $\frac{G}{M}$	O $\frac{G}{C}=q$
1955	3	29.57		4.920			82.2	0.530	0.480	0.088	0.900	0.300	4.021	4.409	4.215	67.7	0.008
	4	65.84		4.978			55.8	1.954	1.500	0.276	0.687	0.229	3.516	4.021	3.768	43.3	0.301
	5	70.24	12.29	3.860	0.3919	1.480	33.7	3.861	2.470	0.454	0.485	0.161	2.901	3.516	3.208	24.7	0.640
	6	47.65		4.260			18.2										
1956	3	26.05		6.631			132.0	0.283	0.250	0.046	0.948	0.316	4.523	4.885	4.704	110.0	0.038
	4	94.29		4.432			92.1	1.663	1.330	0.245	0.723	0.241	4.037	4.523	4.280	72.3	0.300
	5	119.91	19.47	4.686	0.4225	1.526	56.7	4.036	2.550	0.469	0.532	0.177	3.391	4.037	3.714	41.0	0.544
	6	88.84		4.593			29.7										
1957	3	30.15		4.819			133.0	0.326	0.270	0.050	0.944	0.314	4.526	4.890	4.708	111.0	0.056
	4	137.43		4.643			92.4	2.559	1.820	0.335	0.621	0.207	3.984	4.526	4.255	70.5	0.392
	5	156.58	17.09	4.849	0.4379	1.549	53.7	5.915	3.270	0.602	0.318	0.106	3.276	3.984	3.630	37.7	0.674
	6	81.01		4.761			26.5										

TABLE Vb. (i) Correction of the effective fishing effort on age-groups 3–5 (Appendix Table A) for the effect of partial recruitment (3 = 5%, 4 = 31%, 5 = 62%).

Age (years)	1946	1947	1948	1949	1950	1951	1952	1953	1954	1955	1956	1957	1958	1959	1960	1961	1962	1963
3	0.037	0.057	0.086	0.063	0.141	0.226	0.158	0.130	0.124	0.141	0.333	0.347	0.315	0.424	0.308	0.287	0.316	0.315
4	0.215	0.320	0.454	0.347	0.746	1.058	0.872	0.790	0.922	0.967	1.677	1.722	1.702	1.648	1.467	1.537	1.622	1.592
5	0.439	0.655	0.859	0.763	1.317	1.883	1.938	1.450	1.873	2.143	3.137	3.253	2.926	2.820	2.393	2.905	3.006	2.951

(ii) Extrapolation of the log catch per unit effort of partially recruited age-groups, using the regressions from Table IV

Age (years)	1946	1947	1948	1949	1950	1951	1952	1953	1954	1955	1956	1957	1958	1959	1960	1961	1962	1963
3.0	4.729	4.319	4.594	4.242	4.421	5.108	5.385	4.929	4.494	3.827	4.593	5.068				4.962	4.172	2.395
3.5	4.559	4.148	4.421	4.247	4.421	4.927	5.211	4.757	4.321	3.653	4.406	4.877	4.396	4.533	4.408	4.778	3.987	2.048
4.0		4.390	3.978	4.249	3.902	4.074	4.747	5.037	4.585	4.149	3.480	4.220	4.687	4.341	4.223	4.039	4.595	3.802
4.5	4.424	4.200	3.781	4.058	3.698	3.854	4.526	4.827	4.379	3.934	3.232	3.948	4.402	4.025	4.150	3.782	4.333	3.369
5.0	4.241	4.058	4.011	3.585	3.867	3.495	3.634	4.306	4.617	4.173	3.919	2.984	3.777	4.118	3.897	3.645	3.525	4.072
5.5	3.684	3.841	3.782	3.368	3.606	3.235	3.370	4.049	4.346	3.915	3.395	2.659	3.332	3.755	3.495	3.307	3.182	3.566
6.0	3.485	3.287	3.624	3.554	3.151	3.345	2.976	3.106	3.793	4.076	3.657	3.072	2.335	2.988	3.185	2.876	2.970	2.839

(iii) Actual abundance of age-groups, estimated as the antilog of half years of (ii) above

Age (years)	1946	1947	1948	1949	1950	1951	1952	1953	1954	1955	1956	1957	1958	1959	1960	1961	1962	1963
3	95.5	63.3	83.2	58.7	69.9	138.0	183.0	116.0	75.3	38.6	82.0	131.0	67.4	76.8	68.2	119.0	53.9	7.8
4	69.5	66.7	43.9	57.9	40.4	47.2	92.4	125.0	79.8	51.1	25.4	51.8	81.6	43.0	49.3	43.9	76.2	29.1
5	32.6	46.6	43.9	29.0	36.8	25.4	29.1	57.4	77.2	50.2	29.8	14.3	28.0	42.8	24.2	27.3	24.1	35.4

G L, but using the appropriate values of $(q+\frac{M}{f})$ and $\bar{f}_x$ for each year-class. The half-yearly estimates of abundance are also calculated (Table VB(iii)), as in columns M and N of Table VA, so that these three estimates of mean annual catch per effort can be compared with the observed figures.

RESULTS

THE DISTRIBUTION OF FISHING MORTALITY BETWEEN THE THREE FISHERIES SUSTAINED BY THE ARCTO-NORWEGIAN COD

The results of the correlation between total mortality and fishing effort have already been presented in order to define a value of natural mortality for use in the determination of partially recruited age-groups. The derived value of $q = 0.12$ per 10^8 ton-hours can be combined with the fishing effort data to give idealized estimates of fishing mortality. The fishing effort data from Table VA for ages 3–5, and Appendix Table A for ages 6 + are repeated in Table VI, distinguishing between effort in the feeding fisheries of Regions I and IIB, and in the spawning fishery at the Norway coast (Region IIA), and including estimates of effort on older age-groups which are not properly sampled in the feeding fisheries although they are fished in the spawning fishery. The total fishing mortality is shown in Table VII and compared with "virtual population" estimates of fishing mortality using a value $M = 0.3$. The means of these two sets of estimates for age-groups 6–12, both by age-groups and by years, demonstrate the similarity of the results.

The fishing mortality as determined by this catch/effort analysis can be subdivided in a number of ways to demonstrate the effect of the different fisheries. Table VIII gives separately the effects of fishing in the feeding fisheries and the spawning fishery upon the total stock of Arcto-Norwegian cod. It has been noted earlier that the method of computing effort at the Norway coast will have underestimated its true level, and if this is taken into account a greater proportion of the total should be attributed to the spawning fishery. The difference amounts to about 25% of the value of F on the older age-groups in the Norway coast fishery, i.e. rather less than 0.1.

In Table IX the fishing mortality in the Barents Sea and at Bear Island is expressed as a proportion of that part of the Arcto-Norwegian stock which occupies each of the feeding fisheries, i.e. as if each area contained a unit stock. Thus the fishing mortality on the total stock is partitioned between the two feeding fisheries by the ratio of the catches in the two areas, and this is divided by the proportion of the stock actually in one or other of the areas (from Appendix Table A),

$$\text{i.e. } F_{\text{Reg. I}} = \text{Log}_e \left\{ e^{-F_{I+IIB}} \times \frac{Y_{N_x}}{Y_{N_{I+IIB}}} \times \frac{1}{(Y_N/f)_I/(Y_N/f)_{I+IIB}} \right\}.$$

TABLE VI. The effective fishing effort, by age-groups, in different fisheries for the Arcto-Norwegian cod.

Total stock

Age-group	1946	1947	1948	1949	1950	1951	1952	1953	1954	1955	1956	1957	1958	1959	1960	1961	1962	1963
3	0.037	0.057	0.086	0.063	0.141	0.226	0.158	0.130	0.124	0.141	0.333	0.347	0.315	0.424	0.308	0.287	0.316	0.315
4	0.215	0.320	0.454	0.347	0.746	1.058	0.872	0.790	0.922	0.967	1.677	1.722	1.702	1.648	1.467	1.537	1.622	1.592
5	0.439	0.655	0.859	0.763	1.317	1.883	1.938	1.450	1.873	2.143	3.137	3.253	2.926	2.820	2.393	2.905	3.006	2.951
6		0.997	1.422	1.423	2.048	2.699	3.537	2.515	2.833	3.831	4.855	4.866	4.164	3.445	4.176	4.229	4.560	4.761
7		1.004	1.497	1.907	2.286	2.527	2.806	2.428	3.057	4.368	4.964	5.410	4.684	3.220	5.236	4.260	4.778	5.060
8		1.195	1.680	1.960	2.318	3.040	3.078	2.171	3.579	4.782	5.389	4.783	4.558	3.867	5.077	4.929	6.154	6.263
9		1.648	1.922	1.683	2.766	3.278	4.112	3.101	3.073	4.888	5.755	3.264	2.293	5.931	3.546	5.172	6.423	7.901
10		1.924	2.123	1.721	2.698	3.258	4.646	3.076	5.254	4.821	6.408	6.936	5.469	4.427	5.589	12.596	7.935	9.010
11		1.620	1.910	3.442	3.096	5.712	5.485	4.078	4.831	3.965	5.852	6.399	6.435	6.606	5.000	13.190	7.665	12.031
12		2.704	2.528	2.755	4.504	1.990	12.131	3.781	4.957	5.233	5.495	7.570	5.483	4.235	7.796	6.000	8.051	9.659
13		3.767	3.030	2.964	3.000	2.990	4.818	2.370	5.330	4.263	5.461	3.953	2.922	5.536	4.377	6.000	7.500	5.123
14		3.750	3.000	2.996	3.000	3.200	5.000	2.520	5.000	4.500	4.830	4.160	4.000	5.000	5.000	6.000	7.500	7.500
15	3.000	3.750	3.000	2.998	3.000	3.200	5.000	4.000	5.000	4.500	5.000	4.500	4.000	5.000	5.000	6.000	7.500	7.500
16	3.000	3.750	3.000	3.000	3.000	3.200	5.000	4.000	5.000	4.500	5.000	4.500	4.000	5.000	5.000	6.000	7.500	7.500
17+	3.000	3.750	3.000	3.000	3.000	3.200	5.000	4.000	5.000	4.500	5.000	4.500	4.000	5.000	5.000	6.000	7.500	7.500

Region IIA

Age-group	1947	1948	1949	1950	1951	1952	1953	1954	1955	1956	1957	1958	1959	1960	1961	1962	1963
6		0.004	0.002	0.015	0.020	0.004	0.019	0.017	0.007	0.046	0.022	0.022	0.026	0.041	0.031	0.033	0.021
7	0.016	0.011	0.040	0.177	0.056	0.045	0.052	0.093	0.086	0.094	0.099	0.059	0.056	0.071	0.135	0.196	0.146
8	0.088	0.058	0.081	0.562	0.344	0.221	0.163	0.315	0.506	0.759	0.221	0.510	0.484	0.247	0.663	0.697	0.665
9	0.465	0.295	0.217	0.474	1.022	1.209	0.851	0.521	1.154	1.337	1.114	0.674	1.962	1.041	1.825	1.403	1.674
10	0.758	0.538	0.382	0.714	1.484	2.194	1.916	1.595	1.811	1.997	2.222	1.508	2.582	3.040	2.602	2.229	1.575
11	0.503	0.343	1.395	1.679	1.541	3.086	2.556	2.014	1.978	1.780	2.381	3.247	2.324	4.651	3.830	3.486	3.819
12	0.739	0.352	1.795	1.836	0.976	2.994	1.546	2.635	2.810	1.921	2.793	2.392	1.736	4.377		4.428	2.859
13	2.611	0.938	1.143	1.499	2.990	3.474	2.370	3.052	3.922	4.486	3.953	2.922					2.523
14		1.443	2.046	1.996	3.242		2.520		4.263	4.830	4.160						
15			1.969	1.996	3.242												
16																	
17+																	

Regions I + IIB

Age-group	1946	1947	1948	1949	1950	1951	1952	1953	1954	1955	1956	1957	1958	1959	1960	1961	1962	1963
3	0.037	0.057	0.086	0.063	0.141	0.226	0.158	0.130	0.124	0.141	0.333	0.347	0.315	0.424	0.308	0.287	0.316	0.315
4	0.215	0.320	0.454	0.347	0.746	1.058	0.872	0.790	0.922	0.967	1.677	1.722	1.702	1.648	1.467	1.537	1.622	1.592
5	0.439	0.655	0.859	0.763	1.317	1.883	1.938	1.450	1.873	2.143	3.137	3.253	2.926	2.820	2.393	2.905	3.006	2.951
6	0.645	0.997	1.486	1.421	2.048	2.679	3.533	2.496	2.816	3.745	4.809	4.844	4.142	3.419	4.135	4.229	4.560	4.740
7	0.667	0.988	1.642	1.867	2.271	2.471	2.761	2.376	2.964	3.862	4.870	5.311	4.625	3.164	5.165	4.094	4.582	4.914
8	0.734	1.107	1.627	1.879	2.141	2.696	2.857	2.008	3.264	3.734	4.630	4.562	4.048	3.383	5.830	4.266	5.457	5.598
9	0.764	1.183	1.585	1.466	2.204	2.256	2.903	2.250	2.552	2.971	4.418	2.150	1.619	3.969	2.505	4.347	5.020	6.227
10	0.704	1.166	1.529	1.339	2.224	1.774	2.452	1.160	3.659	2.843	4.411	4.714	3.961	3.116	1.815	9.994	5.706	7.435
11	0.700	1.117	1.558	2.047	2.382	4.171	2.399	1.522	2.817	1.155	4.072	4.018	3.188	4.024	—	9.359	4.179	8.212
12	0.737	1.965	1.590	0.960	2.825	1.014	9.137	2.235	2.322	1.311	1.574	4.777	3.091	1.911	3.145		3.623	6.800
13		1.156	1.587	1.821	1.128		1.344		2.278		0.975			3.800				2.600
14				0.950	1.650													
15				1.029	1.512													

TABLE VII. Comparison of fishing mortality determined by catch/effort analysis, and by "virtual population" analysis.

Age-group	1946	1947	1948	1949	1950	1951	1952	1953	1954	1955	1956	1957	1958	1959	1960	1961	1962	1963	Mean values, by years Virtual pop.	Catch effort
(i) Catch/effort analysis																				
3		0.007	0.010	0.008	0.017	0.027	0.019	0.016	0.015	0.017	0.040	0.042	0.038	0.051	0.037	0.034	0.038	0.038		
4		0.04	0.05	0.04	0.09	0.13	0.10	0.09	0.11	0.12	0.20	0.21	0.20	0.20	0.18	0.18	0.19	0.19		
5		0.08	0.10	0.09	0.16	0.23	0.23	0.17	0.22	0.26	0.38	0.39	0.35	0.34	0.29	0.35	0.36	0.35		
6		0.12	0.17	0.17	0.25	0.32	0.42	0.30	0.34	0.53	0.58	0.58	0.50	0.41	0.50	0.51	0.55	0.57	0.34	0.39
7		0.12	0.18	0.23	0.27	0.30	0.34	0.29	0.37	0.46	0.60	0.65	0.56	0.39	0.63	0.51	0.57	0.61	0.38	0.40
8		0.14	0.20	0.24	0.28	0.36	0.37	0.26	0.43	0.52	0.65	0.57	0.55	0.46	0.73	0.59	0.74	0.75	0.39	0.44
9		0.20	0.23	0.20	0.33	0.39	0.49	0.37	0.37	0.59	0.69	0.39	0.28	0.71	0.43	0.74	0.77	0.95	0.45	0.45
10		0.23	0.25	0.21	0.32	0.39	0.56	0.37	0.63	0.57	0.77	0.83	0.66	0.53	0.67	1.51	0.95	1.08	0.55	0.59
11		0.19	0.34	0.41	0.37	0.67	0.66	0.49	0.58	0.58	0.70	0.77	0.77	0.79	0.60	1.58	0.92	1.44	0.68	0.65
12		0.32	0.23	0.33	0.54	0.24	1.46	0.45	0.59	0.48	0.42	0.91	0.66	0.51	0.94		0.97	1.16	0.65	0.60
13		0.45	0.30	0.36	0.36	0.36	0.58	0.28	0.64	0.63	0.66	0.47	0.35	0.66	0.53			0.61		
14			0.36	0.36	0.36	0.38		0.30		0.51	0.58	0.50								
15				0.36	0.36	0.38														
16					0.36	0.38														
(ii) Virtual population analysis																				
4		0.02	0.01	0.04	0.02	0.10	0.11	0.10	0.10	0.07	0.09	0.07	0.20	0.17	0.16	0.20	0.12			
5		0.08	0.05	0.13	0.06	0.19	0.27	0.19	0.21	0.24	0.35	0.20	0.27	0.37	0.31	0.36	0.40			
6		0.09	0.14	0.25	0.19	0.25	0.39	0.27	0.26	0.45	0.55	0.39	0.45	0.37	0.44	0.40	0.58			
7		0.16	0.36	0.28	0.29	0.39	0.43	0.26	0.34	0.46	0.50	0.45	0.43	0.37	0.38	0.44	0.52			
8		0.12	0.25	0.25	0.35	0.36	0.37	0.31	0.36	0.58	0.44	0.54	0.35	0.38	0.35	0.56	0.66			
9		0.35	0.36	0.31	0.38	0.47	0.53	0.39	0.38	0.59	0.28	0.45	0.39	0.49	0.58	0.59	0.85			
10		0.39	0.33	0.38	0.42	0.51	0.68	0.53	0.70	0.82	0.29	0.48	0.60	0.65	0.73	0.72	0.72			
11		0.61	0.46	0.50	0.67	0.66	0.92	0.66	0.88	1.20	0.26	0.61	0.61	0.50	0.70	0.64	0.92			
12		0.36	0.41	0.49	0.69	0.43	1.24	0.67	0.73	1.09	0.15	0.49	0.60	0.54	0.38	0.93	0.81			
13			0.24	0.47	0.60	0.47	0.43	0.61	0.88	0.85	0.14	0.59	0.24	0.27	0.43	0.73	0.80			
14								0.21	0.92		0.09	0.86	0.12	0.24			1.04			
Mean values by age-groups, 6–12 only																				
Catch/effort		0.19	0.23	0.26	0.34	0.38	0.61	0.36	0.47	0.53	0.63	0.67	0.57	0.54	0.64	0.78	0.78			
Virtual population		0.30	0.33	0.35	0.43	0.44	0.65	0.44	0.52	0.74	0.35	0.49	0.49	0.47	0.51	0.61	0.72			

TABLE VIII. The fishing mortality, by age-groups, in different fisheries, as a proportion of the total stock of Arcto-Norwegian cod.

Age-group	Calendar year																
	1947	1948	1949	1950	1951	1952	1953	1954	1955	1956	1957	1958	1959	1960	1961	1962	1963
(i) *Fishing mortality in regions I & IIB*																	
3	0.07	0.01	0.01	0.02	0.03	0.02	0.02	0.01	0.02	0.04	0.04	0.04	0.05	0.04	0.03	0.04	0.04
4	0.04	0.05	0.04	0.09	0.13	0.10	0.09	0.11	0.12	0.20	0.21	0.20	0.20	0.18	0.18	0.19	0.19
5	0.08	0.10	0.09	0.16	0.23	0.23	0.17	0.22	0.26	0.38	0.39	0.35	0.34	0.29	0.35	0.36	0.35
6	0.12	0.17	0.17	0.25	0.318	0.42	0.298	0.338	0.529	0.574	0.577	0.497	0.407	0.495	0.506	0.546	0.567
7	0.118	0.179	0.225	0.268	0.293	0.335	0.284	0.359	0.45	0.589	0.638	0.553	0.383	0.621	0.494	0.546	0.592
8	0.13	0.19	0.23	0.26	0.32	0.34	0.24	0.39	0.46	0.56	0.54	0.49	0.40	0.70	0.51	0.66	0.67
9	0.14	0.19	0.17	0.26	0.27	0.34	0.27	0.31	0.45	0.53	0.26	0.20	0.47	0.31	0.52	0.60	0.75
10	0.14	0.19	0.16	0.26	0.21	0.30	0.14	0.44	0.35	0.53	0.56	0.48	0.37	0.22	1.20	0.68	0.89
11	0.13	0.18	0.24	0.28	0.49	0.29	0.18	0.34	0.34	0.49	0.48	0.38	0.48	0.24	1.12	0.50	0.98
12	0.23	0.19	0.11	0.34	0.12	1.10	0.26	0.27	0.14	0.19	0.57	0.37	0.23	0.38	-	0.44	0.82
13	0.14	0.19	0.22	0.14	-	0.16	-	0.27	0.16	0.12	-	-	0.45	-	-	-	0.31
14			0.11	0.18	-												
15			0.12	0.12	-												
(ii) *Fishing mortality in region IIA*																	
6	0.002	-	0.005	0.002	0.002	0.005	0.002	0.002	0.001	0.006	0.003	0.003	0.003	0.005	0.004	0.004	0.003
7	0.01	0.001	0.01	0.02	0.007	0.03	0.006	0.011	0.010	0.011	0.012	0.007	0.007	0.009	0.016	0.024	0.018
8	0.06	0.01	0.03	0.07	0.04	0.15	0.02	0.04	0.06	0.09	0.03	0.06	0.06	0.03	0.08	0.08	0.08
9	0.09	0.04	0.05	0.06	0.12	0.26	0.10	0.06	0.14	0.16	0.13	0.08	0.24	0.12	0.22	0.17	0.20
10	0.06	0.06	0.17	0.09	0.18	0.37	0.23	0.19	0.22	0.24	0.27	0.18	0.16	0.45	0.31	0.27	0.19
11	0.09	0.16	0.22	0.20	0.12	0.36	0.31	0.24	0.24	0.21	0.29	0.39	0.31	0.36	0.46	0.42	0.46
12	0.31	0.04	0.14	0.22	0.36	0.42	0.19	0.32	0.34	0.23	0.34	0.29	0.28	0.56		0.53	0.34
13		0.11	0.25	0.18	0.39		0.28	0.37	0.47	0.54	0.47	0.35	0.21	0.53			0.30
14		0.17	0.24	0.24	0.39		0.30		0.51	0.58	0.50						
15																	
16																	
17																	

TABLE IX. Fishing mortality of Arcto-Norwegian cod in the feeding fisheries, as a proportion of the stock in each of those regions.

Age-group	1946	1947	1948	1949	1950	1951	1952	1953	1954	1955	1956	1957	1958	1959	1960	1961	1962	1963
(i) Region I																		
6		0.11	0.17	0.19	0.31	0.29	0.46	0.29	0.39	0.61	0.65	0.47	0.42	0.33	0.49	0.40	0.48	0.57
7		0.16	0.18	0.20	0.25	0.23	0.31	0.23	0.27	0.40	0.48	0.50	0.43	0.29	0.60	0.36	0.30	0.46
8		0.13	0.19	0.24	0.25	0.34	0.38	0.25	0.41	0.47	0.55	0.47	0.47	0.35	0.71	0.47	0.61	0.68
9		0.13	0.20	0.17	0.25	0.27	0.50	0.31	0.33	0.46	0.54	0.25	0.19	0.47	0.33	0.51	0.54	0.91
10		0.13	0.20	0.16	0.25	0.21	0.29	0.16	0.45	0.36	0.63	0.50	0.47	0.33	0.24	1.18	0.46	0.83
11		0.12	0.20	0.24	0.25	0.62	0.24	0.21	0.31	0.34	0.52	0.46	0.37	0.45	0.38	1.40	0.43	0.98
12		0.21	0.18	0.21	0.35	0.16	0.90	0.27	0.19	0.13	0.19	0.58	0.37	0.23	0.38		0.44	0.82
13		0.13	0.20	0.22	0.15		0.16		0.28	0.14	0.12			0.45				0.31
14			0.20	0.11	0.20													
15				0.12	0.18													
16																		
17																		
(ii) Region IIB																		
6		0.16	0.16	0.14	0.21	0.33	0.19	0.23	0.17	0.23	0.34	0.75	0.75	0.81	0.36	0.79	0.63	0.44
7		0.15	0.17	0.12	0.24	0.24	0.19	0.21	0.18	0.21	0.53	0.59	0.66	0.64	0.27	0.34	0.48	0.33
8		0.16	0.16	0.16	0.29	0.27	0.21	0.21	0.23	0.31	0.59	1.06	0.56	0.64	0.52	1.16	0.81	0.61
9		0.15	0.16	0.17	0.33	0.27	0.19	0.19	0.20	0.34	0.48	0.29	0.16	0.47	0.17	0.78	1.87	0.38
10		0.16	0.16	0.16	0.45	0.27	0.36	0.09	0.40	0.23	0.63	0.09	0.59	0.83	0.10	1.82	?	?
11		0.16	0.16	0.09	0.13	0.10	0.04	0.10	0.06	0.29	0.32	0.50	0.64	0.78	0.10	0.57	1.98	?
12		0.26	0.13		0.09	0.59		0.12	0.09	0.30								
13		0.17	0.16		0.07				0.24	0.22								
14																		
15																		
16																		
17																		

Calendar year

In general terms this shows that the effect of fishing on the Bear Island–Spitsbergen "stock" has been at least as high as that of fishing on the Barents Sea "stock" but, in terms of the effect on the total stock, the Bear Island fishery has relatively little significance, as can be judged directly from the ratios of total catches in the two areas.

A further important aspect of the data is shown in Table X, where these estimates of *fishing* mortality within each feeding fishery are compared with the estimates of *total* mortality which are obtained directly from the catch per unit effort in each region.

In Table X, column 1 gives the mean value of total mortality (Z) for each pair of age-groups, and columns 2, 3, and 4 sum the fishing mortality which should be included within that value of Z, as shown in Tables VIII and IX and recalculated as

$$\sum \frac{(6^F 50\text{--}62 \ + \ 7^F 51\text{--}63)}{24}.$$

This is then subtracted from the value of Z to give an implied natural mortality in each feeding fishery. The Barents Sea data give an answer close to that estimated from the grouped data for the total stock, but in the Bear Island–Spitsbergen fishery there is a clear increase with age in the apparent natural mortality. The mean value of M for this latter group is not unlike that suggested by Garrod (1964) from conventional catch/effort analysis of the Bear Island fishery alone, and it is entirely consistent with the biological background of some of the "Bear Island" cod returning to the Barents Sea after their spawning, resulting in a net loss of older fish from Bear Island to the Barents Sea grounds (Trout, 1957).

THE ESTIMATION OF DISCARD RATES ABOARD ENGLISH TRAWLERS

In the analysis of the partially recruited age-groups of the Arcto-Norwegian cod (see Methods, section on "The estimation of the abundance and mortality of partially recruited age-groups"), estimates are given of their actual abundance in the stock (Table V). These can be compared with the catch per unit effort in English landings in order to reconstruct the trends in discard rates that have taken place during the post-war period: these are set out in Table XI. The estimated actual abundance of each age-group is first corrected for their availability to the trawl fleet, using the values of 5% of 3-year-olds, 32% of 4-year-olds, and 62% of 5-year-olds (see Table V). This is then compared with the catch per unit effort recorded by English trawlers, the difference between the two sets of figures being an estimate of the discarding of particular age-groups. These have been assessed as a numerical percentage of the total catch per unit effort of all age-groups, and as the percentage by weight, derived from an age–weight relationship for Arcto-Norwegian cod.

These estimates show very high discard rates in the 1950–55 period, with a peak of 40% by numbers and 20% by weight in 1953–54 and a second peak

TABLE X. A comparison of total mortality of Arcto-Norwegian cod deduced from catch/effort analysis in each region, and the fishing mortality in each region deduced from the fishing effort/total stock relationships (see Table IX).

Age-group	1950-51	51-52	52-53	53-54	54-55	55-56	56-57	57-58	58-59	59-60	60-61	61-62	62-63	1 Mean Z	2 I F.	3 IIA F	4 ΣF	5 (M) Z-ΣF
(i) Total mortality in Region I																		
6 – 7	-1.53	0.02	0.71	0.50	0.27	0.74	0.88	0.54	0.58	1.15	0.85	0.60	1.18	0.50	0.36	0.01	0.37	0.13
7 – 8	0.83	1.02	0.85	0.69	0.57	1.09	0.75	0.58	0.63	1.33	0.25	0.86	1.22	0.82	0.38	0.03	0.41	0.41
8 – 9	0.98	1.13	0.90	1.25	0.67	1.57	1.49	1.45	0.94	1.71	0.35	1.16	1.67	1.17	0.39	0.08	0.47	0.70
9 – 10	1.05	0.98	1.07	0.53	0.16	1.19	0.74	0.36	0.40	2.02	0.32	1.74	1.78	0.95	0.38	0.16	0.54	0.41
10 – 11	0.06	0.64	2.30	0.07	0.49	1.73	0.51	0.35	1.13	–	1.54	0.59	0.63	0.77	0.42	0.25	0.67	0.10
11 – 12	–	0.75	1.44	0.40	1.46	2.50	1.14	1.24	1.51	0.97	–	–	2.29	1.05	0.39	0.28	0.67	0.38
Mean:	0.28	0.76	1.21	0.57	0.60	1.47	0.92	0.75	0.86	1.44	0.66	0.99	1.46				Mean:	0.35
(ii) Total mortality in Region IIB																		
6 – 7	0.76	0.69	0.69	1.19	0.91	0.95	1.03	0.67	1.15	0.77	0.67	0.74	1.35	0.89	0.41			0.48
7 – 8	0.79	1.17	0.67	0.83	1.10	0.83	0.83	1.41	1.07	2.16	1.15	1.14	1.46	1.12	0.46			0.66
8 – 9	1.23	0.48	0.55	1.66	1.44	0.86	1.32	1.73	0.91	2.64	0.32	1.13	1.75	1.23	0.49			0.74
9 – 10	1.57	1.19	1.80	1.75	0.40	0.98	1.26	2.07	0.52	2.38	1.05	1.16	1.91	1.39	0.39			1.00
10 – 11	1.79	2.45	1.49	1.81	0.91	1.39	2.02	1.34	0.27	–	0.75	–	–	1.42	0.45			0.97
Mean:	1.23	0.88	1.04	1.45	0.95	1.00	1.29	1.44	0.78	1.99	0.79	1.04	1.62				Mean:	0.77

TABLE XI. The estimation of discarding on English trawlers from the estimated abundance of partially recruited age-groups of Arcto-Norwegian cod (given as numbers/100 ton-hours).

Age	Calendar year																	
	1946	1947	1948	1949	1950	1951	1952	1953	1954	1955	1956	1957	1958	1959	1960	1961	1962	1963
(i) Actual abundance (Table V) corrected for availability																		
3	6.1	4.1	5.3	3.8	4.5	8.8	11.7	7.4	4.8	2.5	5.2	8.4	4.3	4.9	4.4	7.6	3.4	0.5
4	22.9	22.0	14.5	19.1	13.3	15.6	30.5	41.3	26.3	16.9	8.4	17.1	26.9	14.2	16.3	14.5	25.1	9.6
5	20.2	28.9	27.2	18.0	22.8	15.7	18.0	35.6	47.9	31.1	18.5	8.9	17.4	26.5	15.0	16.9	14.9	21.9
(ii) Observed abundance (c.p.e. English Appendix Table A)																		
3	5.3	1.7	0.7	4.4		0.4	0.6	0.6	0.5	0.1		0.1	0.9	1.5	4.4	2.2	1.3	0.5
4	11.1	17.3	2.2	15.6	1.5	3.5	14.9	5.1	11.0	3.8	0.8	2.7	11.5	8.9	21.4	15.6	16.2	10.3
5	14.5	34.3	17.5	30.6	5.6	11.2	18.8	25.4	26.5	28.8	11.8	4.7	13.8	23.7	18.2	22.5	28.0	30.8
(iii) Rejected abundance (i-ii)																		
3	0.8	2.4	4.6	-0.6	4.5	8.4	11.1	6.8	4.3	2.4	5.2	8.3	(3.4)	(3.4)	(-)	5.4	2.1	-0.5
4	11.8	4.7	12.3	3.5	11.8	12.1	15.6	36.2	15.3	13.1	7.6	14.4	15.4	5.3	(-5.1)	(-1.1)	8.9	-0.7
5	5.7	-5.4	9.7	-12.6	17.2	4.5	-0.8	10.2	21.4	2.3	6.7	4.2	3.6	2.8	(-3.2)	(-5.6)	(-13.1)	-8.9
Total (iii)	18.3	1.7	26.6	-9.7	33.5	25.0	25.9	53.2	41.0	17.8	19.5	26.9	22.4	11.5	-8.3	-1.3	-2.1	-10.1
Total stock (i) + 6 yr+	154.8	141.7	145.7	148.5	102.7	91.6	107.2	127.6	132.2	119.1	96.8	77.1	79.5	72.6	63.3	63.7	73.6	57.0
Percentage reject	12	-	18	-6	33	27	24	42	31	15	20	35	28	16	-13	-2	-3	-17
Weight of rejects	0.017		0.024		0.032	0.019	0.017	0.045	0.040	0.014	0.017	0.021	0.014	0.009				
Weight of total stock	0.515		0.364		0.248	0.189	0.187	0.202	0.214	0.225	0.197	0.154	0.138	0.121				
Percentage reject	3		7		13	10	9	22	19	6	9	14	10	7				

in 1957 and 1958. However, the decline in discard rates to an apparently negative value in 1959 is clearly impossible and arises from the original assumption
of the method, that discarding was zero for the 1955, 1956, and 1957 year-
classes. Some discarding must have taken place, and so all the estimates of
rejection must be slightly too low. Bearing this in mind there is a clear overall
decline in discard rates through the period, with peaks during years when
particularly strong year-classes have been amongst the age-groups 3–5. Thus
in 1953 and 1954 the very strong 1948–50 year-classes were only partially
recruited, and in 1957–58 the good 1954 year-class was just entering the fishery.

There are no quantitative records of the discard rates on English trawlers,
and indeed the interpretation put forward above can only be tentative, since
it refers to grouped data for both the Barents Sea and Bear Island regions.
Nevertheless the percentages are of the expected order of magnitude and bear
out available "hearsay" evidence.

RELATIONSHIPS BETWEEN THE COD FISHERY AND THE ENVIRONMENT: SOME
TENTATIVE INTERACTIONS

It is well known that the Arcto-Norwegian cod fishery is influenced by
fluctuations in temperature. For example, Maslov (1960b) has associated year-
to-year changes in the easterly summer migration of Barents Sea cod with
the distribution of temperature. Konstantinov (1964) has outlined temperature-
dependent relationships defining the relative success of the coastal and offshore
fishery areas of the Murman Coast, and Cushing (1959) and Beverton and Lee
(1965) have described various aspects of the distribution of cod and temperature
on the Spitsbergen Shelf. Other workers, e.g. Kislyakov (1961), have considered
the influence of temperature upon the survival of young fish.

Following the broad findings of Konstantinov it might be expected that
fluctuations in temperature should influence the catchability of cod, and hence
the effectiveness of fishing effort, but this cannot be defined by catch/effort
analysis alone because successive annual estimates of mortality, and hence q,
are interdependent. This difficulty can be overcome by using the catch/effort
analysis in conjunction with the "virtual population" analysis.[3] The former
gives estimates of effective fishing effort in each year (f), and the latter gives
independent estimates of fishing mortality (F) in each year. The division F/f
thus gives estimates of q, for each calendar year, which can be reletad to any
particular feature of the environment.

In fact the mean value of q derived in this way is $q = 0.118$ per 10^8 ton-
hours, very close to the value of $q = 0.12$ derived from the catch/effort analysis.
However, the chronological plot of these data as the full line in Fig. 6, indicates
that q may fluctuate widely, and that in the Arcto-Norwegian cod stock the
value of q in the second half of the 1950s was somewhat lower than in the early
years of the decade. However, as pointed out earlier, the English catch per unit

---

[3]This method of virtual population analysis is described and elaborated by Gulland in an
Appendix to the Report of the Arctic Working Group (ICES, 1965), and since it has not been
widely published the method is outlined in some detail in Appendix B of this paper.

effort data for 1946–49 are not very reliable, and over the period since 1950 there is no distinctive trend in q.

Figure 6 also shows the mean monthly anomaly (April–December) of temperatures in the 0–200-m layer at the Kola Meridian. These can be compared with the year-to-year variations of q for the whole Arcto-Norwegian stock. There is some indication that q tends to be low in warm years, e.g. 1950, 1954, and 1959, but this is not consistent and there is no correlation between the two sets of data. Indeed there is no reason to expect that fluctuations in this particular estimate of temperature should account for all the variance of q, and the comparison is given more to illustrate the value of catch and effort analysis when used in conjunction with virtual population analysis to define the variability of q.

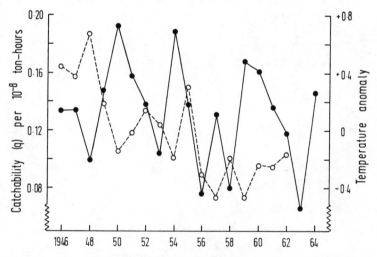

FIG. 6. Chronological changes in the catchability, q, of Arcto-Norwegian cod (full line) and in the temperature of the Kola meridian (broken line). The temperature changes are shown as the mean monthly anomaly (April–December) in the 0–200-m layer.

The increased frequency of low values of q since 1955 has an important bearing upon the interpretation of the catch/effort analysis: the fishing mortality matrices in Tables VII–IX will tend to overestimate fishing mortality during recent years. From the scale of change in q indicated in Fig. 6 the error could amount to F = 0.10 when estimating the real increase in fishing mortality from fishing effort between 1950 and 1963. From the point of view of assessments using the ratio E = F/F+M this implies a value of 0.65 as against a value of 0.70. This is not a very important difference and it can be seen that, even taking into account the environmental influence on the catchability of cod, fishing activity has still been the most important source of mortality in recent years.

One further influence of the environment upon the fishery can be deduced from Appendix Table A. Column 6 shows the proportion of each age-group located in the Bear Island–Spitsbergen stock. Of these, the estimates for 6-year-old cod are the only accurate indication of the partition of each year-class between the two feeding fisheries: estimates of younger age-groups are influenced by the partial recruitment discussed earlier, and estimates for the older fish are biassed by the redistribution of fish following maturity. In Fig. 7 this estimated

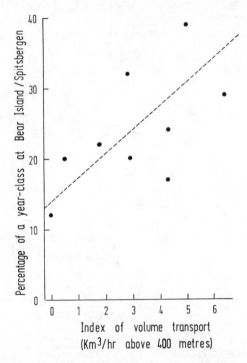

Fig. 7. The proportional distribution of year-classes of Arcto-Norwegian cod between the feeding fisheries, in relation to the volume transport in the West Spirtsbergen current during the year of spawning.

proportion of 6-year-olds of each year-class which is present on the Svalbard Shelf is related to the estimated volume transport in the West Bear Island Section during the summer of the year in which they were spawned (data from Lee, personal communication). The data do not cover the entire period, but those available do show that an increased volume transport leads to relatively more young cod being drifted to the Svalbard Shelf. The analysis has not been carried further, owing to the lack of more relevant data on the differences in volume transport between the West Bear Island and Kola Sections.

THE ESTIMATION OF SPAWNING STOCK SIZE FROM THE ABUNDANCE OF MATURE
FISH IN THE FEEDING FISHERIES, AND THE RELATIONSHIP WITH YEAR-
CLASS STRENGTH

The analysis so far discussed gives a comprehensive interpretation of the
changes observed in the commercial population of Arcto-Norwegian cod, given
a certain recruitment and the observable fishing effort. The information can
also be used to investigate further the recent fluctuations in year-class strength
in relation to the abundance of the spawning stock.

Hitherto this aspect of the dynamics of the stock has been complicated
by the lack of accurate information concerning the abundance of the spawning
stock at the Norway coast. Norwegian estimates have been biassed by unspecified increases in the fishing power of their vessels (Saetersdal and Hylen,
1964), and data from offshore trawl fisheries of England and Germany have been
biassed by fluctuating objectives in the fishery. English skippers have fished
further north, almost on to the overwintering grounds of the immature stock,
during recent years, especially since the extension of territorial limits in 1961,
and German trawlers have concentrated to a greater extent on redfish. All
three sets of data may well be biassed by changes in relative distribution of a
shrinking spawning stock. For example, the catch per effort recorded by Norwegian fishermen may tend to an asymptote, regardless of spawning stock size
if there is some optimum density of fish for spawning purposes: as the stock is
reduced the spawning zone may contract, either geographically or temporally.

The present treatment of the data permits the abundance of the spawning
stock to be estimated from the catch per unit effort of the relevant age-groups
in the feeding fisheries. The calculation is not given in detail, but the procedure
has been to extrapolate the abundance of each year-class at successive age
intervals, using the regressions given in Table IV and the effective effort per
age-group in Table VI. This table also gives estimates of the fishing effort on
age-groups which are not accurately represented by sampling in the feeding
fisheries, but which are presumed to be subject to the same mortality as this
mature age-groups which are fully represented in the spawning fishery. The
permits a reconstruction of the stock in the feeding fisheries, which is then
adjusted by the maturity ogive calculated earlier, to give an estimate of the
spawning stock. There is no direct method of testing the accuracy of this
method, beyond comparing it with other estimates derived from the Norway
coast data.

The errors inherent in the Norwegian catch per unit effort data have already
been mentioned, but nevertheless these should still give a representative estimate of the age composition of the spawning stock, if not of its size, bearing
in mind some bias due to gear selection. Figure 8 compares the per mille age
composition of Norwegian catches (from the original Working Group data)
and the age composition estimated by "reconstruction" from the trawl fishery
data. The similarity between the two offers some verification for the method,
and, most especially, the agreement between partially matured age-groups
supports the assumption that the proportion of mature fish within an age-

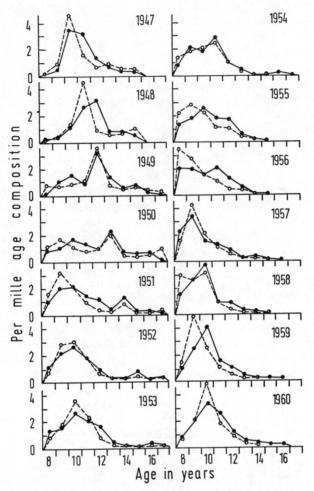

Fig. 8. A comparison between the per mille age composition of spawning cod in the Norway coast fishery, taken from the Norwegian data (broken line) and from the reconstruction of the stock from the estimated abundance of different age-groups in the feeding fisheries (full line).

group has not changed, although the observed mean may have decreased. (This was discussed earlier.) The favourable comparison in Fig. 8 also justifies the use of the mean weight per fish, as indicated by the Norwegian data, to derive an estimate of the weight per effort for this numerical reconstruction of the stock. This series of weights per unit effort has then been referred to the mean value for 1947–63, in order to compare it with the trend indicated by the Norwegian and English data from the Norway coast fishery itself (from ICES, 1965, annex 1, figure 4). The result, set out in Fig. 9, again compares favourably

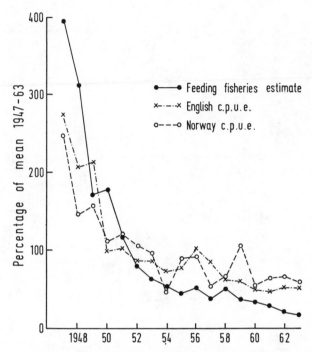

Fig. 9. The decline in the abundance of spawning cod, as deduced from the abundance of mature cod in the feeding fisheries, compared with the decline in the catch per unit effort of Norwegian and English Norway coast fisheries.

with previous data and reflects the progressive overestimate of spawning stock abundance which has taken place in the Norway coast fisheries themselves.

This procedure thus gives estimates of spawning stock which are in accord with the known facts. Now, the fecundity of female cod is proportional to their weight (A. D. Woodhead, personal communication), so that this estimated weight per effort of spawning stock is also an index of egg production (but not necessarily "egg quality") in each year.

Estimates of year-class strength have been derived earlier in the estimation of the abundance of partially recruited age-groups. These are given as logarithms in Table VA and repeated in Table XII as the relative abundance of codling at the beginning of their fourth year, i.e. 3-group. Confirmation of these estimates of year-class strength at 3 years old is shown in Fig. 10 where these data, expressed relative to the mean for 1946–60, are compared with the results of the young cod (age 0–3) surveys carried out by the USSR in the Barents Sea (Baranenkova, 1964). Bearing in mind the possible errors in the present estimates for the 1949 and 1950 year-classes, the agreement between the two sets of data is good (see below).

TABLE XII. Spawning stock size and year-class strength of Arcto-Norwegian cod.

| Year-class | Present analysis | | Gulland (1964) year-class strength | | | | Spawning stock | | Raising factors | | Index of year-class | Index of spawning stock |
| | Year-class at 3 years old (Table V) | Spawning stock weight | Region I | Region IIB | $\dfrac{\text{IIB}}{3.54}$ | Adjusted I + IIB | Number | No. weighted by kilo/fish | A/C | B/D | C × 0.56 1937–1942 | D × 0.197 1937–1946 |
	A	B				C		D				
1937			96	581	164	260	104	597			146	118
1938			29	95	27	56	141	832			31	164
1939			24	25	7	31	154	916			17	180
1940			39	25	7	46	142	873			26	172
1941			62	53	15	77	126	764			43	151
1942			77	92	26	103	117	708			58	139
1943	113		66	116	33	99	106	620	1.14		113	122
1944	75		44	77	22	66	102	580	1.14		75	114
1945	99		69	52	15	84	102	706	1.18		99	139
1946	70		95	26	7	102	122	710	0.69		70	140
1947	83	153	144	48	14	158	104	628	0.53	0.244	83	153
1948	165	121	222	129	36	258	81	513	0.64	0.236	165	121
1949	218	67	279	102	29	308	65	444	0.71	0.151	218	67
1950	138	70	276	324	92	368	66	452	0.37	0.155	138	70
1951	89	45	124	268	76	200			0.44		89	45
1952	46	31	65	87	25	90			0.51		46	31
1953	99	24	105	151	43	148			0.67		99	24
1954	159	21	160	347	98	258			0.62		159	21
1955	81	17	176	136	38	214			0.38		81	17
1956	93	20									93	20
1957	82	15									82	15
1958	143	20									143	20
1959	65	18									65	18
1960	(11)	13									(11)	13
1961		11									11	11
1962		8									8	8
1963		7									7	7
									Mean: 0.56	0.197		
									(excluding 1943–45)			

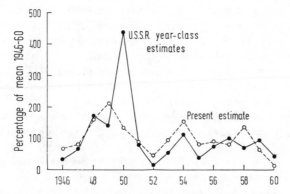

FIG. 10. A comparison of the estimated year-class strength of Arcto-Norwegian cod at 3 years old with USSR estimates from their winter surveys of 0-3 group cod in Region I.

The comparison between the estimate of spawning potential and the year-class subsequently generated from it is set out in Table XII. However, since this analysis has been based upon data from post-war years the earliest estimate of spawning stock is that of 1947, and for year-class strength that of 1943. The data have been extended using Gulland's estimates of spawning stock and year-class strength (Gulland, 1964b). His estimates of year-class strength in the two feeding fisheries are weighted for their different areas before summation to a single value for the whole stock in column C. Columns E and F determine raising factors from the years of overlap of the two series and these are used to unify Gulland's estimates with the data presented above.

The results are plotted in Fig. 11. The interpretation of this figure will be the source of further discussion, but a number of points may be made clear:

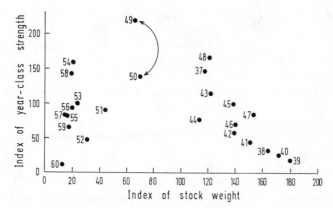

FIG. 11. The relation between spawning stock size of Arcto-Norwegian cod and the subsequent abundance at 3 years old of the resultant year-class.

277

(i) the raising factors are probably not very accurate, but because of the time series implicit in the results the error will have little effect on the general interpretation of Fig. 11. For example, inspection of these factors in Table XII suggests that they may be too low, so that the estimates of both spawning stock and year-class strength may be too low for the period 1937–43. (ii) Gulland's results for earlier years were based upon pre-war data and have not been included in the illustration because the original estimates of the abundance of different age-groups were not reliable, being based (a) upon scale readings — which are known to be difficult for higher age-groups, and (b) upon German length compositions converted to English catch per effort. (Trial plots of each year-class after the method developed here have shown that the fish were not fully recruited to the German landings until 7–8 years of age. This is related to market preferences (Gulland, 1964b; Lundbeck, 1964) and means that the German age composition data cannot be reliably used to estimate the composition of English landings, where year-classes were certainly fully recruited at a lower age.) (iii) The estimate of year-class strength for 1949 may be too high. The reason for this can be seen by reference to Fig. 5 and Appendix Table A. This year-class seems to "disappear" in 1959–60 and the last point in the series has rotated the regression downward in Fig. 5, giving a possible overestimate of the slope $q + \dfrac{M}{f}$ and hence an overestimate of the year-class strength at 3 years old. For the 1950 year-class the reverse is true. It seems possible that the acknowledgment of the 1950 group as an outstanding year-class may have led to the inclusion of some of the 1949 year-class as 1950 class, when the fish reached an age where the otolith margins become difficult to interpret. (iv) For the 1958, 1959, and 1960 year-classes these can only be preliminary estimates, pending further evidence collected since 1963.

DISCUSSION

The principal changes in the primary statistics of the Arcto-Norwegian cod fishery are shown in Fig. 12 (taken from ICES, 1965). Briefly the total catch of all regions has remained fairly steady at about 800,000 tons since the renewal of fishing after World War II, apart from two exceptional years 1955 and 1956. At the same time the proportion of the catch taken from the Norway coast fishery has decreased, and the stock abundance as an average of all areas has decreased to about 20% of its immediate post-war level. This implies that the increase in fishing effort during the same period has been wasted, insofar as it has not led to an increase in total yield. The question then arises as to what extent the trends have been generated by fishing itself, and to what extent the decline in stock abundance can be atributed to natural changes which would have taken place irrespective of changes in the amount of fishing. These questions can be answered from the foregoing analysis, the basic issue being what factors have generated the decline in stock abundance as measured by the catch per unit effort.

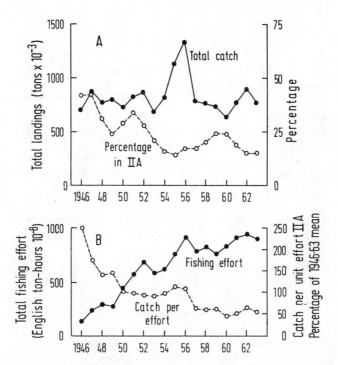

FIG. 12. A. The total catch of Arcto-Norwegian cod and the proportion caught in Region IIA. B. Total effective fishing effort
on the Arcto-Norwegian cod and the trend in mean catch per unit
effort.

In any fishery, changes in any one of five factors may influence the observed
abundance of the stock. These factors are: (i) the fishing effort, and hence
fishing mortality; (ii) the catchability of the stock; (iii) the rate of partial
recruitment to the exploited stock; (iv) the absolute level of recruitment to the
fishery; (v) the natural loss from the stock, either by mortality or emigration.

The changes in fishing mortality as judged from the fishing effort data
are summarized in Table VII. During the war, and before the redevelopment of
fishing in the Barents Sea and at Bear Island, the fishing mortality on age-
groups 3–7 can be assumed to have been negligible, though, as judged from
Table VIII, fishing mortality on older age-groups may have been about $F = 0.1$
as a result of fishing at the Norway coast. In the intervening years 1946–63
the fishing mortality in the feeding fisheries on both immature and mature
cod has risen to about $F = 0.6$ and at the Norway coast to $F = 0.3$.

However, during the post-war period there has been a change in the catchability of Arcto-Norwegian cod as measured by the English trawler fleet. This
was illustrated in Fig. 6. It is possible that since 1956 the value of q has been
lower, though it cannot be shown whether this reflects a change in the biological

characteristics of the stock, or in the efficiency of the English fleet. This change can also be seen in the mean values of fishing mortality of 6- to 12-year-old cod derived from the catch/effort analysis and virtual population analysis and given in Table VII. From this evidence it is concluded that in recent years estimates of fishing mortality derived from English data are slightly too high.

The analysis of mortality in the partially recruited age-groups is obscured by trends in the discard practices aboard English trawlers. It has been suggested that there has been a tendency to fish more for smaller cod during recent years, as the older age-groups have become less abundant, thus increasing the rate of recruitment. However, an assumption that the rate of recruitment of each age-group remained constant throughout the period implies discard rates which are consistent with what is known to have occurred (see Table V). It therefore seems more probable that the apparent increase of fishing on small cod is a change in recruitment to English landings and not to the exploited stock, since the observed changes in the stock can be explained without postulating a change in recruitment to the catches.

On the other hand the effects of (random) variations in recruitment to the stock are particularly evident in the peak landings of 1955 and 1956 of Fig. 12. These were associated with the 1948–50 year classes, which were exceptionally good (see Table XII) and which were fully recruited to the fisheries as 5- to 7-year-olds during the mid 1950s. Subsequent year-classes have on the whole been smaller, but not less than the year-classes of the pre-1948 period. Consequently, though there are indications of complex stock–recruitment relations in Fig. 11, during the time period considered here they have not contributed a great deal to the decline in stock abundance. This may be qualified by the corollary that a decline in abundance has resulted from recent year-classes being less abundant than those of 1948–50, but it is not yet certain to what extent, if any, these good year-classes were themselves generated by fishing. Compared with the pre-1948 period, recruitment to the stock has, if anything, improved.

A comparison between stock abundance in 1948 and in 1963, i.e. outside the period of influence of the 1948–50 year-classes, shows that the recently low abundance is not primarily related to a serious decline in the absolute level of recruitment.

Finally it is noted that it has not been possible to detect any change in natural loss from the stock. The pattern of recapture of fish marked in the Arcto-Norwegian cod area has remained typical of the migrations deduced in earlier years (Cushing, 1966), and there is no clear indication in Fig. 4 that the natural mortality in the stock as a whole should lie on a different regression for more recent years, when high levels of fishing effort have been recorded.

Thus, during the post-war years up to 1960, and excluding 1948–50, recruitment to the Arcto-Norwegian cod stock has remained more or less steady. Similarly there has been no marked change in the rate of recruitment of codling to the exploited stock and no changes in natural loss have been detected apart from the suspected change in catchability discussed above. It is therefore concluded that the changes outlined in Fig. 12 can be almost entirely attributed to the effect to fishing itself.

SUMMARY

1) The population dynamics of the Arcto-Norwegian cod are analysed by the reconstruction of changes that have taken place in the stock as measured in the feeding fisheries of the Barents Sea, and at Bear Island–Spitsbergen.

2) Three modifications of analytic technique are introduced:
    i) the measurement of variations in concentration of fishing upon different age-groups of cod within the same year;
    ii) the adjustment of catch per unit effort and fishing effort to offset differences in fish density and fishing intensity per unit area between the Barents Sea and Bear Island–Spitsbergen fisheries;
    iii) the development of a method for estimating the abundance and effective fishing effort on partially recruited age-groups, and which at the same time gives improved estimates of year-class strength at recruitment.

3) The analysis of catch/effort data gives values ($M = 0.33$ and $q = 0.12$) which are the same as those determined by the alternative virtual population analysis.

4) The catch/effort analysis detects emigration of maturing cod from the Bear Island fishery. Previously this has only been identified in experiments where fish tagged at Bear Island have been returned from the southeast Barents Sea.

5) The determinations of partial recruitment and the comparison of the actual and observed abundance of these age-groups in English data imply certain changes in the discard rates in this fleet. These are shown to be in accordance with available evidence.

6) The combination of results from the catch/effort analysis and "virtual population" analysis permits the determination of changes in catchability between calendar years. Lower values of q have been more frequent since 1956.

7) Evidence is presented to show that, in years of high volume transport in the West Spitsbergen current, a greater proportion of a year-class is drifted on to the Svalbard Shelf.

8) The method of estimating the total abundance of Arcto-Norwegian cod in its feeding fisheries permits the reconstruction of the Norway coast spawning stock. This reconstruction illustrates changes in the spawning stock which, though previously suspected, had not been detected owing to various sources of bias in the data collected in the Norway coast fishery itself.

9) The new estimates of year-class strength, and the reconstruction of the spawning stock, are related to show a time series of changes in both aspects of the stock.

10) Post-war changes in the fisheries are discussed in relation to the implications of the changes in fishing effort, catchability, partial recruitment, and year-class strength which have been demonstrated during the analysis.

REFERENCES

BARANENKOVA, A. S. 1964. Some results of estimation of cod fry in the Barents Sea during 1946–1961. Materialy Rybokhoz. Issled. Sev. Bass., 3: 72–106.

BEVERTON, R. J. H. [ed.] 1960. Second Progress Report of the Working Group on Arctic Fisheries. Copenhagen, 1959. ICES C.M. 1960, Gadoid Fish. Comm.

BEVERTON, R. J. H., AND HOLT, S. J. 1957. On the dynamics of exploited fish populations. Fish. Invest. London, Ser. II, 19: 533 p.

BEVERTON, R. J. H., AND LEE, A. J. 1965. The influence of hydrographic and other factors on the distribution of cod on the Spitzbergen Shelf. Spec. Publ. Intern. Comm. Northwest Atlantic Fish. No. 6, (In press).

BISHOP, Y. M. M. 1959. Errors in estimates of mortality obtained from virtual populations. J. Fish. Res. Bd. Canada, 16: 73–90.

CUSHING, D. H. 1966. The arctic cod. Pergamon Press, Oxford.

CUSHING, D. H. (In I. D. RICHARDSON et al.,) 1959. Echo-sounding experiments in the Barents Sea. Fish. Invest. London, Ser. II, 22(9), 57 p.

FRY, F. E. J. 1949. Statistics of a lake trout fishery. Biometrics, 5(1): 27–67.

GARROD, D. J. 1964. Effective fishing effort and the catchability coefficient q. Rapp. Proc.-Verb. Réun. Conseil Perm. Intern. Exploration Mer, 155: 66–70.

GULLAND, J. A. 1956. On the fishing effort in English demersal fisheries. Fish. Invest. London, Ser. II, 18(9): 46 p.
1961. Fishing and the stocks of fish at Iceland. Ibid., 23(4): 52 p.
1964a. Manual of methods of fish population analyses. F.A.O. Fish. Biol. Tech. Papers, No. 40.
1964b. The abundance of fish stocks in the Barents Sea. Rapp. Proc.-Verb. Réun. Conseil Perm. Intern. Exploration Mer, 155: 126–127.

HJORT, J. 1914. Fluctuations in the great fisheries of northern Europe viewed in the light of biological research. Ibid., 20: 228 p.

INTERNATIONAL COUNCIL FOR THE EXPLORATION OF THE SEA, 1965. Report of the meeting of the Arctic Fisheries Working Group in Hamburg, January 18–23, 1965. Coop. Res. Rept., Ser. B, 1965; Annex I, p. 15–32.

INTERNATIONAL FISHERIES CONVENTION. 1946. Report by the President on the ninth meeting of the Permanent Commission held in Copenhagen, May 1961.

JONES, R. 1964. Estimating population size from commercial statistics when fishing mortality varies with age. Rapp. Proc.-Verb. Réun. Conseil Perm. Intern. Exploration Mer, 155: 210–214.

KISLYAKOV, A. G. 1961. The relation between hydrological conditions and fluctuations in the abundance of cod year-classes. Tr. Soveshch. Ikhtiol. Komis, 13: 260–264.

KONSTANTINOV, K. G. 1964. Effect of water temperature on fish resources for trawling in the Barents Sea. Vopr. Ikhtiol., 4, No. 2 (31): 255–269.

LUNDBECK, J. 1964. Biologisch-Statistiche Utersuchungen über die deutsche Hochseefischerei. Ber. Deut. Wiss. Komm. Meeresforsch., 17: 133-164.

MASLOV, N. A. 1960a. Soviet investigations on the biology of the cod and other demersal fishes of the Barents Sea. Soviet Fish. Invest. N. European Seas, VNIRO/PINRO, Moscow; p. 185–231.
1960b. Changes in geographical distribution of the demersal fishes in the Barents Sea during the last thirty years. ICES C.M. 1960, Gadoid Fish. Comm., Doc. No. 174.

PALOHEIMO, J. E. 1961. Studies in estimation of mortalities. 1. Comparison of a method described by Beverton and Holt and a new linear formula. J. Fish. Res. Bd. Canada, 18(5): 645–662.

PONOMARENKO, V. P. 1964. Natural, fishing and total mortalities of cod of the Barents Sea stock in 1946–1963. Materialy Rybokhoz. Issled. Sev. Bass., 4: 3–8.

RICKER, W. E. 1940. Relation of 'catch per unit effort' to abundance and rate of exploitation. J. Fish. Res. Bd. Canada, 5: 43–70.
1944. Further notes on fishing mortality and effort. Copeia, 1: 23–44.
1958. Handbook of computations for biological statistics of fish populations. Bull. Fish. Res. Bd. Canada, No. 119. 300 p.

ROLLEFSEN, G. 1953. Observations on the cod and cod fisheries of Lofoten. Rapp. Proc.-Verb. Réun. Conseil Perm. Intern. Exploration Mer, 136: 40–47.

SAETERSDAL, G., AND HYLEN, A. 1964. The decline of the skrei fisheries. Fiskeridirektorat. Skrifter Ser. Havunders., 13(7): 56–69.

TROUT, G. C. 1957. The Bear Island cod: migrations and movements. Fish. Invest. London Ser. II, 21(6): 51 p.

APPENDIX A

A COMPARISON OF ENGLISH AND USSR CATCH PER UNIT EFFORT DATA

The key-statistics of the catch/effort analysis are the total international catch of each age-group, by numbers, and the catch per unit effort in the feeding fisheries. In this latter statistic there is a choice between the USSR catch per hour, and the English catch per 100 ton-hours, the fishing by other countries not having been sufficiently representative at both the Barents Sea and at Bear Island–Spitsbergen throughout the period. The analysis has been based upon the English data because it has the additional advantage of having incorporated in it a correction for secular changes in fishing power that have taken place as the size of trawlers has increased. This is not available for USSR data.

In order to check any major discrepancies between the English and USSR data the catches per effort of the two countries for the Barents Sea have been compared in Appendix Table B.

If the USSR and English data were uniform then the ratio for each age-group should remain constant. Looking first at the best sampled age-groups (6–10) this ratio is satisfactorily constant within years, and discrepancies in older age-groups are in part due to the "rounding" of estimates of abundance which are very low. However there are systematic changes between years, particularly during the period 1954–58, as shown by the mean ratio of all groups over 6 years old. This was a period of expansion of the USSR fishing in the Barents Sea, so the trend may be associated with increased fishing power of the newly built trawlers. There is an additional possible source of bias in USSR statistics in that the main objective of their fishery was switched to redfish in the late 1950s, with an apparent relative underestimate of the catch per effort of cod during these years. It would appear from the increase in the ratio at the end of the period that the USSR fleet has again fished more intensively for cod as redfish catches have declined.

It is felt that these trends in fleet expansion and fleet objectives in the USSR data will explain the major discrepancies between the two sets of data.

APPENDIX TABLE A.   The exploited life history of successive year-classes of Arcto-Norwegian cod.

Column: 1 Total catch in numbers, Region I, ×10⁻³
2 Total catch in numbers, Region IIB, ×10⁻³
3 No/100 ton-hours, English, Region I
4 No/100 ton-hours, English (corrected), Region IIB
5 Total No/100 ton-hours (3+4)
6 Proportion of each age-group in Region IIB (col.4/col.5)
7 Fishing effort, ton-hours ×10⁸, Region I
8 Fishing effort, ton-hours ×10⁸, Region IIB
9 Total fishing effort. Calculation as in text

Column: 10 Total catch in numbers, Region IIA, ×10⁻³
11 Fishing effort, ton-hours ×10⁻⁸, Region IIA (for derivation see text)
12 Corrected fishing effort in Region IIA
13 Corrected fishing effort in feeding fisheries (from col. 9)
14 Corrected catch per unit effort (from col. 5)
15 Total fishing effort for pairs of years as $\dfrac{9x + 9_{x+1}}{2} + 12_x$
16 Cumulative fishing effort to the midpoint of each year (col. 14)
17 Log$_e$ catch per unit effort (col. 14)

Year-class	Sample year	1-	2	3	4	5	6	7	8	9	10	11	12	13	14	15	16	17
1934	1946	634	1 036	0.95	1.32	2.27	0.58	0.670	0.785	0.737	5 927	2.611	2.611	0.737	2.27	3.557	0.368	0.820
	1947	1 330	671	1.23	0.50	1.73	0.29	1.081	1.342	1.156	2 497	1.443	1.443	1.156	1.73	2.804	3.925	0.548
	1948	940	298	0.56	0.22	0.78	0.28	1.679	1.354	1.587	1 536	1.969	1.969	1.587	0.78	3.277	6.729	-0.248
	1949	623	68	0.44	0.05	0.49	0.10	1.416	1.360	1.410	1 335	2.725	1.996	1.029	0.67	10.310	10.006	-0.400
	1950	156	—	0.01	—	0.01	—	15.600	—	15.600	606							
1935	1946	2 426	1 579	3.62	2.10	5.72	0.37	0.670	0.752	0.700	4 229	0.739	0.739	0.700	5.72	2.071	0.350	1.744
	1947	2 055	2 151	1.90	1.61	3.51	0.46	1.082	1.336	1.198	2 007	0.572	0.938	1.965	2.14	2.715	2.421	0.761
	1948	787	230	0.47	0.17	0.64	0.27	1.675	1.353	1.590	1 309	2.046	2.046	1.590	0.64	3.316	5.136	-0.446
	1949	575	—	0.23	—	0.23	—	2.500	—	2.500	1 221	5.309	3.242	0.950	0.61	3.227	8.452	-0.494
	1950	263	36	0.04	—	0.04	—	6.575	—	6.575				1.512	0.17		11.679	-1.772
1936	1946	8 146	5 236	12.15	6.84	18.99	0.36	0.670	0.766	0.704	9 558	0.503	0.503	0.704	18.99	1.413	0.352	2.943
	1947	9 555	1 778	8.83	1.33	10.16	0.13	1.083	1.337	1.117	3 576	0.352	0.352	1.117	10.16	1.689	1.765	2.313
	1948	1 732	901	1.03	0.66	1.69	0.39	1.682	1.365	1.558	1 931	1.143	1.143	1.558	1.69	2.832	3.454	0.525
	1949	1 438	10	0.79	—	0.79	—	1.821	—	1.821	1 184	1.499	1.499	1.821	0.79	3.234	6.286	-0.236
	1950	265	47	0.06	0.01	0.07	0.14	4.418	4.700	4.460				1.650	0.19		9.520	-1.661
1937	1946	11 199	16 921	16.72	20.08	36.80	0.55	0.670	0.843	0.764	27 884	0.758	0.758	0.764	36.80	1.723	0.382	3.605
	1947	9 083	5 553	8.38	4.17	12.55	0.33	1.084	1.332	1.166	16 853	1.343	1.343	1.166	12.55	2.690	2.105	2.529
	1948	4 579	3 309	2.73	2.43	5.16	0.47	1.677	1.362	1.529	9 262	1.795	1.795	1.529	5.16	3.039	4.795	1.641
	1949	2 674	101	1.02	0.05	1.07	0.47	2.622	2.020	2.594	5 309	4.962	1.836	0.960	2.89	2.880	7.834	1.061
	1950	939	76	0.16	0.02	0.18	0.11	5.867	3.800	5.638				1.128	0.90		10.714	-0.105
	1951	291																
1938	1946	5 422	3 551	8.10	4.14	12.24	0.34	0.670	0.858	0.734	5 694	0.465	0.465	0.734	12.24	1.423	0.367	2.501
	1947	6 168	5 110	5.70	3.83	9.53	0.40	1.082	1.334	1.183	5 129	0.538	0.538	1.183	9.53	1.922	1.790	2.254
	1948	2 539	901	1.51	0.66	2.17	0.30	1.682	1.365	1.585	3 027	1.395	1.395	1.585	2.17	3.211	3.712	0.775
	1949	3 275	31	1.60	—	1.60	—	2.047	—	2.047	2 687	1.679	2.990	2.047	1.60	4.115	6.923	0.470
	1950	768	140	0.17	0.01	0.18	0.06	4.517	14.000	5.044	961	5.339		2.825	0.32		11.038	-1.139
	1951	25	22	—														

(Continued)

1939																	
1946	5 859	2 187	9.48	2.59	12.07	0.21	0.618	0.844	0.667	1 067	0.088	0.088	0.667	12.07	0.975	0.333	2.491
1947	5 194	671	4.80	0.50	5.30	0.09	1.082	1.342	1.107	1 562	0.295	0.295	1.107	5.30	1.662	1.308	1.668
1948	3 927	1 508	5.34	1.10	2.99	0.17	1.680	1.371	1.627	2 459	0.382	0.382	1.627	6.44	1.865	2.970	1.863
1949	2 630	76	2.94	0.05	1.23	0.02	1.336	1.520	1.339	2 135	0.714	0.714	1.339	2.99	2.574	4.835	1.095
1950	147	300	1.20	0.03	0.03	0.02	2.192	10.000	2.382	1 201	0.976	0.976	2.382	1.23	2.674	7.409	0.207
1951	64	22	0.02	0.01	0.02	0.33	7.350	2.200	5.633	579	19.298	3.474	1.014	0.16	4.653	10.083	−1.833
1952	52	13	0.02	—			3.200		3.200				1.344	0.05		14.736	−2.996
1953		1															
1940																	
1946	7 748	3 548	13.01	4.47	17.48	0.26	0.596	0.794	0.645	280	0.016	0.016	0.645	17.48	0.832	0.322	2.862
1947	17 750	588	18.10	0.44	18.54	0.24	0.980	1.337	0.988	1 075	0.058	0.058	0.988	18.54	1.363	1.154	2.918
1948	9 726	2 111	8.47	1.54	10.01	0.15	1.668	1.371	1.622	173	0.217	0.217	1.622	10.01	1.761	2.517	2.303
1949	2 645	565	2.09	0.38	6.96	0.05	1.478	1.250	1.466	3 302	0.474	0.474	1.466	6.96	2.319	4.278	1.940
1950	1 438	491	1.97	0.15	2.24	0.07	2.114	3.767	2.224	3 452	1.541	1.541	2.224	2.24	4.738	6.597	0.806
1951	94	126	0.93	0.03	2.00	0.01	1.343	1.833	1.350	1 939	0.969	0.969	1.350	0.64	9.648	11.335	−0.446
1952	72		0.04		0.93		1.546	—	1.546	373	0.401	0.401	1.546	0.16		20.983	−1.833
1953		14			0.04		2.350	—	2.350	168	4.200		2.350				
1941																	
1946	4 573	5 897	7.84	6.95	14.79	0.47	0.584	0.849	0.708	304	0.011	0.011	0.708	14.79	0.852	0.354	2.695
1947	21 361	5 332	22.79	4.00	26.79	0.15	0.238	1.333	0.997	272	0.081	0.081	0.997	26.79	1.252	1.206	3.288
1948	52 607	7 605	35.00	5.57	40.57	0.14	1.505	1.365	1.486	5 740	0.562	0.562	1.486	40.57	1.763	2.458	3.704
1949	697	491	9.08	1.13	10.21	0.11	1.948	1.320	1.879	8 681	1.484	1.484	1.879	10.21	2.603	4.221	2.322
1950	3 069	2 050	5.06	0.79	5.85	0.13	2.143	2.595	2.204	9 957	3.086	3.086	2.204	5.85	3.473	6.824	1.766
1951	1 900	356	1.77	0.16	1.93	0.08	1.734	2.225	1.774	453	1.546	1.546	1.774	1.93	5.172	10.297	0.658
1952	504	356	0.93	0.01	0.94	0.04	2.042	3.000	2.399	702	3.052	3.052	2.399	0.94	4.543	15.469	−0.062
1953	143	10	0.22	0.01	0.23	0.04	2.291	—	2.235	348	8.700	8.700	2.235	0.23	4.557	19.332	−1.470
1954	18	15	0.03	0.01	0.08	0.25	2.350	4.300	2.278				2.278	0.08	5.738	24.640	−2.528
1955							4.767		4.650								
1942																	
1946	4 562	2 639	7.93	3.14	11.07	0.28	0.575	0.841	0.651	141	0.004	0.004	0.651	11.07	0.853	0.325	2.407
1947	20 818	15 444	24.84	11.56	34.35	0.34	0.914	1.336	1.056	239	0.040	0.040	1.056	34.35	1.241	1.178	3.535
1948	35 586	8 810	20.08	7.46	31.29	0.27	1.431	1.226	1.418	4 863	0.177	0.177	1.418	31.29	1.682	2.419	3.444
1949	42 269	9 150	11.33	2.78	27.54	0.21	2.105	2.346	1.867	14 003	1.022	1.022	1.867	27.54	2.181	4.101	3.314
1950	23 062	6 265	4.13	0.24	13.70	0.19	2.091	2.208	2.141	10 769	2.194	2.194	2.141	13.70	2.256	9.502	2.617
1951	3 354	1 722	1.55	0.05	4.91	0.16	2.265	2.738	2.256	817	2.556	2.556	2.452	4.91	4.548	14.050	1.591
1952	732	657	0.10	0.01	1.79	0.13	2.408	2.900	2.452	1 817	2.635	2.556	1.522	1.79	4.543	18.593	0.582
1953	905	145	0.15	0.11	0.52	0.25	6.034	15.000	5.250	799	3.922	7.085	2.322	0.52	4.557	23.150	−0.654
1954	323	150	0.10	0.04	0.18	0.09	3.230	—	4.300	420	4.830	7.264	1.311	0.18	5.738	28.888	−1.715
1955	76	38	0.03	0.02	0.07	0.25	2.533		2.850			10.500		0.07			−2.659
1956	27		0.02				1.350		1.350								
1943																	
1946	2 845	220	5.00	0.27	5.27	0.05	0.569	0.815	0.582	39	0.002	0.002	0.582	5.27	0.774	0.291	1.662
1947	13 161	893	14.63	2.67	17.30	0.15	0.899	1.335	0.966	825	0.015	0.015	0.966	17.30	1.176	1.065	2.851
1948	16 283	3 928	11.67	5.79	17.46	0.33	1.395	1.369	1.386	566	0.344	0.344	1.386	17.46	1.406	2.241	2.862
1949	54 116	21 635	34.48	18.46	53.30	0.35	1.553	1.173	1.421	7 566	1.209	1.209	1.421	53.30	1.861	3.647	3.976
1950	34 829	15 163	15.26	6.74	22.00	0.31	2.281	2.250	2.271	11 749	1.916	1.916	2.271	22.00	2.827	5.508	3.091
1951	19 076	7 121	6.66	3.06	4.05	0.31	2.865	2.327	2.696	762	2.014	1.916	2.696	4.05	3.006	8.335	2.274
1952	8 451	3 280	2.16	1.89	0.74	0.47	3.925	1.735	2.903	405	2.014	2.014	2.903	0.74	3.947	16.290	1.399
1953	2 053	484	0.69	0.31	0.25	0.30	2.774	1.561	2.416	2 080	2.810	2.810	2.817	0.74	4.002	20.292	0.784
1954	1 785	301	0.02	0.05	0.17	0.07	2.586	6.020	2.817	1 207	4.486	7.010	1.555	0.25	4.796	25.088	−0.301
1955	270	37		0.01	0.02	0.06	1.687	3.700	1.805				0.975	0.08	5.551	30.639	−1.386
1956	75			—			3.750		3.750								−2.526
1957	9	7															

APPENDIX TABLE A.   The exploited life history of successive year-classes of Arcto-Norwegian cod. — (Concluded)

Year-class	Sample year	1	2	3	4	5	6	7	8	9	10	11	12	13	14	15	16	17
1944	1947	1 484	22	1.67	0.77	1.67	0.35	0.889	—	0.889	879	0.056	0.056	0.889	1.67	1.131	0.444	0.513
	1948	1 937	162	1.41	12.24	2.18	0.40	1.374	1.370	1.373	3 500	0.221	0.221	1.373	2.18	1.301	1.575	0.779
	1949	24 200	13 469	18.39	9.07	30.63	0.58	1.316	1.100	1.230	4 682	0.851	0.851	1.230	30.63	1.639	2.876	3.421
	1950	17 878	16 935	6.49	4.26	15.56	0.27	2.300	1.868	2.048	3 924	1.595	1.595	2.048	15.56	2.315	4.515	2.747
	1951	28 214	2 571	11.58	1.32	15.84	0.24	2.435	2.572	2.857	2 235	1.978	1.978	2.471	15.84	2.885	6.830	2.760
	1952	13 147	1 246	4.18	0.76	5.50	0.31	3.145	1.948	2.250	1 268	1.921	1.921	2.857	5.50	3.404	9.715	1.705
	1953	4 289	289	1.70	0.13	2.46	0.12	2.523	1.640	3.659	1 324	6.480	3.953	2.250	1.13	4.549	13.119	0.900
	1954	3 689	446	1.00	0.05	1.13	0.08	3.689	3.430	2.843				3.659	0.66	5.229	17.668	0.122
	1955	1 743	134	0.61		0.66	—	2.580	2.680	2.580				1.574	0.08	4.129	22.897	-0.416
	1956	129	12	0.05		0.05			—								23.864	-1.238
	1957	42	4														27.026	-2.526
1945	1948	889	6	0.66	2.68	0.66	0.17	1.347	1.133	1.347	113	0.020	0.020	1.347	0.66	1.197	0.673	-0.416
	1949	13 307	3 036	12.93	4.24	15.61	0.76	1.029	2.037	1.047	824	0.045	0.045	1.047	15.61	1.586	1.870	2.747
	1950	9 263	13 682	13.66	4.77	5.60	0.26	2.613	2.868	2.125	587	0.163	0.163	2.125	5.60	2.422	3.456	1.723
	1951	35 698	4 341	13.42	2.43	18.43	0.15	2.938	1.786	2.679	3 650	0.521	0.521	2.679	18.43	2.765	5.878	2.912
	1952	39 438	2 283	5.76	1.25	15.85	0.18	2.047	1.826	2.761	4 962	2.625	1.811	2.761	15.85	2.547	8.643	2.760
	1953	11 789	633	1.65	0.24	7.01	0.13	3.852	2.638	2.008	2 795	1.780	1.780	2.008	7.01	2.801	11.190	1.947
	1954	6 356	341	1.41	0.16	1.89	0.10	3.067	2.131	2.971	810	2.793	2.793	2.971	2.74	4.572	13.991	1.008
	1955	4 325	118	0.25	0.04	1.57	0.14	4.252	2.950	4.072	263	2.922	2.922	4.072	1.57	5.301	18.563	0.451
	1956	1 063	41	0.08	0.01	0.29	0.11	4.862	4.100	4.777				4.777	0.29	7.217	23.864	-1.238
	1957	389				0.09									0.09		31.081	-2.408
	1958	45																
1946	1949	4 351	3	4.43	0.90	4.43	0.61	0.982	2.161	0.982	6	0.004	0.004	0.982	4.43	1.618	0.491	1.488
	1950	3 936	1 945	0.58	2.53	1.48	0.23	2.400	3.139	2.255	42	0.004	0.004	2.255	1.48	2.648	2.109	0.392
	1951	48 273	7 941	8.66	2.18	11.19	0.12	3.000	1.759	3.533	968	0.052	0.052	3.033	18.77	3.287	4.757	2.415
	1952	62 459	3 835	16.59	1.09	18.77	0.12	3.766	2.199	2.376	921	0.315	0.315	3.533	9.28	3.006	8.044	2.228
	1953	19 636	1 104	8.19	0.48	9.28	0.11	2.398	2.300	3.264	5 238	1.154	1.154	3.264	4.54	3.135	14.185	1.513
	1954	13 714	315	4.06	0.11	4.54	0.05	3.378	2.864	3.734	4 595	1.997	1.997	3.734	2.19	4.653	18.838	0.784
	1955	7 861	189	2.08	0.04	2.19	0.06	3.780	4.725	4.018	933	2.381	2.381	4.411	0.67	6.069	24.907	-0.400
	1956	2 766	116	0.63	0.01	0.67	0.03	4.390	11.600	3.091	191	2.392	2.392	4.018	0.39	6.595	31.502	-0.942
	1957	1 451	3	0.11		0.39		3.091		3.800		1.736	1.736	3.091	0.11	5.946	37.448	-2.207
	1958	340				0.11		3.800						3.800				
	1959	76																
1947	1950	304	407	1.42	0.05	0.05	1.00	3.200	2.200	2.200	360	0.019	0.019	2.200	0.05	2.698	1.100	-2.996
	1951	54 032	6 580	12.08	2.06	3.48	0.59	3.900	3.194	3.197	2 007	0.093	0.093	3.197	3.48	3.161	3.798	1.247
	1952	85 563	11 596	12.23	6.71	21.64	0.36	2.600	2.089	2.496	9 953	0.506	0.506	3.126	18.79	2.830	6.959	1.934
	1953	44 830	9 209	10.43	4.41	11.77	0.20	3.067	2.764	2.964	8 459	1.337	1.337	2.964	21.64	2.823	9.789	3.073
	1954	31 985	2 907	5.88	1.34	6.33	0.11	3.945	4.021	3.862	3 133	2.222	2.222	3.862	11.77	3.919	12.612	1.845
	1955	23 192	1 244	1.22	0.45	1.41	0.07	4.480	9.740	4.418	2 046	3.247	3.247	4.418	1.41	5.477	16.531	0.344
	1956	2 465	764	0.58	0.19	0.63	0.13	4.282	3.900	4.714	976	2.324	2.324	4.714	0.63	6.798	22.008	-0.462
	1957	2 484	487	0.41	0.05	0.42	0.08	3.171		3.188	394	4.377	4.377	3.188	0.42	7.198	28.796	-0.867
	1958	1 300	39	0.09	0.01	0.09	0.02	1.911		1.911				1.911	0.09	4.873	35.994	-2.408
	1959	172	91														40.867	
	1960	3	2															

1948	1951	24 543	1 379	0.04	0.39	0.04	0.43	0.91	3.400	3.536	3.524	429	0.017	0.017		3.524	0.43	3.079	1.762	−0.844
	1952	78 245	19 822	4.75	10.17	4.75	14.92	0.68	4.100	1.950	2.634	3 898	0.086	0.086		2.634	14.92	2.486	4.841	2.701
	1953	87 383	21 814	14.46	10.94	19.35	25.40	0.43	2.600	1.994	2.339	15 941	0.759	0.759		2.339	25.40	2.594	7.327	3.235
	1954	74 167	16 215	22.12	9.96	28.99	32.08	0.31	3.352	2.357	3.745	8 322	1.114	1.114		3.745	32.08	3.366	9.921	3.469
	1955	69 163	9 473	16.97	4.02	6.54	20.99	0.19	4.074	4.835	4.630	6 987	1.508	1.508		4.630	20.99	3.946	13.287	3.045
	1956	26 077	8 510	1.29	1.76	1.54	7.47	0.24	5.387	6.399	5.657					4.150	4.63	4.504	18.233	2.011
	1957	6 949	3 007	0.90	0.47	1.01	1.76	0.27	5.943	4.218	5.961	2 478	2.582	2.582		3.961	0.96	4.563	22.737	1.533
	1958	3 549	253	0.29	0.06		0.96	0.06	3.779	5.800	4.024	1 535	4.651	4.651		4.024	0.33	6.574	27.300	−0.041
	1959	1 096	232	0.11	0.04		0.33	0.12	3.145		3.145	542	4.927	4.927		3.145	0.11	8.235	33.874	−1.109
	1960	346	18	0.14			0.11		1.636		1.636					1.636	0.14	7.317	42.109	−2.207
	1961	244	41				0.14													
1949	1952	17 809	1 658	0.16	0.44	0.16	0.60	0.73	4.300	1.800	2.467	8	0.002	0.002		2.467	0.60	2.426	1.233	−0.511
	1953	100 082	6 538	2.12	3.02	2.35	5.14	0.59	2.700	2.165	2.386	177	0.007	0.007		2.386	5.14	2.704	3.659	1.637
	1954	118 507	12 365	19.35	7.18	28.99	26.53	0.27	3.500	1.722	3.019	3 419	0.094	0.094		3.019	26.53	3.720	6.363	3.277
	1955	144 162	14 994	28.99	7.47	13.80	36.46	0.20	4.974	2.217	4.408	3 692	0.094	0.094		4.408	36.46	4.733	10.083	3.597
	1956	66 212	8 932	13.80	2.88	6.54	16.68	0.17	4.800	5.206	4.870	5 260	0.221	0.221		4.562	16.68	4.937	14.816	2.815
	1957	26 649	879	6.54	1.26	1.54	7.80	0.16	4.075	7.089	4.562	6 236	0.674	0.674		1.619	7.80	3.764	19.753	2.054
	1958	2 852	700	1.54	0.22	1.01	1.76	0.13	4.431	3.995	4.377	3 466	1.311	3.563		3.116	1.76	3.678	23.517	1.560
	1959	661	13	1.01	0.13		1.14	0.11	2.824	5.384	3.116	1 471		3.040			1.14		27.195	0.131
	1960	229	44									307		5.118						
	1961		151																	
	1962			0.06		0.06	0.06		6.701		6.701		5.118			6.701	0.06			
1950	1953	42 211	4 562	0.02	0.60	0.02	0.62	0.97	2.800	2.000	2.026	1 311	0.046	0.046		2.026	0.62	2.405	1.013	−0.478
	1954	126 472	11 050	5.42	5.55	5.42	10.97	0.51	3.600	1.991	2.784	3 751	0.099	0.099		2.784	10.97	3.120	3.418	2.397
	1955	164 145	37 332	11.71	17.05	19.68	28.76	0.59	4.900	4.665	3.457	7 606	0.510	0.510		3.457	28.76	4.179	6.538	3.360
	1956	170 904	66 744	23.25	14.72	9.68	37.97	0.39	5.001	5.884	4.809	13 896	1.962	1.962		4.809	37.97	5.159	10.717	2.701
	1957	48 409	30 839	9.68	5.24	5.45	14.92	0.36	3.929	4.555	5.311	7 203	3.774	15.326		5.311	14.92	5.190	15.876	1.907
	1958	21 412	5 830	5.45	1.28	2.12	6.73	0.14	3.934	4.109	4.048	8 203	3.831			3.969	6.73	5.970	21.066	0.971
	1959	8 339	2 137	2.12	0.52	0.28	2.64	0.20	7.563	3.320	3.969	9 962	4.428			1.815	2.64	6.666	27.036	0.278
	1960	2 230	166	0.28	0.05	0.06	0.33	0.15	5.071	3.850	7.259	1 476	2.523			1.359	1.32	9.418	33.702	−1.109
	1961	402	77	0.06	0.02	0.04	0.13	0.25	2.600		4.666	328				3.623	0.33	10.919	43.120	−2.040
	1962	471	50	0.04			0.04				3.623					2.600	0.13	5.634	54.039	−3.219
	1963	104									2.600						0.04		59.673	
1951	1954	9 686	770	0.08	0.47	0.08	0.55	0.85	3.700	1.639	1.939	263	0.022	0.022		1.939	0.55	2.429	0.969	−0.598
	1955	31 728	807	0.79	3.05	6.40	3.84	0.79	5.000	2.200	2.920	839	0.059	0.059		2.920	3.84	3.990	3.398	1.345
	1956	95 695	823	6.40	5.38	9.79	11.78	0.46	5.700	5.135	5.060	3 896	0.484	0.484		5.060	11.78	4.974	7.388	2.468
	1957	41 546	791	9.79	4.52	5.73	14.31	0.32	4.242	6.147	4.844	3 997	1.041	1.041		4.625	14.31	4.793	12.362	2.660
	1958	24 018	226	5.73	2.32	3.04	8.05	0.29	4.191	5.700	4.625	3 969	2.602	6.505		3.383	8.05	4.488	17.155	2.086
	1959	9 154	837	3.04	0.80	0.55	3.84	0.21	3.011	4.796	3.383	2 033	3.486	4.841		2.505	3.84	7.349	21.643	1.345
	1960	3 600	221	0.55	0.06	0.40	0.61	0.21	6.545	3.683	6.262	686	2.859	2.859		6.990	1.52	9.070	25.628	0.419
	1961	3 042	188	0.40	0.02	0.22	0.42	0.10	9.047	9.250	9.065					4.179	0.53	8.348	32.977	−0.635
	1962	818	185	0.22	0.02	0.02	0.24	0.05	3.718		4.197					3.623	0.24		42.047	−1.427
	1963	136		0.02			0.02	0.08	6.800		6.800					6.800	0.02		50.395	−3.912
1952	1955	3 692	316	0.51	0.10		0.10	1.00	6.100	2.200	2.200	105	0.022	0.022		2.200	0.10	2.642	1.100	−2.302
	1956	21 127	2 307	3.39	0.27		0.78	0.35	5.100	5.000	5.065	557	0.056	0.056		5.065	0.78	5.156	3.742	−0.248
	1957	16 795	8 734	8.19	1.32		4.71	0.28	4.555	5.819	5.247	1 263	0.247	0.247		5.247	4.71	4.716	8.898	1.550
	1958	31 064	9 951	4.57	1.71		9.90	0.17	3.793	5.550	4.142	2 318	1.825	1.825		4.142	9.90	3.709	13.614	1.293
	1959	13 173	997	1.21	0.54		5.11	0.11	3.882	5.876	5.830	2 006	2.229	2.229		5.830	5.11	4.744	17.323	1.631
	1960	7 110	294	0.85	0.06		1.27	0.05	5.876	4.256	4.347	611	3.819	3.819		4.347	1.27	6.913	22.067	0.239
	1961	3 642	295	0.15	0.05		0.90	0.06	4.256	4.900	5.706					5.706	0.90	7.255	28.035	−0.105
	1962	657	271	0.08	0.01		0.16	0.06	4.280	5.900	8.212					8.212	0.16	10.778	36.235	−1.833
	1963						0.08		8.212	27.100							0.08		47.013	−2.526

APPENDIX TABLE A. The exploited life history of successive year-classes of Arcto-Norwegian cod. — (*Concluded*)

Year-class	Sample year	1	2	3	4	5	6	7	8	9	10	11	12	13	14	15	16	17
1953	1956	10 430	146	0.02		0.02				5.200				5.200	0.02		2.600	-3.912
	1957	20 852	5 977	1.98	0.75	2.73	0.27	5.200	5.000	5.202				5.202	2.73		7.801	1.004
	1958	42 735	22 462	9.83	3.98	13.81	0.29	4.346	6.000	4.720				4.720	13.81	5.201	12.762	2.625
	1959	37 941	10 070	12.43	1.62	14.05	0.12	3.054	6.216	3.419	363	0.026	0.026	3.419	14.05	4.961	16.857	2.639
	1960	21 805	2 260	3.04	0.75	4.66	0.16	5.578	3.013	5.165	995	0.071	0.071	5.165	4.66	4.095	21.220	1.539
	1961	12 374	1 617	0.95	0.24	3.28	0.07	4.070	6.738	4.266	3 090	0.663	0.663	4.266	3.28	4.363	26.598	1.188
	1962	4 403	766	0.16	0.08	1.03	0.08	4.636	9.575	5.020	4 601	1.403	1.403	5.020	1.03	6.046	32.644	0.030
	1963	1 135	129		0.01	0.17	0.06	7.094	12.900	7.435	1 622	1.575	1.575	7.435	0.17	7.802	40.446	-1.772
1954	1957	6 110	7 595	0.08	0.06	0.14	0.43	5.000	6.000	5.429				5.429	0.14		2.714	-1.966
	1958	48 585	77 030	4.38	7.11	11.49	0.62	4.400	6.800	5.143				5.143	11.49	5.286	8.000	2.442
	1959	59 197	69 004	14.84	8.90	23.74	0.37	3.200	3.531	4.549				4.549	23.74	4.846	12.846	3.167
	1960	57 119	18 465	13.06	5.23	18.29	0.29	4.375	3.922	4.135	972	0.041	0.041	4.135	18.29	4.383	17.229	2.907
	1961	23 432	5 511	5.61	2.68	8.29	0.32	4.178	6.441	4.094	2 467	0.135	0.135	4.094	8.29	4.249	21.478	2.115
	1962	12 141	5 538	2.38	0.86	3.24	0.27	5.102	3.860	5.457	5 779	0.697	0.697	5.457	3.24	5.472	26.950	1.176
	1963	3 157	579	0.45	0.15	0.60	0.25	7.015		6.227	5 424	1.674	1.674	6.227	0.60	7.516	34.466	-0.511
1955	1958	15 277	14 291	0.55	0.34	0.89	0.38	4.500	4.500	4.920				4.920	0.89		2.461	-0.117
	1959	30 051	35 788	5.33	3.55	8.88	0.40	3.300	7.500	4.978				4.978	8.88	4.950	7.411	2.184
	1960	56 535	13 701	14.38	3.81	18.19	0.21	3.929	5.998	3.860				3.860	18.19	4.420	11.831	2.901
	1961	32 537	15 117	8.75	2.52	11.27	0.22	3.539	6.294	4.229	571	0.031	0.031	4.229	11.27	4.077	15.908	2.422
	1962	19 938	7 553	4.80	1.20	6.00	0.20	4.153	5.024	4.582	2 214	0.196	0.196	4.582	6.00	4.601	20.509	1.792
	1963	8 112	1 407	1.42	0.28	1.70	0.16	5.712		5.598	3 990	0.665	0.665	5.598	1.70	5.755	26.264	0.531
1956	1959	10 328	15 719	0.50	1.03	1.53	0.67	3.400	8.200	6.631				6.631	1.53		3.315	0.425
	1960	79 100	15 825	16.75	4.67	21.42	0.22	4.722	3.388	4.432				4.432	21.42	5.531	8.846	3.063
	1961	73 854	46 055	14.88	7.66	22.54	0.34	4.010	5.196	4.684	93	0.004	0.004	4.684	22.54	4.561	13.407	3.114
	1962	64 364	24 477	4.56	4.71	19.47	0.24	4.357	4.096	4.560	745	0.033	0.033	4.560	19.47	4.654	18.061	2.970
	1963	23 396	4 956		1.21	5.77	0.21	5.131		4.914	2 839	0.146	0.146	4.914	5.77	4.883	22.944	1.753
1957	1960	27 949	2 202	3.80	0.60	4.40	0.14	5.000	3.670	4.819				4.819	4.40		2.409	1.482
	1961	88 144	49 289	10.60	4.96	15.56	0.32	5.000	6.000	4.638	20	0.005	0.005	4.638	15.56	4.733	7.142	2.745
	1962	120 746	35 837	21.20	6.77	27.97	0.24	4.700	5.293	4.843	94	0.006	0.006	4.843	27.97	4.746	11.888	3.320
	1963	67 383	13 632	13.75	3.34	17.09	0.20	4.899	4.081	4.740	598	0.021	0.021	4.740	17.09	4.812	16.700	2.839
1958	1961	46 667	17 220	1.69	0.54	2.23	0.24	4.000	6.000	4.484				4.484	2.23		2.242	0.802
	1962	133 685	26 570	12.17	4.07	16.24	0.25	4.800	5.200	4.901				4.901	16.24	4.692	6.934	2.787
	1963	157 346	22 725	24.74	6.03	30.77	0.20	5.000	3.768	4.759				4.759	30.77	4.830	11.764	3.428
1959	1962	48 233	6 298	1.13	0.20	1.33	0.15	4.900	5.200	4.945				4.945	1.33		2.472	0.285
	1963	99 563	9 497	8.64	1.65	10.29	0.16	5.000	3.800	4.810				4.810	10.29	4.877	7.349	2.332
1960	1963	17 084	1 829	0.47	0.03	0.50	0.06	5.000	3.800	4.928				4.928	0.50		2.464	-0.693

APPENDIX B

GULLAND'S METHOD OF VIRTUAL POPULATION ANALYSIS

The population (r) of a year-class at the end of a year can be expressed as a proportion of the catch during the year:

$$_xr_n = \frac{_xN_{n+1}}{_xC_n} = \frac{_xN_ne^{-(F+M)}}{_xN_n \ (F/(F+M)) \ (1-e^{-(F+M)})} \tag{A}$$

where F represents the fishing mortality on year-class x at t years old in year n. Thus if the expression

$$\frac{(F+M)e^{-(F+M)}}{F(1-e^{-(F+M)})}$$

is tabulated at a known value of M for a range of values of F, $_tF_n$ can be determined immediately once $_xr_n$ is known.

The virtual population (V) of a year-class can be defined as

$$_xV_{n+1} = {_xE_{n+1}} \ {_xN_{n+1}} \tag{B}$$

so, from A,

$$_xr_n = \frac{_xV_{n+1}}{_xE_{n+1} \ _xC_n} = \frac{1}{_xE_{n+1}} \left(\frac{_xV_{n+1}}{_xV_n - _xV_{n+1}}\right) = \frac{1}{_xE_{n+1}} \left(\frac{_xS_n}{1 - _xS_n}\right) \tag{C}$$

Thus $_xr_n$ is a simple fraction of the apparent survival during year n (as estimated from virtual populations) and the exploitation ratio $_xE_{n+1}$ applicable to the fish of the x year-class alive at the end of year n.

The exploitation ratio $_xE_n$ applicable to fish at the beginning of the year n will be the sum of the proportions of fish alive at the beginning of the year which are caught during that year and in subsequent years;

$$_xE_n = \frac{_tF_n}{_tF_n + M} \ (1-e^{-(_tF_n+M)}) + e^{-(_tF_n+M)} \ _xE_{n+1} \tag{D}$$

Thus, from the determined value of M, and an assumed value of $_xE_{n+1}$ for the final year in which a year-class contributes to the catches, successive values of $_xr_n$, $_tF_n$, $_xE_n$, $_xr_{n-1}$, $_{t-1}F_{n-1}$, etc. can be estimated.

APPENDIX TABLE B. Comparison of USSR and English catch per unit effort data for the Barents Sea cod fishery.

Calendar year

Age	1946	1947	1948	1949	1950	1951	1952	1953	1954	1955	1956	1957	1958	1959	1960	1961	1962	1963
USSR Numbers/1 hr fishing																		
6	48.21	70.30	93.14	113.99	60.40	79.35	116.64	82.34	151.02	251.42	233.36	58.68	35.32	53.74	55.09	30.69	60.45	71.70
7	35.12	55.82	131.24	117.53	89.40	47.86	68.10	26.38	58.35	94.80	81.44	56.58	25.66	17.35	21.79	18.98	17.91	23.04
8	30.01	14.81	31.75	41.19	46.81	20.78	16.44	15.19	22.31	36.27	27.73	35.63	28.49	10.91	8.71	10.30	9.05	7.55
9	61.93	17.57	20.04	15.63	20.54	7.56	7.83	4.00	8.58	12.37	5.78	6.55	8.95	7.56	3.11	2.99	3.86	2.75
10	45.02	25.84	5.65	4.26	9.36	2.52	2.35	1.60	5.15	6.60	2.89	3.14	5.18	3.64	1.87	1.47	0.60	1.02
11	13.41	27.22	10.23	6.75	6.34	1.89	1.56	0.80	2.57	2.47	1.16	2.10	2.12	1.12	0.93	0.21	0.35	0.61
12	3.51	5.86	3.88	3.91	1.81		0.78	0.80				0.52	0.47	0.28	0.31	0.23	0.49	0.17
13	1.28	3.79	1.77	2.13	1.21													0.11
14	0.63	2.07	2.11	1.06	0.61											0.21		0.26
15	1.28				0.30											0.27		0.05
English Numbers/100 ton-hours — See Appendix Table A.																		
Ratio USSR/English																		
6	3.7	3.1	3.8	3.4	9.2	5.8	7.1	4.8	6.8	8.6	10.1	6.0	4.3	4.4	4.2	3.5	4.1	5.2
7	3.7	3.1	3.8	5.8	5.9	4.2	5.1	3.2	5.6	5.6	5.9	6.0	4.5	3.9	5.6	3.4	3.7	5.0
8	3.7	3.1	3.8	4.5	4.2	3.1	3.9	2.7	5.6	6.1	4.9	5.4	5.3	3.6	7.3	3.3	3.9	5.4
9	3.7	3.1	3.8	2.3	4.1	1.8	3.7	2.4	3.7	5.9	4.8	1.9	2.1	3.6	2.2	3.3	4.3	6.9
10	3.7	3.1	3.8	1.5	4.7	1.5	1.5	1.1	5.1	4.7	4.8	5.2	5.8	3.6	1.7	2.9	3.0	5.0
11	3.7	3.1	3.8	4.2	5.3	3.1	1.7	2.0	3.9	3.5	3.9	5.2	5.3	3.7	2.3	1.0	1.7	6.1
12	3.7	3.1	3.8	2.8	6.0		3.9	4.0				5.2	4.7	2.8	3.1	2.3	4.9	—
13	3.7	3.1	3.8	2.7	1.5													
14	3.7	3.1	3.8	1.8	3.0													
15		3.1	3.8		1.5													
Mean ratio:	3.7	3.1	3.8	3.2	4.5	3.3	3.8	2.9	5.1	5.7	5.7	5.0	4.6	3.7	3.8	2.8	3.7	5.6

An investigation of accuracy of virtual population analysis using cohort analysis

J.G.Pope

Research Bulletin of the International Commission for the Northwest Atlantic Fisheries, **9**, 65-74, 1972.

An Investigation of the Accuracy of Virtual Population Analysis Using Cohort Analysis

By J. G. Pope[1]

Abstract

Cohort analysis is a simplified, approximate form of Gulland's virtual population analysis. As such it may be used to obtain estimates of the instanteous rate of fishing mortality and the population surviving for each age of a year-class, given the catch-at-age data, and an estimate of the instanteous rate of natural mortality and an estimate of the fishing mortality at the final age of exploitation. More importantly, the simplicity of cohort analysis makes it possible to investigate the errors generated in such estimates by the arbitrary choice of the rate of fishing mortality on the last age exploited and by the sampling errors of the catch-at-age data.

Introduction

Gulland's virtual population analysis (Gulland, 1965) is an extremely useful technique when assessing a fishery, because it enables estimates of population at age and fishing mortality to be made independently of the measurement of effort. These estimates are however subject to various errors which might adversely affect an assessment. What causes these errors and how can their magnitude be calculated?

Section 1. Cohort Analysis as an Approximation to Virtual Population Analysis

In the text and appendices the following symbols are used:
M is the instantaneous coefficient of natural mortality;
F is the instantenous coefficient of fishing mortality;
Z is the instantaneous coefficient of total mortality;
N_i is the population of a year-class at the ith birthday;
C_i is the catch of a year-class at age i;

t is the last age of a year-class for which catch data are available;
exp is the exponential function.

Virtual population analysis is an ingenious step-wise procedure developed by Gulland to calculate, for a year-class, the instanteous fishing mortality and population at each age, given a knowledge of the catch at each age and the natural mortality. Details of the method are given in Appendix A but basically it is based on the solution of the formula:

$$\frac{N_{i+1}}{C_i} = \frac{(F+M)\exp\{-(F+M)\}}{F\left(1-\exp\{-(F+M)\}\right)} \tag{1.1}$$

Since this formula does not yield an analytical solution for F it has to be solved numerically, either by reference to tables or by using an iterative procedure. Either method makes calculation of F and N by hand somewhat laborious. More importantly, the lack of an explicit formula for F makes it difficult to comprehend the effect that errors in the input information cause in the results.

Cohort analysis is a new form of virtual population analysis developed by the author to overcome these problems. It is in fact an approximation to Gulland's virtual population analysis which is usable at least up to values of $M = 0.3$ and $F = 1.2$. A detailed explanation of the method is given in Appendix B. The method is based on the approximate formula:

$$N_i = C_i \exp\{M/2\} + N_{i+1} \exp\{M\}. \tag{1.2}$$

Thus, using 1.2 as a recurrence relationship,

$$N_i = \left(C_i \exp\{M/2\}\right) + \left(C_{i+1}\exp\{3M/2\}\right) + \tag{1.3}$$
$$+ \left(C_{i+2}\exp\{5M/2\}\right) + \ldots + \left(N_t \exp\{(t-i)M\}\right)$$

As with Gulland's virtual population analysis, N_t has two possible forms. The first form is when C_t refers to the

[1] Ministry of Agriculture, Fisheries and Food, Fisheries Laboratory, Lowestoft, England.

catch in year t only, which is the case with the last year's catch of a year-class that is still being fished.

In this case

$$N_t = \frac{C_t Z_t}{F_t (1 - \exp\{-Z_t\})} \qquad (1.4)$$

and consequently

$$N_i = \left(C_i \exp\{M/2\}\right) + \left(C_{i+1} \exp\{3M/2\}\right) + \qquad (1.5)$$
$$+ \left(C_{i+2} \exp\{5M/2\}\right) + \cdots + \left(\frac{C_t Z_t \exp\{(t-i)M\}}{F_t (1 - \exp\{-Z_t\})}\right)$$

The second form is when C_t refers to the catch in year t and all subsequent years. This is usually the case with a completely fished year-class. In this case

$$N_t = \frac{C_t Z_t}{F_t} \qquad (1.6)$$

and consequently

$$N_i = \left(C_i \exp\{M/2\}\right) + \left(C_{i+1} \exp\{3M/2\}\right) + \qquad (1.7)$$
$$+ \left(C_{i+2} \exp\{5M/2\}\right) + \cdots + \left(\frac{C_t Z_t \exp\{(t-i)M\}}{F_t}\right)$$

In either case

$$F_i = \log_e\{N_i / N_{i+1}\} - M . \qquad (1.8)$$

The closeness with which these formulae approximate the results of virtual population analysis can be judged from Table 1, where results of both methods are compared. It can be seen that in no case do the estimates given by the two methods differ by more than 2%. Consequently an investigation of the errors of cohort analysis is an approximate investigation of the errors of Gulland's virtual population analysis. It can be seen from equations 1.5 and 1.7 that errors in N_i, and consequently errors in F_i, can be introduced by the incorrect choice of F_t and by the sampling errors in the C_i. These two sources of error are investigated in the next two sections. Errors in M can also cause errors in N_i and F_i, but for the purpose of this document M will be considered as fixed.

TABLE 1. Comparison of the results of virtual population analysis and cohort analysis. Arcto-Norwegian cod, 1956 year-class, M = 0.3.

Age (years)	Fishing mortality, F_n			Population, $N_i \times 10^{-6}$		
	Virtual population analysis	Cohort analysis	% error	Virtual population analysis	Cohort analysis	% error
12	0.8000[a]	0.8000[a]		0.2	0.2	
11	1.3400	1.3670	2	1.1	1.1	0
10	0.7826	0.7806	–	3.1	3.2	2
9	0.6768	0.6747	–	8.3	8.5	2
8	0.6582	0.6570	–	21.7	22.2	2
7	0.8636	0.8657	–	69.6	71.2	2
6	0.7341	0.7333	–	195.6	200.1	2
5	0.4289	0.4261	1	405.5	413.6	2
4	0.1874	0.1854	1	660.2	672.0	2
3	0.0411	0.0405	1	928.5	944.7	2
2	0.0024	0.0024	–	1256.4	1278.2	2
1	0.0007	0.0007	–	1697.1	1726.6	2

[a]Assumed.

293

Section 2. Error in Cohort Analysis Due to the Incorrect Choice of F_t

If an incorrect value F_t is chosen for the terminal fishing mortality when its true value is $\overline{F}_t$, then the proportional error in N_t, $\rho(N_t)$, is given as follows in the case when C_t is the catch in year t only:

$$\rho(N_t) = \frac{Z_t \, \overline{F}_t \, (1 - \exp\{-\overline{Z}_t\})}{\overline{Z}_t \, F_t \, (1 - \exp\{-Z_t\})} - 1 \qquad (2.1)$$

Since

$$\rho(N_i) = \rho(N_{i+1}) \exp\{-F_i\} \qquad (2.2)$$

it follows that

$$\rho(N_i) = \left(\frac{Z_t \, \overline{F}_t \, (1 - \exp\{-\overline{Z}_t\})}{\overline{Z}_t \, F_t \, (1 - \exp\{-Z_t\})} - 1 \right) \times$$
$$\times \left(\exp\{-F_i - F_{i+1} \ldots - F_{t-1}\} \right) \qquad (2.3)$$

For small values of Z_t this is approximately given by

$$\rho(N_i) \simeq \left(\frac{\overline{F}_t - F_t}{F_t} \right) \times$$
$$\times \left(\exp\{-F_i - F_{i+1} \ldots - F_{t-1}\} \right) \qquad (2.4)$$

while for larger values of Z this formula tends to overstate the error and is therefore still of some value.

A formula similar to 2.4 gives $\rho'(N_i)$, the proportional error in N_i when C_t is the catch in year t and all subsequent years. In this case

$$\rho'(N_i) = \frac{M}{\overline{Z}_t} \left(\frac{\overline{F}_t - F_t}{F_t} \right) \times$$
$$\times \left(\exp\{-F_i \ldots - F_{t-1}\} \right) \qquad (2.5)$$

and therefore

$$\rho'(N_i) = \frac{M}{\overline{Z}_t} \, \rho(N_i). \qquad (2.6)$$

It is therefore simple to convert a table of $\rho(N_i)$ into a table of $\rho'(N_i)$. The porportional error of F_i, $\rho(F_i)$, is given approximately by a formula due to Agger et al., (1971):

$$\rho(F_i) = \rho(N_i) \left(\frac{1 - \exp\{F_i\}}{F_i} \right) \qquad (2.7)$$

This form supersedes a cruder formula in Pope (1971). However, for the purpose of discussion 2.7 may be simplified to the assumption that $\rho(F_i)$ is less than twice the numerical value of $\rho(N_i)$ and of opposite sign provided F_i is less 1.2 (the viable range of cohort analysis) since within this range

$$-2 < \left(\frac{1 - \exp\{F_i\}}{F_i} \right) \leqslant -1. \qquad (2.8)$$

Figure 1 shows graphs of $\rho(N_i)$ plotted against the sum of the fishing mortality from year i to year $t-1$ (cumulative fishing mortality). Figure 1 is based on formula 2.4 which increasingly overestimates $\rho(N_i)$ as $\overline{Z}_t$ increases. Consequently the value of $\rho(F_i)$ is given approximately by:

$$\rho(F_i) \approx - \{\text{value of } \rho(N_i) \text{ on Figure 1}\} \qquad (2.9)$$

provided $F_i \leqslant \overline{F}_t$. It can be seen that the underestimation of F_t results in estimates of N_i which are too large and estimates of F_i which are too small, whereas overestimating F_t has the reverse effect. It can also be seen that as the cumulative fishing mortality increases, both types of error decrease. As an example, if F_t was overestimated by 100% for a year-class and the cumulative fishing mortality from year i to year $t-1$ was 2.0, then the percentage error in N_i would at most be ·7% and the percentage error in F_i would be about +7% (see 2.9). If, however, F_t was underestimated by 50% and the cumulative fishing mortality was equal to 2.0, then the percentage error in N_i would be at the most 14% and the percentage error in F_i would be about -14% (see 2.9).

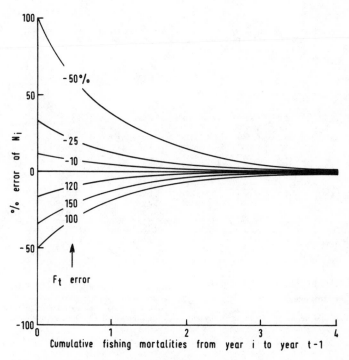

Fig. 1. Graphs of the percentage error in N_i due to incorrect values of F_t plotted against the cumulative fishing mortalities from year i to year $t - 1$.

Thus, provided that F_t can be estimated within this range and provided that the cumulative fishing mortality is greater than 2.0, the error in in the estimates of N_i and F_i should be small enough for many uses. If, however, the cumulative fishing mortality is small, which is the case when the number of recruits to a year-class is estimated from the catches of partially recruited age groups, then the accurate estimation of N_i and F_i will require the accurate choice of F_t. It should also be realized that since the cumulative fishing mortality is the sum of the fishing mortalities from age i to age $t - 1$ it must, for a particular year-class, be a monotonically decreasing function of age. Hence the bias in F_i caused by the incorrect choice of F_t will be greatest amongst the oldest age groups, and this may upset estimates of selectivity with age. Table 2 shows the results of a cohort analysis for the 1956 year-class of the Arcto-Norwegian cod. This assumes that the true values of M and F_t are 0.3 and 0.8 respectively and shows the percentage errors in N_i and F_i when F_t is overestimated by 100% or underestimated by 50%. These errors were computed by rerunning the data with the appropriate value of F_t and are therefore precise. It can be seen that these percentage errors are similar but, in general, smaller than their estimates in Fig. 1.

TABLE 2. The percentage in N_i and F_i when F_t is overestimated by 100% and when F_t is underestimated by 50% for the 1956 year-class of the Arcto-Norwegian cod, with $M = 0.3$ and when the true value of $F_t = 0.8$.

Age (years)	$N_i \times 10^{-6}$	F_i	Cumulative F_i	% error when F_t is taken as 0.4 in N_i	% error when F_t is taken as 0.4 in F_i	% error when F_t is taken as 1.6 in N_i	% error when F_t is taken as 1.6 in F_i
12	0.2	0.8000[a]	–	+ 68.66		- 32.27	
11	1.1	1.3670	1.3670	+ 17.50	- 26.44	- 8.22	+ 22.21
10	3.2	0.7806	2.1475	+ 8.02	- 10.77	- 3.77	+ 6.07
9	8.5	0.6747	2.8222	+ 4.08	- 5.50	- 1.92	+ 2.82
8	22.2	0.6570	3.4792	+ 2.12	- 2.91	- 0.99	+ 1.43
7	71.2	0.8657	4.3449	+ 0.89	- 1.40	- 0.42	+ 0.67
6	200.1	0.7333	5.0782	+ 0.43	- 0.63	- 0.20	+ 0.29
5	413.6	0.4261	5.5043	+ 0.28	- 0.35	- 0.13	+ 0.16
4	672.0	0.1854	5.6897	+ 0.23	- 0.22	- 0.11	+ 0.16
3	944.7	0.0405	5.7302	+ 0.22	- 0.25	- 0.10	+ 0.25
2	1278.2	0.0024	5.7326	+ 0.22	- 0.00	- 0.10	+ 0.00
1	1726.6	0.0007	5.7333	+ 0.22	- 0.00	- 0.10	+ 0.00

[a] Assumed.

Section 3. Error in Cohort Analysis Due to the Sampling Error of C_i

Unlike the estimate of F_t, which is usually an arbitrary choice, each estimate of catch at age can be assigned a variance, although this is seldom available, due to the heavy work involved in its computation (see Gulland, 1955). Assuming such variances to be available it is a simple matter to compute the resulting variance of N_i and F_i, since

$$\text{var}(N_i) = \Big(\text{var}(C_i)\exp\{M\}\Big) + \Big(\text{var}(N_{i+1})\exp\{2M\}\Big), \tag{3.1}$$

and this may be used as a recurrence relationship to obtain

$$\text{var}(N_i) = \Big(\text{var}(C_i)\exp\{M\}\Big) + \Big(\text{var}(C_{i+1})\exp\{3M\}\Big) + \ldots + \left(\text{var}(C_t)\frac{\exp\{2(t-i)M\}(F_t+M)^2}{F_t^2(1-\exp\{-F_t-M\})^2}\right) \tag{3.2}$$

which is a very similar formula to 1.5.

The equivalent variance of F_i can be approximated, since

$$F_i = \log_e\{N_i/N_{i+1}\} - M \tag{3.3}$$

which yields approximately

$$\text{Var}(F_i) \approx \left(\frac{\text{var}(N_i)}{N_i^2}\right) - \left(\frac{2\,\text{var}(N_{i+1})\exp\{M\}}{N_i\,N_{i+1}}\right) + \left(\frac{\text{var}(N_{i+1})}{N_{i+1}^2}\right) \tag{3.4}$$

Equations 3.2 and 3.4 should be used to calculate the respective variances of N_i and F_i in a particular case, but in order to appreciate the approximate magnitude of these variances the following approximate formulae are useful:

$$\left(\text{var ratio } (N_i)\right)^2 \approx \left(\text{var ratio } (C_i)\right)^2 \left(1 - \exp\{-F_i\}\right)^2 + \left(\text{var ratio } (N_{i+1})\right)^2 \left(\exp\{-F_i\}\right)^2 \tag{3.5}$$

$$\left(\text{var ratio } (F_i)\right)^2 \approx \frac{\left(1 - \exp\{-F_i\}\right)^2}{F_i^2} \left[\left(\text{var ratio } (C_i)\right)^2 + \left(\text{var ratio } (N_{i+1})\right)^2\right] \tag{3.6}$$

Details of the derivation of these formulae are given in Appendix C.

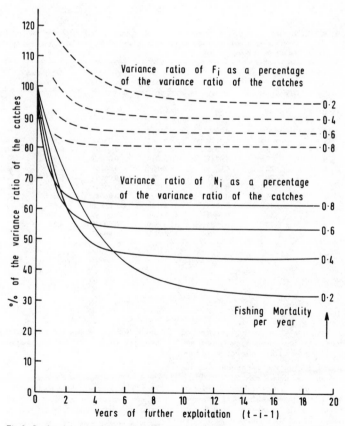

Fig. 2. Graphs of the percentage variance ratio of F_i and of N_i for various constant levels of fishing mortality plotted against the years of further exploitation.

Figure 2 shows graphs of these formulae for each year from the final year, that is for the number of years from the estimate in question to the final year. The graphs are given for the case when the variance ratio of the catch-at-age data is constant, and when the fishing mortality is constant throughout the life of the fish. Although these conditions are unrealistic, the rapid convergence of the graphs to asymptotic values does suggest that the graphs would indicate the approximate value of the variance ratio of the estimates of N_i and F_i, even when F_i is not constant from year to year. As an example of the use of the graphs, the estimate of N_5 (for a year-class with an oldest age group of 12 years old, experiencing a fishing mortality of 0.6 per year) would have a variance ratio of approximately 54% of the variance ratio of the catch data. Similarly, the estimate of F_5 would have a variance ratio of approximately 85% of the variance ratio of the catch data. Hence, if the variance ratio of the catch data was 10%, then the variance ratios of N_5 and F_5 would be 5.4% and 8.5% respectively. As a result the approximate 95% confidence limits for the estimates would be ± 10.8% of the estimate of N_5 and ± 17.0% of the estimate of F_5.

TABLE 3. Standard deviations and variance ratios of N_i and F_i calculated for the 1956 Arcto-Norwegian cod, assuming that the variance ratio for the catch at each age was 10% and that M = 0.3 and F_t = 0.8.

Age (years)	$N_i \times 10^{-6}$	F_i	Standard deviation $N_i \times 10^{-6}$	F_i	Variance ratio (%) N_i	F_i
12	0.2	0.8000[a]	0.01449			
11	1.1	1.3670	0.08361	0.09192	7.60	6.72
10	3.2	0.7806	0.20763	0.06801	6.49	8.71
9	8.5	0.6747	0.50349	0.05865	5.92	7.51
8	22.2	0.6570	1.26666	0.05601	5.71	8.53
7	71.2	0.8657	4.46487	0.06669	6.27	7.70
6	200.1	0.7333	12.01878	0.06134	6.01	8.36
5	413.6	0.4261	21.65851	0.04047	5.24	9.50
4	672.0	0.1854	31.37061	0.01910	4.67	10.30
3	944.7	0.0405	42.51186	0.00439	4.50	10.83
2	1278.2	0.0024	57.38580	0.00026	4.49	10.77
1	1726.6	0.0007	77.46281	0.00007	4.49	10.59

[a]Assumed.

Table 3 shows the 1956 Arcto-Norwegian cod results, together with the standard deviations and variance ratios of N_i and F_i. These were computed from equations 3.2 and 3.4 on the assumption that the variance ratio of the catch data at each age was 10%. It can be seen that the variance ratios of these estimates are not very different from those which would have been predicted by entering the graphs of Fig. 2 with appropriate values of F_i at the asymptotic parts of the graphs. Thus Fig. 2 should prove to be of some value in providing quick estimates of the variance ratios of N_i and F_i for any year-class which has catch data which have approximately constant variance ratios.

Summary

This document provides formulae for calculating the error introduced in cohort analysis (and therefore virtual population analysis) by errors in F_t and by the sampling error of catch data. It also provides some quick estimates of the likely size of such errors. These estimates suggest that such errors converge to fairly small values, but they also suggest that a knowledge of the approximate value of these errors will always be a safeguard against misinterpretation of data!

References

AGGER, P., I. BOËTIUS, and H. LASSEN. 1971. On errors in the virtual population analysis. *ICES C.M.* 1971 Doc. No. H: 16 (mimeographed).

GULLAND, J. A. 1955. Estimation of growth and mortality in commercial fish populations. *Fish. Invest., Lond. (2),* **18** (9).

⎯⎯⎯ 1965. Estimation of mortality rates. Annex to Arctic Fisheries Working Group Report (meeting in Hamburg, January 1965). *ICES, C.M.* 1965, Doc. No. 3 (mimeographed).

POPE, J. G. 1971. An investigation of the accurancy of virtual population analysis. *Annu. Meet. int. Comm. Northw. Atlant. Fish, 1971,* Res. Doc. No. 116, Serial No. 2606 (mimeographed).

Appendix A.

Gulland's Virtual Population Analysis

Gulland's virtual population analysis is based on the following two equations:

$$N_{i+1} = N_i \exp\{-(F_i + M)\} \tag{A.1}$$

$$C_i = N_i \frac{F_i \left(1 - \exp\{-(F_i + M)\}\right)}{F_i + M}. \tag{A.2}$$

It follows therefore that

$$\frac{N_{i+1}}{C_i} = \frac{(F_i + M) \exp\{-(F_i + M)\}}{F_i\left(1 - \exp\{-(F_i + M)\}\right)}. \tag{A.3}$$

Hence, if N_{i+1}, C_i, and M are known, then (as was explained in Section 1) it is possible to solve A.3/1.1 for F_i and then to use A.1 to obtain N_i. Then A.3/1.1 may be used to obtain F_{i-1} and so on. To start the procedure it is necessary to estimate N_t, the population size of the oldest age at which the year-class was fished. This is done by taking the best estimate (or guess) of F_t available and using A.2 to solve for N_t. It is shown in Section 2 that in general the accurate choice of F_t is not crucial to the accuracy of F or N for other ages.

Appendix B

Cohort Analysis

The basic equation of cohort analysis is obtained by rewriting A.1 as

$$N_{i+1} \exp\{+M\} = N_i \exp\{-F_i\}, \tag{B.1}$$

which can be expressed as

$$N_{i+1} \exp\{+M\} = N_i - N_i(1 - \exp\{-F_i\}). \tag{B.2}$$

Substituting A.2 in B.2,

$$N_{i+1} \exp\{+M\} = N_i - C_i \left(\frac{(F_i + M)(1 - \exp\{-F_i\})}{F_i(1 - \exp\{-(F_i + M)\})}\right) \tag{B.3}$$

However, within the range $M < 0.3$, $F_i < 1.2$ the function $\left(\dfrac{(F_i + M)(1 - \exp\{-F_i\})}{F_i(1 - \exp\{-(F_i + M)\})}\right)$ can be approximated by $\exp\{+M/2\}$ with an error that is always less than 4%. Thus, with little loss of accurancy, B.3 can be rewritten as:

$$N_i = N_{i+1} \exp\{+M\} + C_i \exp\{+M/2\}. \quad (B.4)$$

This is the basic equation of cohort analysis. Given a knowledge of N_{i+1}, C_i and M, it is possible to use B.4 (or 1.2) to obtain N_i and then to use B.1 to obtain F_i. Having found N_i, equation B.4 (or 1.2) can be used to obtain N_{i-1} and so on. The value of N_t is obtained as for virtual population analysis.

Appendix C

The Derivation of the Formulae of Section 3

If ΔN_i, ΔN_{i+1} and ΔC_i denote random errors in N_i, N_{i+1} and C_i that are a result of the sampling errors of the C_i, then making a Taylor expansion of 1.2 gives $\Delta N_i = \Delta C_i \exp\{M/2\} + \Delta N_{i+1} \exp\{M\}$. \quad (C.1)

Squaring and summing gives

$$\text{var}(N_i) = \left(\text{var}(C_i) \exp\{M\}\right) + \left(\text{var}(N_{i+1}) \exp\{2M\}\right) + \left(\text{covar}(C_i, N_{i+1}) \exp\{3M/2\}\right). \quad (C.2)$$

Since C_i and N_{i+1} are unrelated, the covariance term will be equal to zero and C.2 will yield 3.1, that is:

$$\text{var}(N_i) = \left(\text{var}(C_i) \exp\{M\}\right) + \left(\text{var}(N_{i+1}) \exp\{2M\}\right).$$

Dividing both sides of this by N_i^2 gives, after substituting A.1 and A.2,

$$\frac{\text{var}(N_i)}{N_i^2} = \left(\text{var}(C_i) \exp\{M\}\right) \left(\frac{F_i(1 - \exp\{-(F_i+M)\})}{(F_i+M)C_i}\right)^2 + \left(\frac{\text{var}(N_{i+1}) \exp\{2M\}}{N_{i+1} \exp\{F_i+M\}^2}\right) \quad (C.3)$$

Using the same approximation as in Appendix B, this simplifies to:

$$\left(\text{var ratio}(N_i)\right)^2 = \left(\text{var ratio}(C_i)\right)^2 \left(1 - \exp\{-F_i\}\right)^2 + \left(\text{var ratio}(N_{i+1})\right)^2 \left(\exp\{-F_i\}\right)^2$$

which is the form of equation 3.5.

If ΔF_i is the error in F_i due to the sampling errors in the C_i, then expanding 3.3 as a Taylor series gives

approximately: $$\Delta F_i = \frac{\Delta N_i}{N_i} - \frac{\Delta N_{i+1}}{N_{i+1}} \quad (C.4)$$

Squaring and summing this gives:

$$\text{var}(F_i) = \frac{\text{var}(N_i)}{N_i^2} - \frac{2\,\text{covar}(N_i, N_{i+1})}{N_i\,N_{i+1}} + \frac{\text{var}(N_{i+1})}{N_{i+1}^2} \tag{C.5}$$

Multiplying C.1 by ΔN_{i+1} and summing gives:

$$\text{covar}(N_i, N_{i+1}) = \Big(\text{covar}(C_i, N_{i+1})\exp\{M/2\}\Big) + \Big(\text{var}(N_{i+1})\exp\{M\}\Big). \tag{C.6}$$

And since there is no relation between C_i and N_{i+1}, C.6 becomes:

$$\text{covar}(N_i, N_{i+1}) = \text{var}(N_{i+1})\exp\{M\}. \tag{C.7}$$

Substituting C.7 in C.5 gives:

$$\text{var}(F_i) = \frac{\text{var}(N_i)}{N_i^2} - \frac{2\,\text{var}(N_{i+1})\exp\{M\}}{N_i\,N_{i+1}} + \frac{\text{var}(N_{i+1})}{N_{i+1}^2} \tag{C.8}$$

which is of course the same as 3.4. This can be further simplified by dividing both sides by F_i^2 and substituting in A.1, which gives:

$$\left(\text{var ratio}(F_i)\right)^2 = \frac{\left(\text{var ratio}(N_i)\right)^2 + \left(\text{var ratio}(N_{i+1})\right)^2 \left(1 - 2\exp\{-F_i\}\right)}{F_i^2} \tag{C.9}$$

Hence, using 3.5, this gives:

$$\left(\text{var ratio}(F_i)\right)^2 = \frac{\left(1 - \exp\{F_i\}\right)^2 \left[\left(\text{var ratio}(C_i)\right)^2 + \left(\text{var ratio}(N_{i+1})\right)^2\right]}{F_i^2} \tag{C.10}$$

which is the same form as 3.6.

A generalized stock production model

J.J.Pella and P.K.Tomlinson

Bulletin, Inter-American Tropical Tuna Commission, **13,** 421-458, 1969.

A GENERALIZED STOCK PRODUCTION MODEL

by

Jerome J. Pella and Patrick K. Tomlinson[1]

ABSTRACT

A generalization of the Schaefer model is described which allows for skewness of the stock production curve relating production with population size. A fitting scheme is developed by which the stock production curve can be determined for an exploited population using only the catch and effort history of the fishery. Because of the extensive computations required in estimating the parameters, a computer program for use in the calculations is included. Examples are provided which demonstrate the suitability of the model for describing the dynamics of certain fish populations. In particular the catch and effort history for the yellowfin tuna fishery in the eastern Pacific Ocean is analyzed.

ACKNOWLEDGMENTS

We are grateful to Mr. John A. Gulland, Food and Agriculture Organization of the United Nations, Dr. James Joseph, Inter-American Tropical Tuna Commission, and Dr. M. B. Schaefer, Institute of Marine Resources of the University of California, for their comments on the manuscript. We owe Dr. G. J. Paulik, College of Fisheries, University of Washington, a special debt of gratitude for his extensive and constructive review.

INTRODUCTION

Stock production models represent an attempt by fisheries biologists to assess directly the relationship between the sustainable yield from a stock (or population) and the stock size. The earliest analytic approach of this type is due to Graham (1935), who utilized the logistic model to estimate the yield which might be expected from fish stocks of the North Sea. In the estimation procedure he adopted it is necessary that the fishery be stable at least once in its history.

Feller (1940) developed a modified form of the logistic model to describe the growth of some colonies of Infusoria from experiments by Gause. In these experiments, as described by Feller, portions of colonies were being continuously removed from the cultures. He developed estimators of the parameters of the logistic model, assuming the rates of removal were known and that periods occurred when the Infusoria colonies were in equilibrium with the rates of removal.

Schaefer (1954) independently developed essentially the same model as that of Feller (now commonly referred to in fisheries applications as

---

[1]Operations Research Branch, California Department of Fish and Game, Terminal Island, California.

the Schaefer model) and described a technique to estimate the logistic parameters under non-equilibrium exploitation conditions. The information
required is the catch and effort history for the stock, together with an
independent estimate of the catchability coefficient. Schaefer (1957) extended his estimation scheme so that the catchability coefficient could
also be estimated from the catch and effort history.

The logistic assumption is justly criticized, since it results in the curve
of equilibrium yield as related to stock size (hereafter called the stock
production curve) having its maximum at a stock size of exactly one-half
of the theoretical maximum stock size. Ricker (1958) and Schaefer and
Beverton (1963), among others, have remarked that for many fish populations the curve may well be skewed, with the maximum occurring at stock
sizes less than one-half the maximum stock size. These authors have suggested models which would account for this skewness. Neither of the suggestions is accompanied by an estimation scheme using only the catch and
effort history as a data base. The model we now propose permits positive
or negative skewness of the stock production curve, and the estimation
technique permits us to determine the magnitude and direction of skewness
from the catch and effort information alone.

DESCRIPTION OF THE MODEL

We hypothesize that the stock or population under consideration has
at each level P (which may be measured in terms of either numbers or
biomass) a certain potential for growth. If at time t, the population is at
size $P(t)$, then we assume that the instantaneous rate of growth of the
population at time t is

$$\frac{dP(t)}{dt} = HP^m(t) - KP(t) \qquad (1)$$

where H, K, and m are constants. $m \geq 0$. If the population is limited in
growth to an absolute maximum, say P_{max}, it is easy to show that H and K
must be negative if $m > 1$ and positive if $m < 1$. We defer the discussion
of the case when $m = 1$ momentarily. Equation (1) is a special case of
Bernoulli's equation, and has been previously discussed in a different context (Richards 1959; Chapman 1960).

If the population described by (1) is permitted to increase without
external interference, then upon integrating we have

$$P(t) = [P_{max}^{1-m} - (P_{max}^{1-m} - P(0)^{1-m}) e^{-K(1-m)t}]^{\frac{1}{1-m}} \qquad (2)$$

where

$P(0)$ is the population size at time 0 and

$$P_{max} = (\frac{K}{H})^{\frac{1}{m-1}}. \tag{3}$$

On the other hand, if individuals are removed from this population during the interval $(0, t)$, the rate of growth will be altered. If a fishery of $f(t)$ units of effort is operating on the population at time t, the instantaneous catch rate is

$$\frac{dC(t)}{dt} = qf(t)P(t) \tag{4}$$

provided the units of effort operate independently. The constant q is called the catchability coefficient in fisheries literature. With a fishery present the instantaneous rate of growth of the population given by (1) is decreased by the instantaneous catch rate at (4), resulting in a modified growth rate

$$\frac{dP(t)}{dt} = HP^m(t) - KP(t) - qf(t)P(t). \tag{5}$$

We may indicate at this juncture that with $m = 2$, (5) becomes the Schaefer model

$$\frac{dP(t)}{dt} = k_1P(t) \ (L - P(t)) - k_2f(t)P(t) \tag{6}$$

(Schaefer 1954, 1957) where our parameters are defined in terms of his notation as

$$H = -k_1$$
$$K = -k_1L$$
$$q = k_2.$$

If the fishing effort remains constant during the time interval $(0, t)$ the population size at time t is obtained by integrating (5) so as to obtain

$$P(t) = [\frac{H}{K+qf} - (\frac{H}{K+qf} - P(0)^{1-m})e^{-(K+qf)(1-m)t}]^{\frac{1}{1-m}} \tag{7}$$

where f is the constant effort. If $f = (HP(0)^{m-1} - K)/q$, the population will be in equilibrium with the fishery and $P = P(0)$ at any time t. Hereafter we shall drop the variable, time, from the notation to emphasize equilibrium conditions when they occur. If $f < [(HP(0)^{m-1} - K)/q]$, the population will increase, and given t sufficiently large re-establish a new equilibrium at

$$P = (\frac{qf + K}{H})^{\frac{1}{m-1}}.$$

If $[(HP(0)^{m-1} - K)/q] < f < \infty$ and $0 < m < 1$, the population will decline and given t sufficiently large re-establish itself in equilibrium at a positive population size given by the equation above. The implication here is that the population cannot be fished to extinction. If $[(HP(0)^{m-1} - K)/q] < f < (-K/q)$, and $m > 1$, the population again given sufficient time will decline to a positive equilibrium population size given by the equation above, but if $f \geq (-K/q)$, the population will eventually be fished to extinction. Practically speaking, we have little confidence in prognostications of survival or extinction of a population under extreme exploitation, but this does not diminish the utility of the model.

If the fishery is of such a magnitude as to exactly, remove the production by the population, the equilibrium catch per unit of time, say 1 year, is

$$C = HP^m - KP = qfP. \tag{8}$$

We observe from (8) that the equilibrium annual catch, C, is obtained by the equilibrium effort which maintains the population in equilibrium at size P.

Our generalized production model results in a family of stock production curves indexed by the parameter m (Figure 1). The previous objection to the Schaefer model on the basis of skewness of the stock production curve is easily overcome by permitting $m < 2$.

Figure 1 includes the curve for $m = 1$. The curve plotted there was obtained by interpolating between the curves for $m = 0.999$ and $m = 1.001$, which for practical purposes were indistinguishable. When $m \to 1$ and P_{max} is finite, (1) can be shown to correspond to the Gompertz growth model (see Richards 1959). The coefficients H and K in this case become equal and infinitely great. It shall be obvious from the discussion on fitting of the model that this difficulty is circumvented. Clearly we cannot allow $m = 1$ but may allow $m \to 1$.

The relation between equilibrium yield and effort for any choice of m can be obtained from (8) and is

$$C = qf \left(\frac{qf + K}{H}\right)^{\frac{1}{m-1}} \tag{9}$$

These curves are skewed in the same direction of the stock production curves for any m (Figure 2). If the stock production curve is positively skewed, as has been suggested, fishing beyond the maximum of the yield curve at (9) will result in a less pronounced decline in yield than would be predicted by the Schaefer model or models with $m > 2$. As m decreases to 1, the curves intersect the abscissa at $f = 0$ and $f = -K/q$. With $0 < m < 1$ the curves rise to a maximum and then decline with increasing effort, becoming asymptotic to the abscissa. With $0 < m < 2$ the curves have an inflection point to the right of the maximum at

$$f = \frac{2K(1 - m)}{qm}.$$

When $m = 0$, the curve is asymptotic to the maximum equilibrium catch.

The equilibrium catch per unit of effort, as related to effort, is obtained from (9) by dividing by f to obtain

$$U = q(\frac{qf + K}{H})^{\frac{1}{m-1}} \tag{10}$$

These curves are plotted for a variety of m values (Figure 3). Some plots of catch per unit of effort against effort in fisheries literature (*e.g.* FAO, 1968) are suggestive of these curves with $m < 2$. The curvature of these lines, even with values of m substantially less than 2, is slight even over broad ranges of effort, so that it is quite difficult to reject the logistic assumption from such plots by a test for linearity unless m is greatly different from 2, effort has been observed over a wide range, or large numbers of observations are available. Even then simple plots of catch per unit of effort against effort are not strictly appropriate in attempting to discriminate between different m values since the curves at (10) are under equilibrium conditions. Attempts to adjust the data to equilibrium conditions are generally unsatisfactory for reasons which will later be demonstrated.

Several characteristics of the population model at (5) which are of particular interest in management of a fishery are the maximum of the equilibrium yield curve (C_{max}), the population size at which this catch can be taken (P_{opt}), and the fishing effort required to maintain the population at this level (F_{opt}). In terms of the constants of the model these parameters are as follows:

$$C_{max} = H(\frac{K}{mH})^{\frac{m}{m-1}} - K(\frac{K}{mH})^{\frac{1}{m-1}} \tag{11}$$

$$P_{opt} = (\frac{K}{mH})^{\frac{1}{m-1}} \tag{12}$$

$$F_{opt} = \frac{K(1 - m)}{mq} . \tag{13}$$

Before commencing with the estimation section, we wish to qualify the conditions under which (5) might validly approximate the dynamics of an exploited fish population. Some of this discussion stems from earlier comments by Schaefer (1957) and Schaefer and Beverton (1963).

ASSUMPTIONS
Random variation in production and catching rates

We have delayed the discussion of the problem of random variation in the production and catching rates in (5) to this point. At the present there is little to offer concerning the stochastic nature of the population under exploitation. We recognize the process is not deterministic as we have so far represented it, so that perhaps a more realistic representation at (5), say, would be

$$\frac{dP(t)}{dt} = \eta_1(t)\,[HP^m(t) - KP(t)] - \eta_2(t)\,qf(t)P(t) \qquad (14)$$

where $\eta_1\,(t)$ and $\eta_2\,(t)$ are time-varying random variables. The variable η_1 represents random variation in the rate of production from the stock due to changes in recruitment, growth, and natural mortality caused by the environment. Variation in the catching rate due to random changes in availability and catchability is represented by the variable η_2. It is assumed that these variables are distributed so that (8), (9), and (10) represent regression curves, and (5) describes the rate of population change under average environmental and fishing conditions. These sweeping assumptions are necessitated by the difficulty encountered in attempting to develop stochastic analogues of (1). Complete results are available only for the case $m = 0$. In this instance, Ahuja and Nash (1965) have developed a stochastic analogue of (1) in terms of a discrete population growth process, derived expressions for the mean and variance of the process, and found maximum likelihood estimators of the parameters for the situation where the population can be counted at points in time while it grows toward the maximum. Some studies on stochastic analogues of the logistic model are available (Feller 1939; Kendall 1949; Bartlett 1957; and Leslie 1958). These investigations demonstrate the difficulty of working with probability analogues of (1) with $m \neq 0$ since neither of the moments of the logistic process nor probabilistic estimators of the parameters could be determined. The situation would not be simplified by adding dynamic exploitation to the problem.

Equilibrium size and age structure

The population is assumed to tend to a stable size and age distribution at each level of fishing effort. Thus at each level of effort there is an associated equilibrium population size with a corresponding stable age distribution. During transition periods between changes in fishing effort and concomitant changes in population size, it is assumed that the age structure of the population continuously adjusts to the stable age distribution corresponding to the population levels encountered. We then interpret the differential equation at (5) as describing the growth of the population under average environmental and fishing conditions when the age distribution continuously assumes the stable age distribution with changing population size.

In reality, during transition periods the actual age distribution will lag the stable age distribution in time as the population size changes and hence (5) will not exactly describe the rate of change of population size under the average conditions. However, if the changes in the fishery are gradual, the transient age distribution at any moment should not differ greatly from the associated stable age distribution so that (5) will give a good approximation to the actual rate of change. Large and rapid changes in effort will result in (5) providing a poorer representation of the actual rate of population change. Species with relatively short generation times should most closely meet these conditions concerning population size and age structure.

Other time lags

It is assumed that the population size at any instant is the primary determinant of the potential for increase of the population at that instant. Reduced population sizes are favorable for increase, while large population sizes are less favorable. At low population levels the growth rate of individuals may be accelerated, the natural mortality rate may be reduced, and perhaps reproductive success is improved. At high levels, competition for limited quantities of the species' niche can affect some or all of these components of the rate of population growth so as to reduce the potential for increase. The model requires that these primary factors of individual growth, natural mortality, and reproduction respond immediately to changes in the population size, whereas in reality these responses may well be lagged. Reproduction, for instance, is necessarily lagged, and whatever the relationship among population size, age distribution, and reproduction might be for the stock, the reproduction during transition periods should deviate from that at stability. However, if the changes are gradual, this discrepancy is diminished.

Closed population

The population must be a distinct self-sustaining unit. Significant changes in the population size caused by persistent immigration and/or emigration unrelated to the population size preclude the use of this model to describe the dynamics of the population.

Constant catchability

Each unit of effort expended should remove, on the average, a constant fraction of the population. When a variable portion of the population is subject to exploitation during the history of the fishery, this assumption will be violated. A developing fishery, expanding over the area inhabited by a population that mixes slowly through its range provides a likely example. Temporal changes in gear efficiency such as have been observed in the yellowfin fishery (Pella 1969) also violate the assumption unless the effort is adjusted to a reference level of efficiency.

PARAMETER ESTIMATION

We introduce in this paper a technique for estimating the parameters of the production model from the catch and effort history of the stock.

We begin by indicating the motivation for our method by describing some techniques based on a linearization approach currently employed for the special case of the Schaefer model. After indicating certain shortcomings of these methods, we describe our technique, which avoids some of the difficulties in the linearization methods.

Linearization approach

Were we to follow the approach taken by Schaefer (1954, 1957) we would develop a linear model from (5) by integrating over an interval of time, say 1 year, to obtain

$$\Delta P = C_e - C \tag{15}$$

$$\Delta P = \int_{P(0)}^{P(1)} dP(t) \tag{16}$$

$$C_e = H \int_0^1 P^m(t)\,dt - K \int_0^1 P(t)\,dt \tag{17}$$

$$C = qf \int_0^1 P(t)\,dt . \tag{18}$$

We have assumed fishing effort is constant during the year. Equation (15) is a mathematical formulation of the obvious fact that the change in stock size during the year, ΔP, is the difference between the equilibrium catch, C_e, and C, the actual catch.

Further paralleling the development by Schaefer we would set

$$C_e = H\bar{P}^m - K\bar{P} \tag{19}$$

where

$$\bar{P} = \int_0^1 P(t)\,dt . $$

A minor objection can be raised here in that equation (19) is only an approximation since in general

$$\int_0^1 P^m(t)\,dt \neq \left[\int_0^1 P(t)\,dt \right]^m . \tag{20}$$

With $m > 1$ it can be shown by Hölder's inequality that

$$\int_0^1 P^m(t)\,dt > \left[\int_0^1 P(t)\,dt \right]^m \tag{21}$$

except when the stock is in equilibrium. Under equilibrium conditions the two sides of the inequality at (21) are equal.

Therefore, for the Schaefer model

$$\Delta P = C_e - C - B \qquad (22)$$

where C_e is given at (19) with $m = 2$ and B, the resultant difference between the equilibrium catch as calculated at (19), minus that given by (17), is nonnegative with its magnitude depending on H, K, q, f, and $P(0)$.

We would then return to (4) and integrate over a year to obtain the catch

$$C = qf\bar{P} \qquad (23)$$

$$C = \int_0^1 dC(t) .$$

We let $\bar{U} = C/f = q\bar{P}$. $\bar{U}$ is called the annual catch per unit of effort, and is directly proportional to the average stock size during the year.

We may then estimate $\bar{P}$ by

$$\hat{\bar{P}} = \bar{U}/q \qquad (24)$$

and rewrite (15) as

$$\Delta P = H(\bar{U}/q)^m - K(\bar{U}/q) - C \qquad (25)$$

in which the variables $\bar{U}$ and C are observable. The change in stock size in (25) remains to be estimated.

Schaefer suggested that the stock size at the beginning of year i, say $P_i(0)$, might be estimated by linearly interpolating between the average stock sizes in years $i-1$ and i, viz.,

$$\hat{P}_i(0) = (\bar{P}_{i-1} + \bar{P}_i)/2 . \qquad (26)$$

The stock size at the end of the year would be estimated by

$$\hat{P}_i(1) = (\bar{P}_i + \bar{P}_{i+1})/2 . \qquad (27)$$

The difference between these two estimates can be used to estimate the change in stock size during the year

$$\Delta\hat{P}_i = (\bar{P}_{i+1} - \bar{P}_{i-1})/2 . \qquad (28)$$

In terms of the observable catch per effort in these years we would estimate the change in stock size as

$$\Delta\hat{P}_i = (\bar{U}_{i+1} - \bar{U}_{i-1})/2q = \Delta U_i/q . \qquad (29)$$

The substitution of this latter expression into (25) for the change in stock size, the addition of appropriate subscripts to identify the year, and some manipulation provides us with the linear model

$$\Delta U_i = \frac{H}{q^{m-1}} \bar{U}_i{}^m - K\bar{U}_i - qC_i \ . \tag{30}$$

Up to this stage of the development in this section we have made the implicit assumption that the population growth process and catching process are deterministic. The more feasible version of the model at (14) requires that when we form the linear expression at (15) we include an error term

$$\gamma = \int_0^1 [\eta_1(t) - 1] \ [HP^m(t) - KP(t)] \ dt - \int_0^1 [\eta_2(t) - 1]qfP(t)dt \ .$$

Additional error in the expression (30) is due to the assumptions at (19) and (29). We shall represent all these sources, some random and some not random, by a single error term ϵ and write (30) as

$$\Delta U_i = \frac{H}{q^{m-1}} \bar{U}_i{}^m - K\bar{U}_i - qC_i + \epsilon_i \ . \tag{31}$$

Equation (31) can now be used to develop a variety of estimation schemes for *fixed m*. The first technique one might use is least squares. The model is linear, and hence usual regression methods could be used to estimate the coefficients, *i.e.*, if we let

$$Y_i = \Delta U_i$$
$$X_{1i} = \bar{U}_i{}^m$$
$$X_{2i} = \bar{U}_i \qquad\qquad \beta_1 = H\!\!\Big/\!_{q^{m-1}}$$
$$X_{3i} = C_i \qquad\qquad \beta_2 = -K$$
$$\beta_3 = -q$$

then (31) can be written in the more familiar form

$$Y_i = \beta_1 X_{1i} + \beta_2 X_{2i} + \beta_3 X_{3i} + \epsilon_i \ . \tag{32}$$

Given the estimates of the β_i, $i = 1, 2, 3$, and m, we can obtain the corresponding estimates of H, K, and q.

On the other hand, we could continue to follow Schaefer's development by dividing (31) by $\bar{U}_i$ to obtain

$$\frac{\Delta U_i}{\bar{U}_i} = \frac{H}{q^{m-1}} \bar{U}_i{}^{m-1} - K - qf_i + \epsilon'_i \ . \tag{33}$$

If our data consist of the annual catches and efforts observed for a series of n years, we might partition our observations into two subsets, placing the data for the first $n*$ years in one subset and the remaining observations in the other. Schaefer nowhere to our knowledge suggests a criterion for partitioning, and thus the scheme lacks uniqueness in the sense that two individuals analyzing the same set of data may partition the observations differently and so obtain different estimates for the parameters.

313

As we point out in the sequel, the data should be partitioned into subsets corresponding to intervals of years over which population change is thought to be great.

Once the data are partitioned a system of three linear equations in three unknowns may be formed:

$$\sum_{i=1}^{n*} \frac{\Delta U_i}{U_i} = \frac{H}{q^{m-1}} \sum_{i=1}^{n*} U_i^{m-1} - n*K - q \sum_{i=1}^{n*} f_i$$

$$\sum_{i=n*+1}^{n} \frac{\Delta U_i}{U_i} = \frac{H}{q^{m-1}} \sum_{i=n*+1}^{n} U_i^{m-1} - (n - n*)K - q \sum_{i=n*+1}^{n} f_i \qquad (34)$$

$$\sum_{i=1}^{n} \left| \frac{\Delta U_i}{U_i} \right| = \sum_{[i:\Delta U_i > 0]} \left[\frac{H}{q^{m-1}} U_i^{m-1} - K - qf_i \right] -$$

$$\sum_{[i:\Delta U_i < 0]} \left[\frac{H}{q^{m-1}} U_i^{m-1} - K - qf_i \right]$$

where we have arbitrarily set

$$\sum_{i=1}^{n*} \epsilon'_i = \sum_{i=n*+1}^{n} \epsilon'_i = 0$$

$$\sum_{[i:\Delta U_i > 0]} \epsilon'_i - \sum_{[i:\Delta U_i < 0]} \epsilon'_i = 0$$

which may be reasonable approximations. The system is easily solved for fixed m and the estimates of H, K, and q are again readily obtained.

To this juncture two schemes have been described which might be used to estimate the parameters of our model for fixed m. It is straightforward to develop others based on this linearization tack, but these will suffice.

In addition to our previous minor criticisms, the linearization approach as presented here depends on estimates of changes in stock sizes which can be shown to be quite seriously in error under the assumed system of a stock under exploitation defined at (5). If we approximate the differential equation at (5) by a finite difference equation, we may compute the stock size at discrete instants by

$$P(t + \Delta t) = P(t) + [HP^m(t) - KP(t) - qf(t)P(t)]\Delta t \qquad (35)$$

and the catch resulting from the fishing between t and $t + \Delta t$ by

$$C(t, t + \Delta t) = qf(t)P(t)\Delta t. \tag{36}$$

Given values for H, K, m, q, the initial population size $P(0)$, and the effort function, $f(t)$, a catch history can be computed using these two equations. Pella (1967) has computed such a catch history for the Schaefer model. He set

$$
\begin{aligned}
m &= 2 \\
H &= -8.814 \times 10^{-3} \\
K &= -2.600 \\
q &= 3.8 \times 10^{-2} \\
P(0) &= 295 \\
\Delta t &= 0.1
\end{aligned}
$$

These parameters, with the exception of Δt, were chosen to correspond to the estimates obtained for the Schaefer model for the yellowfin tuna fishery in the eastern Pacific Ocean (Schaefer 1957). The stock is measured in millions of pounds and the effort in thousands of boat days (standardized to bait vessels of 201 to 300 short tons capacity). The approximation by the finite difference equation can be shown to be excellent through a comparison with the integrated analogues of (35) and (36). In general, it is not possible to obtain the integrated form corresponding to (36). The effort function was taken to be a step function, changing annually, but remaining constant within years. The effort used was that observed in the yellowfin tuna fishery for 1935 through 1964 (Table 6). The statistics generated by this simulation are presented in Table 1. The interesting point here is the comparison between the observed changes in population size and the estimates obtained by linear interpolation with the catchability coefficient known (Table 1; Figure 4). Clearly the estimates in the situation described are poor, generally underestimating the true changes. The absolute values of the errors are generally only slightly smaller than the corresponding absolute values of the stock changes themselves. It is easy to demonstrate graphically how these errors occur (Figure 5). This plot illustrates the stock size under a fishing effort history of overfishing, followed by underfishing, succeeded by overfishing. By overfishing and underfishing we simply mean the fishing effort level is such that the stock will decrease and increase, respectively. We observe that by linearly interpolating between $\bar{P}_1$ and $\bar{P}_2$ we have an estimate of $P(1)$ with a large positive error. The interpolation between $\bar{P}_2$ and $\bar{P}_3$ results in an estimate of $P(2)$ with a large negative error. We would estimate $P(2) - P(1)$, a large positive change by $\hat{P}(2) - \hat{P}(1)$, a small negative change. The opposite fishing pattern — underfishing, overfishing, underfishing — produces the same result.

The catch and effort history for the yellowfin tuna fishery substantiates our claim that the linear interpolation formula does not estimate population changes in this fishery satisfactorily. We draw this conclusion from the following points which will immediately be demonstrated for the yellowfin fishery from the data of 1934 through 1965 (Table 6): (1) the linear regression model (31) explains little of the variability in ΔU (or equivalently scaled estimates of population changes); (2) the Schaefer model with parameters estimated by the Schaefer method (34) explains over 80 percent of the variation in the catches; and (3) the Schaefer method results in much cancellation of errors of estimates of population change.

When we fit the linear regression equation (31) by least squares to the catch and effort data of the yellowfin tuna fishery for say 1934 through 1965, we find that less than 11 percent of the total sum of squares can be accounted for by the model. The fit is so poor one might reasonably argue that the logistic model is inappropriate. (Certain parameter estimates obtained by this least squares fit with $m = 2$ are presented in Table 2.) However, it is patent that the errors in estimating stock changes contribute a great deal to the residual errors of the fitted model. For if we assume the initial stock size at the beginning of 1934 was approximately proportional to the catch per unit of effort of that year, set the constants of the Schaefer model equal to the estimates obtained by use of the Schaefer method (Table 2) for this fishery, and compute the catches predicted by the model as was done in computing the statistics of Table 1, we find that over 80 percent of the variation in the catches can be accounted for by the model, $i.e.,$

$$R = \frac{\sum_i (C_i - \bar{C})^2 - \sum_i (C_i - \hat{C}_i)^2}{\sum_i (C_i - \bar{C})^2} > .80 \qquad (37)$$

where $\hat{C}_i$ = predicted catch for year i
C_i = actual catch for year i

and the summation extends from 1935 through 1964. We compute R from 1935 through 1964 rather than from 1934 through 1965 since the least squares technique fitted these years, $i.e.,$ due to estimating the change in population size by the interpolation formula, the 2 years at the extremes are lost.

The statistic R here is an analogue of the correlation coefficient of regression theory. It is a measure of the improvement in the fit of the catch history by the Schaefer model over the arithmetic mean. So if the Schaefer model fits the catches no better than the mean, $R = 0$. If the model fits the catches perfectly, $R = 1$. If the mean gives a better fit, $R < 0$. Certainly the Schaefer model fits the catch history of the yellowfin fishery remarkably well when viewed in this respect.

The Schaefer estimation scheme partially circumvents the errors in estimating changes in population size during individual years by forming sums of weighted estimates of population changes over periods of years, *viz.*, over n^* and n-n^* years in (34). A great deal of cancellation takes place in this summing process. If we form an unweighted sum of estimates of population change over k years to estimate the total change over those years, ΔP total (k), we have

$$\hat{\Delta P} \text{ total } (k) = \sum_{i=1}^{k} \hat{\Delta P}_i = (\frac{\bar{U}_k + \bar{U}_{k+1}}{2q}) - (\frac{U_0 + U_1}{2q}) \qquad (38)$$

where all the intermediate terms cancel. We therefore estimate the change in stock size over the k years by the difference of the means of the estimates of mean population sizes during the last 2 years and the first 2 years of $k + 2$ years of data. If the change is large over the k years, the error in estimating stock change as above will be relatively small as compared to the actual stock change, whereas the relative errors for the more erratic and smaller changes during individual years may be quite large. A cursory examination of the yellowfin data indicates a large change in population size has occurred since 1934, but that the decline has often been interrupted by periods of growth. The weighting by the inverse of the catch per effort used by Schaefer makes our cancellation of errors argument less cogent, but undoubtedly much of the error in the individual estimates cancels. The terms which cancel in (38) are separated by a single term. In the weighted estimate the denominators of the corresponding terms are the catches per effort in alternating years. Provided the catch per unit of effort in these years is not greatly dissimilar, the cancelling should occur.

So far our results are evincive of the superiority of the Schaefer method over the least squares technique. We would certainly suggest to anyone who wishes to use his technique that the data be partitioned at points in time between which large stock changes are thought to have occurred. If such a partitioning is possible, it should eliminate the absurd estimates reported by Pella (1967) by using the Schaefer method and by Southward (1968) who used a slight variation of the method.

Pella (1967) developed some additional techniques as variants of the linearization approach. In computing the estimates of the parameters for the yellowfin fishery using the data for 1934 through 1965, he obtained values for the parameters ranging between those of Table 2. Some of the estimates could not be considered as inferior to those obtained by the Schaefer method on the basis that they were unrealistic, as was done for the least squares estimates. Thus a variety of estimates is possible from any set of catch and effort data by the linearization approach.

We propose now a criterion by which estimates of the parameters by different techniques can be compared for goodness of fit of the catch and effort data, and we develop a scheme to determine the parameter estimates which are best as defined in our sense. The estimation technique will further permit us to estimate m, which until now has been assumed fixed by the investigator. With an estimate of m we will have an idea of the degree and direction of skewness of the stock production curve.

Minimum S—criterion

We now propose a reasonable measure of the goodness of fit of our production model to a set of catch and effort data based on the catch and effort observations themselves. The introduction of extraneous error to the observations through attempting to estimate stock changes is avoided. We shall presume that the fishery has been under observation for some length of time during which the catch and effort are known for each of n subintervals of duration Δt_i, $i = 1, \ldots n$. The subintervals, of course, need not be equal. Suppose that the effort sequence $[E] = (E_1, E_2, \ldots, E_n)$ is observed during the study. Here E_i is the actual number of units of effort expended during the ith time interval. Our earlier discussion dealt with intervals of unit length, but now to permit variable time intervals we introduce the E notation. Presuming the effort to be constant during the ith interval, we would convert to the earlier effort measure by scaling the E_i by the inverse of Δt_i, viz., $f_i = E_i/\Delta t_i$. Let $[C] = (C_1, C_2, \ldots, C_n)$ be the corresponding sequence of observed catches. Let $[C(H, K, q, m, P(0))] = (\hat{C}_1, \hat{C}_2, \ldots, \hat{C}_n)$ be the sequence of catches predicted by the integration of the differential equation (4) with $P(t)$ given by (7) using the parameters H, K, q, and m when the initial stock size at the beginning of the first time interval was $P(0)$ and the effort sequence $[E]$. Let $\beta = [(H, K, q, m, P(0)]$. If

$$S(H, K, q, m, P(0)) = \sum_{i=1}^{n} (C_i - \hat{C}_i)^2 \qquad (39)$$

then $(\hat{H}, \hat{K}, \hat{q}, \hat{m}, \hat{P(0)})$ is here defined as the best estimate of the generalized production model parameters provided that for all feasible points in β, S is minimal at $(\hat{H}, \hat{K}, \hat{q}, \hat{m}, \hat{P(0)})$. We remark in passing that

$$R = \frac{\sum_{i=1}^{n} (C_i - \bar{C})^2 - \sum_{i=1}^{n} (C_i - \hat{C}_i)^2}{\sum_{i=1}^{n} (C_i - \bar{C})^2} \qquad (40)$$

is maximized at the best estimate. Hence, we have chosen as the best

318

estimate of the generalized production model parameters that point in the parameter space by which the maximum proportion of the variation in the catches is explained by the model in our specially defined sense.

The surface S in the space β cannot in general be expressed in algebraic form since, as we mentioned earlier, an explicit expression for the integral of (4) with $P(t)$ given by (7) cannot generally be found. We can, as we shall presently show, obtain a good numerical approximation of the integral. Therefore, in order to obtain the best estimate of the parameters of the model we must conduct a search over the surface S in the space β by numerically approximating S at points selected in such a manner as to lead us to the minimizing point. The searching routine we are about to describe is subject to the usual pitfalls in examining portions of a surface to locate an extremum. The routine may terminate its search at a relative minimum if the surface has relative minima in addition to the absolute minimum. Therefore estimates generated by the procedure should be treated circumspectly. In our experience with several sets of fisheries data, the surface was well-behaved with a clear minimum and no relative minima.

Searching routine

The computations required to search over S are so lengthy as to make the method impractical without the availability of a high-speed electronic computer. We now describe a computer program, hereafter referred to as GENPROD, which will locate, for fixed m, the minimizing point in a subspace of β, $\beta^* = [(H, K, q, P(0))]$, with as much precision as desired. The computer program permits the investigator to vary m over a likely range of values and in turn find $\hat{m}$ with as much precision as he deems necessary, thus finding the best estimate in β.

We have actually written the program to search over a transformation of β^*, $\beta^{**} = [(F_{opt}, q, r, U_{max})]$, for fixed m. The transformation from β^* to β^{**} is defined as follows:

$$F_{opt} = \frac{K(1 - m)}{mq}$$

$$q = q$$

$$r = P(0)/P_{max} = P(0)/(K/H)^{\frac{1}{m-1}} \tag{41}$$

$$U_{max} = q(K/H)^{\frac{1}{m-1}} .$$

Here F_{opt}, as defined earlier, is the effort required to maintain the population at its optimum size and to harvest the maximum sustainable yield. The parameter q is of course the catchability coefficient. The parameter

r is the ratio of the stock size at the time when the fishery first comes under observation to the maximum stock size. U_{max} is the maximum catch per unit of effort which would be observed when the stock is at its maximum. The inverse transformation is as follows:

$$H = (\frac{m}{1-m})\ (\frac{q}{U_{max}})^{m-1}\ qF_{opt}$$

$$K = (\frac{m}{1-m})qF_{opt} \tag{42}$$

$$q = q$$

$$P(0) = rU_{max}/q.$$

The transformation is one-to-one if only feasible parameter values are permitted. Clearly, searching through β^* or β^{**} for the minimum of S for fixed m is equivalent in the sense that once the minimizing point is located in either subspace, the minimizing point in the other is determined by the appropriate transformation. Our purpose in making the transformation is to facilitate the guessing of initial values for the model parameters which are required to begin the search. The investigator undoubtedly will find his intuition stronger in β^{**} than β^*.

Program GENPROD begins by computing H, K, and $P(0)$ from guesses of F_{opt}, q, r, and U_{max} for a given m. The program then computes the predicted catches, $[\hat{C}]$, by approximating the integration of equation (4) with $P(t)$ given by (7). Each of the n time intervals is partitioned into N subintervals (the length of the subintervals in the ith time interval is $\Delta t_i/N$). The user specifies the value of N. The predicted catch during the ith interval is approximated by

$$\hat{C}_i = \int_{t_{i-1}}^{t_i} qf_i\, P(t)\ dt \cong qf_i \sum_{j=1}^{N} \frac{1}{2}(P_{ij} + P_{ij+1})\ \frac{\Delta t_i}{N}$$

where P_{ij} is the predicted population size at the beginning of the jth subinterval in the ith time interval and $f_i = E_i/\Delta t_i$ is the fishing effort per unit time. The P_{ij} are computed by equation (7). The f_i are computed internally in GENPROD. *The user provides the observed catches $[C]$, the observed efforts $[E]$, and the lengths of the time periods $[\Delta t]$.* For most situations, the change in population size during any time interval can be reasonably represented by a straight line and N is set equal to 1. The error of estimation is reduced by increasing N.

GENPROD now begins to modify the guesses and search for the best estimates in β^{**}. Each time the guesses are modified, $[\hat{C}]$ is recalculated until the minimum S is found. It is necessary for the user to provide one or more values of m and initial guesses of F_{opt}, q, r, and U_{max} when using GENPROD. GENPROD further requires the user to provide step intervals for F_{opt}, q, r, and U_{max} which are here designated as ΔF, Δq, Δr, and ΔU, respectively, and the user must supply lower and upper bounds for each of the four parameters. The step intervals are used in the search across the range of possible solutions for each parameter as next described.

The searching technique was originally programmed by Beisinger and Bell (1963). The initial guess supplied by the user is taken as a base point, (F_{opt}, q, r, U_{max}), for searching, and the routine begins an exploratory phase. It evaluates S at the point $(F_{opt} + \Delta F, q, r, U_{max})$ and if S at $(F_{opt} + \Delta F, q, r, U_{max})$ is less than S at $(F_{op.}, q, r, U_{max})$, then $(F_{opt} + \Delta F, q, r, U_{max})$ is taken as a new base point and the routine explores in the q direction. On the other hand, if S is less at the initial base point, the routine tries the point $(F_{opt} - \Delta F, q, r, U_{max})$, and if S is less at this choice than at (F_{opt}, q, r, U_{max}), then $(F_{opt} - \Delta F, q, r, U_{max})$ is taken as the new base point; if not, it retains the old base point, and in either event, begins exploring in the q direction. In a similar manner each of the remaining axes of the parameter space corresponding to r and U_{max} is explored and at the termination of this exploration a check is made to see if the base point has been changed. If the base point has changed, a pattern phase is entered; if not, the routine re-enters the exploratory phase with the step intervals for each parameter divided by 10. In the pattern phase the routine tries a pattern step which is a combination of all the successful moves in the previous exploratory phase. This pattern step is repeated until it no longer results in a decrease in the value of the function S, at which time the routine returns, with the step intervals for the parameters unchanged, to the exploratory phase. The routine is halted when the exploratory phase, with the step intervals divided by 10^{KK} (KK is specified by the user) fails to change the latest base point. During the entire process, whenever a step in the search moves outside the lower or upper bounds specified by the user, a test senses this and the trial point is drawn back to the boundary. The best estimates of the parameters in β^{**} are those corresponding to the minimum value of S found.

Since guesses are required when using GENPROD, some hints for evaluating these are appropriate. The general situation is depicted as one in which data (catch and effort) are distributed over a range of stock sizes, including the optimum. The technique suggested is to choose F_{opt} equal to the mean of the observed efforts; choose U_{max} equal to the maximum observed catch per unit of effort; choose P_{max} equal to 4 times the maximum observed catch; and set q equal to U_{max} divided by P_{max}; choose r equal to 0.8. The lower bounds of F_{opt} and U_{max} are set at $1/10$ of the guesses and the upper bounds are set at 10 times the guesses. The bounds of q should be more liberal, say $1/100$ and 100 times. The bounds of r are obviously 0 and 1. The values for the step intervals are simply set equal to the guesses. These guesses are based on the assumption that all $\Delta t_i = 1$.

Of course, if it is known that the catch-effort data were obtained from a segment of the range of stock sizes, or some $\Delta t_i \neq 1$, then the guessing process must be modified. If serious doubt exists, one should make guesses as suggested and at the same time set very wide bounds and utilize relatively large step intervals for a quick search across the range. If any of the final estimates equal a bound, the data should be rerun with wider bounds if the user does not consider solutions beyond the bounds as infeasible. As for the values of m, it is appropriate to begin the search by choosing a range of values greater than 0 but lesss than 4 ($m = 1$ must be excluded). If little is known about the shape of the production curve, try 0.4, 0.8, 1.2, 1.6, 2.0, 2.4, 2.8, and 3.2 for a first run, then try additional values when the approximate range is determined by examining the values of S.

Several factors will influence the amount of computer time necessary to find a solution. The two most important are the values of N and KK, since each of these automatically increases the amount of computation as they become larger. Doubling N will roughly double computation. Increasing KK decreases the rate at which the search occurs and unless the search is near the estimate, convergence can be very slow. It would be advisable on initial runs, to set $N = 1$, $KK = 2$ and use values of the step intervals which will search across the range with few steps. It should be pointed out that if independent information exists about one or more of the parameters, this can be incorporated by setting very narrow bounds. This will speed up the computation time. A CDC 1604 computer used 2 minutes, 14 seconds execution time with $N = 1$, $KK = 3$, and $n = 34$ for six values of m.

The instructions for setting up the input data cards are given in the Appendix. The computer output from GENPROD is given in Table 7, which relates the program terminology to the terminology in this paper. The output should be relatively self-explanatory, except the times (AT TIME) given are the times at the end of the interval and the population sizes given are at the end of each interval. The effort and catch are those during the interval.

Data for Fitting

Assuming the model is appropriate for a given species, the only data necessary to estimate the parameters are catch, effort, and time. For a given total catch over some length of time, the data must be presented in terms of intervals of time. The usual case would be catch by year for some number of years. In each interval of time, the number of time units (Δt_i), the catch (C_i), and the effort (E_i) must be available. The values of C_i and E_i may be zero, such as in the case of closed seasons. In fisheries applications, the usual assumption would be that the growth of the stock in biomass could be approximated by the model. Therefore, the catch should be in terms of weight. The unit of effort is arbitrary, but it must be standardized to conform to the assumption that q is constant. Since the model has five parameters, at least six different population sizes must be represented by the catch-effort data. Of course if one is willing to assume a value for m, the numbers of parameters and required data points are reduced by one.

At times, the catch and effort for each time interval are not directly available, but can be estimated from other sources of data. It is straightforward to estimate the effort (E_i) if the catches (C_i) and average catch per unit of effort ($\overline{U}_i$) are known. For some fisheries, the instantaneous fishing mortality is estimable without knowledge of effort or catch per unit of effort (*e.g.* Murphy 1966). The efforts are estimated as being proportional to the instantaneous fishing mortalities.

EXAMPLE PROBLEMS

Examples for stocks of two species of fish have been worked out to explore the use of GENPROD. The first example comes from experiments on exploiting guppies (*Lebistes reticulatus*) by Silliman and Gutsell (1958). The guppy data are particularly useful in exposing deficiencies of the model and showing the utility of the fitting scheme. The second example comes from the catch statistics (Bayliff 1967) of the yellowfin tuna (*Thunnus albacares*) in the eastern Pacific Ocean. An attempt will be made in the latter case to establish why the predicted catches sometimes vary substantially from the observed catches.

Guppies

Silliman and Gutsell (1958) maintained four guppy populations under similar ecological conditions, except two of the populations (A and B) were exposed to varying fishing rates while the other two (C and D) were not fished. The fishing technique used on the guppies involved systematic removal of every n^{th} fish at points in time separated by 3-week intervals, with n depending on the fishing rate. The removal rates for numbers of fish were 10, 25, 50, and 75 percent each 3-week period. The harvesting scheme would imply that every exploitable fish had an equal chance of being in the catch at a time when removals were made, and therefore approximately every n^{th} gram in the population should have entered the catch.

Silliman and Gutsell artificially constructed an effort history for the guppy experiments by assuming the percent removals in each triweekly period were proportional to the effort measures. They equated the 10-percent rate to 1 unit of effort, the 25-percent rate to 2.5 units of effort, the 50-percent rate to 5 units of effort, and the 75-percent rate to 7.5 units of effort. This would imply that $q = 0.1$. Since we decided to fit the model in terms of biomass rather than numbers, we computed the actual percent removals in weight for each time period. Effort values were constructed along the lines of Silliman and Gutsell by scaling the percent removals in weight by one tenth, implying again that $q = 0.1$. Hence the catches $[C]$ we used correspond exactly to those in Silliman and Gutsell's Table 8. The efforts $[E]$ we used correspond closely to those given in that same table, but we actually computed them from their Table 4.

Since the removals were made at discrete points in time separated by 3-week intervals, we have attempted to affect the harvesting process by dividing each time interval into closed and open fishing periods. We assumed that the guppies were exposed to fishing during only the last one hundredth of each triweekly period. During the closed fishing period of each triweekly interval the catch and effort were set equal to 0. The catches $[C]$ actually used in the fitting of the model then consist of those reported by Silliman and Gutsell, each preceded by a 0 catch. There is similarly an augmented effort history $[E]$. A more exact fitting of the guppy data would necessarily treat the removals as true point removals. Our purpose here, however, is to demonstrate both the utility of the general production model and the fitting program GENPROD. Furthermore we doubt that a more refined treatment of these data in fitting the model would significantly alter the parameter estimates we are about to present. In fact much of the fitting about to be described was carried out a second time with the assumption of continuous fishing during the triweekly periods. The parameter estimates resulting from these fits agreed fairly well with those we now report which were computed under the more realistic discontinuous fishing pattern we adopted.

Some results from the initial fitting of the guppy data of populations A and B are illustrated by Figures 6 and 7. The lower panels of Figures 6 and 7 show the correspondence between the observed catch histories for each population and the best-fitting predicted histories. The observed and predicted population sizes generated from these fits are shown in the upper panels of Figures 6 and 7. In spite of the obviously excellent fits to the catch data for both populations, the population biomass was overestimated in both cases, especially during the first two thirds of the experiment. The trends in the actual population size histories were adequately described by the model.

In seeking to explain the overestimation of the population biomass, it was noticed that q was underestimated (Table 3; parameters unrestricted and minimum S). We next restricted q to 0.1 (the lower and upper bounds of q were set equal to 0.1) and recalculated the remaining parameters for population A (Table 3; q restricted). The results of this new fit (Fig. 8) illustrate a considerable improvement in the correspondence between the predicted and observed population size histories for population A.

The estimated maximum biomass (P_{max}) of 40 to 50 grams with q unrestricted is an overestimate for these guppy populations under the experimental conditions. The two unexploited control populations appeared to oscillate around a value between 30 to 36 grams. Thus the estimate of P_{max} for population A with q restricted seems in line.

Estimates of C_{max} and F_{opt} from the different fits are fairly consistent. The C_{max} estimates compare well with the empirical conclusion of Silliman and Gutsell (see their Figure 14), but the F_{opt} estimates are less than that suggested by Figures 12 and 13 of Silliman and Gutsell.

The empirical evidence from the combined data for populations A and B (Figure 14 of Silliman and Gutsell) suggests a stock production curve skewed to the right. The fit to population A indicated such a curve skewed to the right, but population B appeared to have a symmetrical stock production curve at least to the level of precision ($1.9 < m < 2.1$) we chose. It is particularly noteworthy that in the case of population A, the curve of which apparently is skewed, the assumption that the curve is symmetric does not seriously alter the estimates of the important parameters, C_{max} and F_{opt} (Table 3; population A, parameters either restricted or unrestricted, $m = 2$).

The fitting procedure seems to be most sensitive to estimating C_{max}, F_{opt}, and U_{max}, while least sensitive to estimating m, q, and P_{max}. It is almost certain that the overestimation of population biomass in the upper panels of Figures 6 and 7 was caused by poor estimation of q. But even with q restricted, it is apparent that the model does not represent the population changes exactly though we believe the representation is very good. There are times (Figure 8 upper panel) when the population does not respond as rapidly as the model predicts (resulting in overestimation of the catches) and times when the population does better than predicted (resulting in underestimation of the catches). The exploited guppy populations did not appear to have a stable age distribution at any of the fishing rates, and this might account for the growth rate not responding as predicted.

As a final exercise with the guppy data, the data for population A were divided into two parts. The first part contained catch data for times when the population size was always greater than P_{opt} (week 40 to week 109) and the second part contained catches for times when the population size was generally less than P_{opt} (week 110 to week 172). Curiously the fits to the halves of the data suggested the stock production curve either was symmetric or skewed in the opposite direction from that computed by the fit to the entire data (Table 4). By fitting only half of the data at a time, observations were taken primarily from either the left ascending limb or right descending limb of the stock production curve. Apparently observations are required from both limbs if the skew parameter m is to be determined accurately. In any case, neither of the fits to the halves indicated the curve was strongly skewed.

Both fits to the halves of the data produced estimates of F_{opt} greater

325

than and P_{max} less than the corresponding estimates from the entire data with the parameters unrestricted. It would appear, considering the fit to the entire data and the empirical conclusions of Silliman and Gutsell, that F_{opt}, C_{max}, P_{max}, and q were better estimated from the first than the second half of the data. As a partial explanation, it was noted earlier that the population which is not changing very rapidly is most likely to provide the best data. Population A underwent a rapid change during the time represented by the second half of the data.

Yellowfin tuna of the eastern Pacific Ocean

The yellowfin tuna data for 1934 through 1967 (Table 6) were fitted for a range of m values with the parameters unrestricted (Table 5). The predicted and observed catch histories for $m = 1.4$, the best estimate of the skew parameter, are plotted in Figure 9. The curves of equilibrium catch versus effort were obtained from (9) for $m = 2$ and $m = 1.4$ using the parameter estimates of Table 5 (Figure 10). While it appears that the stock production curve for yellowfin may be skewed, the assumption that the Schaefer model with its symmetric curve is appropriate does not appreciably change either the estimates of maximum sustainable yield or optimum fishing effort. If the curve is skewed, it would be expected that overfishing would not produce as noticeable a decline in catch per unit of effort as would be predicted by the Schaefer model.

The fits of the yellowfin data with the catchability coefficient unrestricted resulted in estimates of that parameter which were unreasonably large. We then restricted the catchability coefficient to a more realistic range of values, but found little difference in either the quality of the fit or the estimates of the other parameters (Table 5).

Clearly the fit to the yellowfin data is not of the quality of that for the guppies, but we did not expect it to be. There are sources of variation inherent in the yellowfin population dynamics which are absent or less influential in the guppy experiments. The guppies were kept in a relatively homogeneous environment where variation in reproduction, growth of individuals, or survival, induced by environmental perturbations, should be of less importance than in the case of the yellowfin tuna. Secondly, we developed an effort measure which perfectly satisfied the assumption of a constant catchability coefficient. The development of such a perfect effort measure for an actual commercial fishery of the complexity of the yellowfin fishery would not be possible.

Some discussion of the fit to the catch history will now be undertaken. In Figure 9 we see there are two periods of significant underestimation (1948 through 1951 and 1960 through 1961) and two periods of consistent and considerable overestimation (1952 through 1954 and 1956 through 1959). Such discrepancies between the observed and predicted catch histories may have been caused by unusually successful or unsuccessful reproduction, since each year class influences the catch for several consecutive years. Alternatively, environmental features either favoring or reducing production by the stock could be correlated in time. In any case, periods of consistent overestimation and underestimation of the catch history are not unexpected.

The remainder of the scattering about the predicted catch history could arise from several causes. Naturally productivity of the stock in different years will be subject to some variation caused by environmental changes and so unexplainable by the model. Furthermore, variation in the catchability coefficient could also be a cause. The age composition of the population has varied over these years, and since certain ages are more susceptible to capture than others (Davidoff 1969), the catchability coefficient must have varied. Further variation in the catchability coefficient could arise from changes in availability of the tuna due to, say, environmental features (*e.g.* Broadhead and Barrett 1964).

These comments have indicated some of the probable causes for the discrepancies between the observed and fitted catch histories. Considering the many potential sources of variation which are not taken into account by the model, together with the high proportion of the variation in the catches explained by the model ($R > 0.80$), it appears to be remarkably suitable for describing the dynamics of the yellowfin tuna in the eastern Pacific Ocean.

As a final note, the yellowfin catches generally came from stock sizes greater than P_{opt}. Whether or not the parameter estimates would change if data were generated from allowing the population to go below P_{opt} is left to speculation, but it is not likely that much additional information will be forthcoming if the population goes roughly into equilibrium under current regulation.

CONCLUSION

The problem investigated in this paper is the determination of the sustainable yield from fish stocks which can be anticipated under different levels of exploitation through an examination of only historic catch and effort information. The yield predictions from the model we discussed should be reasonably accurate provided the harvesting techniques are the same as those used to generate the data base from which the parameter estimates are made. A change in size selection by the fishery or in the time of the year when fishing occurs (*e.g.,* compression of fishing seasons due to catch restrictions) could modify the stock production curve. Still in these cases the analysis of catch and effort information on the basis of the generalized production model should provide a bench mark from which refinements in yield estimates can be made on the basis of more detailed studies of the dynamics of the stock.

We have dealt strictly with physical yield from the stock. The formulations would suggest that ordinarily an attempt should be made to maintain the size of the population, through catch or effort restrictions, at some intermediate level (P_{opt}) between the maximum population size and extinction. At such a level the population should be most productive, generating, on the average, a maximum annual sustainable catch (C_{max}). As has been pointed out by a series of authors (Scott 1954; Crutchfield and Zellner 1962; Turvey 1964; and Smith 1968, among others) such a management strategy based only on physical yield and ignoring economics may be a bit shortsighted. These authors suggest that profit maximization should be the goal of fisheries management. In the event that profit is chosen as the criterion, additional information on fishing costs are required to determine the optimum rate of exploitation. Whether physical yield or profit maximization is chosen as the objective, the determination of the physical yield as related to population size and fishing effort is essential in making management decisions. Viewed from this respect, the present study is fundamental to the management of fisheries, and we hope it will be of value to decision makers in the future.

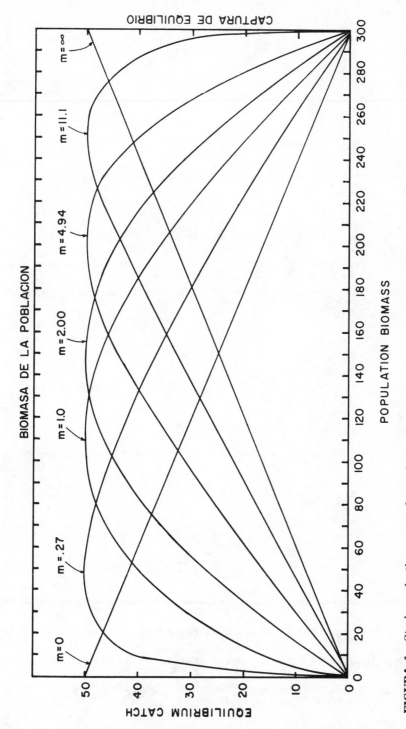

FIGURA 1. Stock production curves for various values of m. $C_{max} = 50$, $P_{max} = 300$, and $q = 0.001$.

FIGURA 1. Curvas de producción del stock para varios valores de m. $C_{max} = 50$, $P_{max} = 300$, y $q = 0.001$.

329

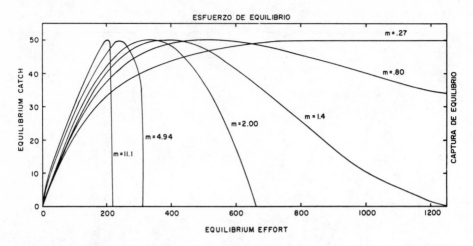

FIGURE 2. Equilibrium yield as related to effort for various values of m. $C_{max} = 50$, $P_{max} = 300$, and $q = 0.001$.

FIGURA 2. Rendimiento de equilibrio con relación al esfuerzo para varios valores de m. $C_{max} = 50$, $P_{max} = 300$, y $q = 0.001$.

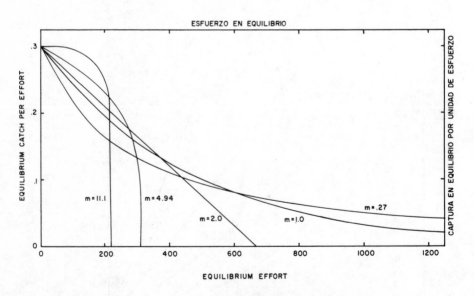

FIGURE 3. Equilibrium catch per unit of effort as related to effort for various values of m. $C_{max} = 50$, $P_{max} = 300$, and $q = 0.001$.

FIGURA 3. Captura de equilibrio por unidad de csfuerzo con relación al esfuerzo para varios valores de m. $C_{max} = 50$, $P_{max} = 300$, y $q = 0.001$.

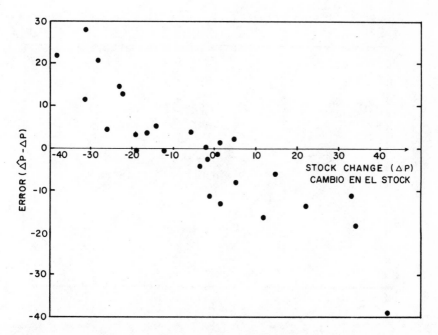

FIGURE 4. Plot of the errors in estimating stock changes by the linear interpolation formula versus the stock changes for a simulated logistic
stock under exploitation. The data are from Table 1.

FIGURA 4. Gráfico de los errores en la estimación de cambios del stock mediante
la fórmula de interpolación lineal en comparación con los cambios del
stock en un stock logístico simulado, bajo explotación. La información
proviene de la Tabla 1.

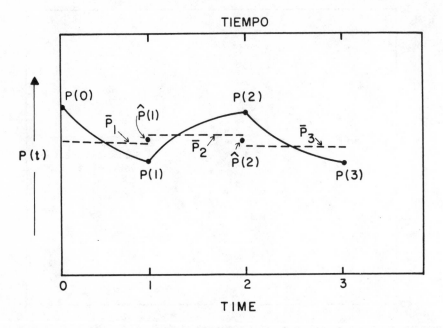

FIGURE 5. Stock size $P(t)$ when overfishing occurs in the first and third time units and underfishing occurs in the second unit.

FIGURA 5. Tamaño del stock $P(t)$ cuando ocurre la sobrepesca en la primera y tercera unidad de tiempo y en el caso de subpesca en la segunda mitad.

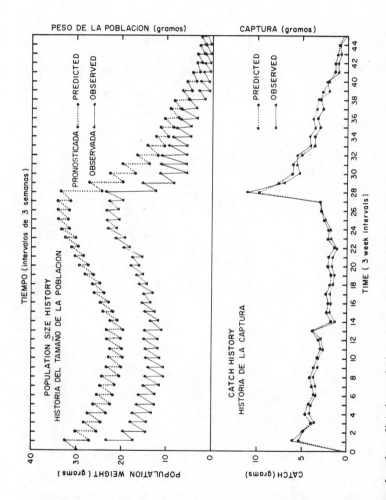

FIGURE 6. Observed and predicted population size and catch histories for guppy population A of Silliman and Gutsell (1963). The parameter values were unrestricted in fitting the catch data.

FIGURA 6. Tamaño observado y pronosticado de la población y cronología de captura de la población A de gambusinos de Silliman y Gutsell (1963). Los valores de los parámetros fueron irrestrictos al ajustar la información de captura.

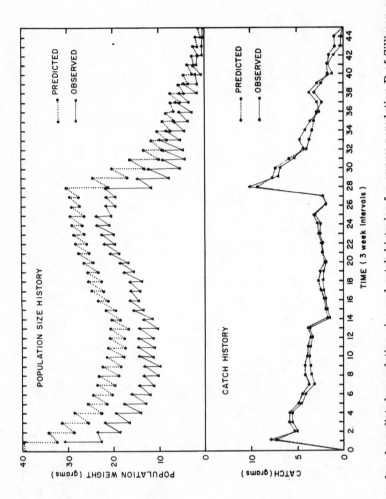

FIGURE 7. Observed and predicted population size and catch histories for guppy population B of Silliman and Gutsell (1963). The parameter values were unrestricted in fitting the catch data.

FIGURA 7. Tamaño observado y pronosticado de la población y cronología de captura de la población B de gambusinos de Silliman y Gutsell (1963). Los valores de los parámetros fueron irrestrictos al ajustar la información de captura.

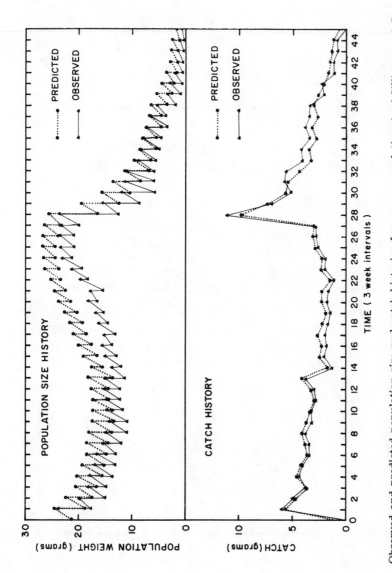

FIGURE 8. Observed and predicted population size and catch histories for guppy population A of Silliman and Gutsell (1963). The catchability coefficient was restricted in fitting the catch data.

FIGURA 8. Tamaño observado y pronosticado de la población y cronología de captura de la población A de gambusinos de Silliman y Gutsell (1963). El coeficiente de capturabilidad fue restringido al ajustar la información de captura.

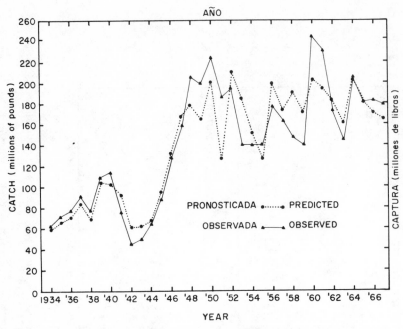

FIGURE 9. Observed and predicted catch histories when $m = 1.4$ and q is unrestricted for the yellowfin tuna in the eastern Pacific Ocean.

FIGURA 9. Cronología de captura observada y pronosticada cuando $m = 1.4$ y q es irrestricta para el atún aleta amarilla en el Océano Pacífico Oriental.

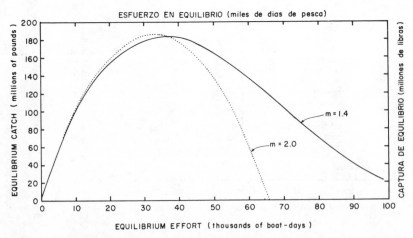

FIGURE 10. Comparison of equilibrium catch curves for $m = 1.4$ and $m = 2.0$ with q unrestricted for yellowfin tuna of the eastern Pacific Ocean.

FIGURA 10. Comparación de las curvas de captura equilibradas para $m = 1.4$ y $m = 2.0$ con q irrestricta para el atún aleta amarilla del Océano Pacífico Oriental.

TABLE 1. Statistics generated from a simulation of a logistic stock under exploitation. The headings for the columns are as follows: t, time in years; P, stock size at the beginning or end of each year; C, annual catch; f, effort during the year; $\bar{U}$, average catch per unit of effort for the year; $\hat{\Delta P}$, predicted change in stock size during the year; ΔP, observed change in stock size during the year. The parameter values were as follows: $k_1 = 8.814 \times 10^{-3}$, $L = 295$, and $k_2 = 3.8 \times 10^{-2}$. The estimated change in stock size was computed by the interpolation method with q known.

TABLA 1. Estadísticas hechas según la simulación de un stock logístico bajo explotación. Los títulos de las columnas son los siguientes: t, tiempo en años; P, tamaño del stock al principio o fin de cada año; C, captura anual; f, esfuerzo durante el año; $\bar{U}$, captura promedio por unidad de esfuerzo para el año; $\hat{\Delta P}$, cambio predicho en el tamaño del stock, durante el año; ΔP, cambio observado en el tamaño del stock durante el año. Los valores de los parámetros fueron los siguientes: $k_1 = 8.814 \times 10^{-3}$, $L = 295$, y $k_2 = 3.8 \times 10^{-2}$. El cambio estimado en el tamaño del stock se computó por el método de interpolación con q conocida.

t	P	C	f	$C/f = \bar{U}$	$\hat{\Delta P}$	ΔP
0	295.00	66.56	6.3	10.565		-25.53
1	269.47	69.04	6.8	10.153	$-\ 7.94$	$-\ 3.53$
2	265.94	81.69	8.2	9.962	$-\ 1.87$	$-\ 5.84$
3	260.10	68.07	6.8	10.011	$-\ 3.07$	$+\ 5.18$
4	265.29	102.15	10.5	9.729	$-\ 8.98$	-14.35
5	250.94	102.39	10.8	9.480	$-\ 2.23$	$-\ 2.29$
6	248.65	91.76	9.6	9.559	$+\ 6.60$	$+\ 4.56$
7	253.21	59.89	6.0	9.981	$+\ 8.70$	$+14.78$
8	267.99	60.30	5.9	10.220	$+\ 2.77$	$+\ 1.47$
9	269.46	65.23	6.4	10.192	$-\ 4.62$	$-\ 1.91$
10	267.55	92.77	9.4	9.869	-12.44	-12.10
11	255.44	129.45	14.0	9.247	-19.76	-18.94
12	236.50	170.69	20.4	8.367	-21.38	-26.05
13	210.45	182.93	24.0	7.622	-12.57	-16.22
14	194.23	170.47	23.0	7.412	-11.83	$+\ 1.36$
15	195.60	214.47	31.9	6.723	$-\ 3.36$	-31.13
16	164.46	133.82	18.7	7.156	$+\ 2.84$	$+41.80$
17	206.26	218.59	31.5	6.939	-17.01	-38.93
18	167.33	213.42	36.4	5.863	$-\ 9.62$	-22.48
19	144.85	155.20	25.0	6.208	$+21.93$	$+33.42$
20	178.27	134.04	17.8	7.530	$+15.87$	$+34.13$
21	212.40	199.44	26.9	7.414	$-\ 7.43$	-28.32
22	184.08	181.79	26.1	6.965	-12.97	$-\ 1.33$
23	182.75	210.21	32.7	6.428	$-\ 8.39$	-23.02
24	159.73	176.54	27.9	6.328	$-\ 4.34$	$+11.92$
25	171.65	209.17	34.3	6.098	-16.00	-19.19
26	152.46	220.83	43.2	5.112	-19.50	-31.34
27	121.12	185.12	40.1	4.616	$-\ 0.45$	$+\ 0.68$
28	121.79	167.57	33.0	5.078	$+\ 8.43$	$+22.10$
29	143.89	200.83	38.2	5.257		$-\ 9.89$
30	134.01					

TABLE 2. Estimates of the Schaefer model parameters for the yellowfin tuna fishery by the least squares method and Schaefer's method. The effort is measured in boat days and the catch in millions of pounds.

TABLA 2. Estimaciones de los parámetros del modelo de Schaefer para la pesquería del atún de aleta amarilla mediante el método de los mínimos cuadrados y el método de Schaefer. El esfuerzo se avalúa por días de operación de los barcos, y la captura en millones de libras.

Method Método	k_1	L	k_2	F_{opt}	C_{max}
Least squares mínimos cuadrados	9.181×10^{-5}	2,313	5.412×10^{-6}	19,620	122.8
Schaefer	2.136×10^{-2}	184	62.90×10^{-6}	31,190	180.8

TABLE 3. Various estimates of parameters for the guppies of Silliman and Gutsell. C_{max} and P_{max} are expressed in grams.

TABLA 3. Varias estimaciones de los parámetros para los gambusinos (*Lebistes reticulatus*) de Silliman y Gutsell. C_{max} y P_{max} se expresan en gramos.

Population Población	m	F_{opt}	C_{max}	P_{max}	U_{max}	q	S	R
A*	1.7***	1.95	3.20	49.3	3.50	0.071	6.95	0.983
A*	2.0	2.02	3.23	46.4	3.20	0.069	7.14	0.982
B*	2.0***	2.18	3.36	39.5	3.08	0.078	8.78	0.980
A**	1.4***	2.23	3.49	36.3	3.63	0.100	9.61	0.975
A**	2.0	2.47	3.79	30.7	3.07	0.100	10.73	0.973

* parameters unrestricted — parámetros irrestrictos
** q restricted — q restrictos
*** minimum S — S mínima

TABLE 4. Comparison of best fits to population A for different sets of catch data. C_{max} and P_{max} are expressed in grams.

TABLA 4. Comparación de los mejores ajustes para la población A de las diferentes series de los datos de captura. C_{max} y P_{max} se expresan en gramos.

Population Población	m	F_{opt}	C_{max}	P_{max}	U_{max}	q	S	R
A (1st half) (1ra mitad)	2.6	2.19	3.00	35.5	2.48	0.070	0.32	0.998
A (2nd half) (2da mitad)	2.0	2.61	4.85	26.6	3.72	0.140	3.41	0.987
A (total)	1.7	1.95	3.20	49.3	3.50	0.071	6.95	0.983

TABLE 5. Comparison of parameter estimates from several fits for the yellowfin tuna. C_{max} is in millions of pounds, F_{opt} in boat days, and U_{max} in pounds per boat-day.

TABLA 5. Comparación de las estimaciones de los parámetros según varios ajustes para el atún aleta amarilla. C_{max} está en millones de libras, F_{opt} en días por barco, y U_{max} en libras diarias por barco.

m	F_{opt}	C_{max}	U_{max}	q	S	R
2.0	32,600	185.8	11,400	9.9×10^{-4}	1.86×10^{10}	0.822
1.4*	35,300	182.6	12,000	4.5×10^{-4}	1.78×10^{10}	0.829
1.3**	36,750	181.5	11,850	5.0×10^{-5}	1.82×10^{10}	0.825

* minimum S with parameters unrestricted — S mínima con parámetros irrestrictos
** minimum S with q restricted — S mínima con q restricta

TABLE 6. Catch, effort, and catch per unit of effort for yellowfin tuna from the eastern Pacific Ocean during 1934 through 1967. The data from 1934 through 1965 were obtained from Bayliff (1967), except that the effort and catch per unit of effort for 1962 through 1965 were corrected for changes in efficiency of the fishing gear.

TABLA 6. Captura, esfuerzo y captura por unidad de esfuerzo para el atún aleta amarilla del Océano Pacífico oriental durante 1934 hasta 1967. Los datos de 1934 hasta 1965 fueron obtenidos de Bayliff (1967), excepto que el esfuerzo y la captura por unidad de esfuerzo para 1962 hasta 1965 fueron corregidos debido a los cambios en la eficiencia de las artes de pesca.

Year / Año	Catch (thousands of pounds) / Captura (miles de libras)	Effort (boat days) / Esfuerzo (días/barco)	Catch per unit of effort (pounds per boat day) / Captura por unidad de esfuerzo (libras diarias por barco)
1934	60,913	5,879	10,361
1935	72,294	6,295	11,484
1936	78,353	6,771	11,571
1937	91,522	8,233	11,116
1938	78,288	6,830	11,463
1939	110,418	10,488	10,528
1940	114,590	10,801	10,609
1941	76,841	9,584	8,018
1942	41,965	5,961	7,040
1943	50,058	5,930	8,441
1944	64,869	6,475	10,019
1945	89,194	9,377	9,512
1946	129,701	13,958	9,292
1947	160,151	20,383	7,857
1948	206,993	24,781	8,353
1949	200,070	23,923	8,363
1950	224,810	31,856	7,057
1951	186,015	18,403	10,108
1952	195,277	34,834	5,606
1953	140,042	36,356	3,852
1954	140,033	26,228	5,339
1955	140,865	17,198	8,191
1956	177,026	27,205	6,507
1957	163,020	26,768	6,090
1958	148,450	31,135	4,768
1959	140,484	28,198*	4,982*
1960	244,331	35,841	6,817
1961	230,886	41,646	5,544
1962	174,063	42,248	4,120
1963	145,469	33,303	4,368
1964	203,882	42,090	4,844
1965	180,086	43,228	4,166
1966	182,294	40,393	4,513
1967	178,944	33,814	5,292

* These statistics differ from those in IATTC (1965, 1966) since they are based on catch rather than landings in 1959.

* Estas estadísticas difieren de las de la CIAT (1965, 1966) ya que están basadas en la captura en vez de los desembarques de 1959.

TABLE 7. Computer program, GENPROD, output with symbology.

TABLA 7. Programa de cómputo GENPROD, producción con simbología.

Text Texto	Program output Producción del programa
m	M
F_{opt}	FOPT
U_{max}	UMAX
q	Q
r	P(0)/L
C_{max}	CMAX
P_{opt}	POPT
P_{max}	PMAX
H	H
K	K
S	SUM OF SQUARES
R	R
$\displaystyle\sum_{}^{n} C_i$	TOTAL CATCH
$\displaystyle\sum_{}^{n} \frac{1}{n} C_i$	AVERAGE CATCH
t	AT TIME
P_i	POPULATION SIZE
E_i	APPLIED EFFORT
C_i	OBSERVED CATCH
$\hat{C}_i$	EXPECTED CATCH
U_i	OBSERVED CATCH/EFFORT
$\dot{U}_i$	EXPECTED CATCH/EFFORT

The effect of fishing on the marine finfish biomass in the Northwest Atlantic from the Gulf of Maine to Cape Hatteras

B.E.Brown, J.A.Brennan, M.D.Grosslein, E.G.Heyerdahl and R.C.Hennemuth

Research Bulletin of the International Commission for the Northwest Atlantic Fisheries, 12, 49-68, 1976.

The Effect of Fishing on the Marine Finfish Biomass in the Northwest Atlantic from the Gulf of Maine to Cape Hatteras

B. E. Brown,[1] J. A. Brennan,[1] M. D. Grosslein,[1] E. G. Heyerdahl,[1] and R. C. Hennemuth[1]

Abstract

Relationships between fishing effort, total finfish community biomass, and yield are determined using data on finfish biomass measured by research vessel trawl surveys and commercial catch and effort statistics. Combined individual species stock assessments and the Schaefer (1954) equilibrium yield model are utilized to estimate potential yields. The fishing effort of different gear types are combined to provide a standardized index of fishing effort in terms of days fished as reported to the International Commission for the Northwest Atlantic Fisheries (ICNAF). A multiplicative learning function is applied as a correction factor to reported days fished of newly developing fisheries, *i.e.* those deployed in areas and on stocks not previously fished. This correction factor adjusts data of fleets entering a fishery to the level of efficiency achieved by the third year in the fishery. These analyses demonstrate a six-fold increase in fishing intensity, and a 55% decline in finfish abundance during the period 1961-72. Plots of yield versus standardized fishing effort indicate that fishing mortality since 1968 has exceeded that which would result in sustaining a maximum yield from the stocks under equilibrium conditions. The projected maximum sustainable yield (MSY) from Schaefer yield curves is 900,000 tons[2], while the sum of the MSY's from individual assessment studies is 1,300,000 tons. It is suggested that because of species interactions summing the MSY's from individual assessments may be an overestimate of the total MSY.

Introduction

Historically, fisheries management has been stimulated by changes in the development of a fishery. New participants increase competition and may force changes in the distribution of the catches. New fisheries develop in areas and on species theretofore not fished; in the face of marked and rapid increases in fishing effort, serious doubts are often raised about the ability of the fish stocks to sustain their full potential productivity, especially when the catch per fishing unit begins to drop.

Such has been the case in the Northwest Atlantic fishery on the continental shelf off the northeast coast of the USA in ICNAF Subarea 5 and Statistical Area 6[3] (Fig. 1). Prior to 1960, almost all of the fishing in the area was carried out by USA vessels less than 300 GRT (gross registered tons), as the fishing grounds were close to the home ports where the landings were processed on shore. After 1960, the distant water fleets of USSR, Poland, Japan, FRG (Federal Republic of Germany), and other countries began fishing in the area. These fleets of large, highly mobile vessels steadily increased both in number and total tonnage (Table 1), and resulted in enlarging the scope of the total fishery with respect to species, area fished and intensity of fishing. While historically the USA fishery had concentrated on selected groundfish species (cod, haddock, redfish and flounders), greater catches of an increasing number of species (Table 2) have been reported since 1960 (ICNAF, 1962a-74a).

The Standing Committee on Research and Statistics of ICNAF (STACRES), which has been evaluating the effects of fishing on the resources in this area (ICNAF, 1961a, 1962b-74b), has on several occasions advised the Commission that the overall fishing intensity was fast approaching that which could not be supported by the stocks. This was emphasized at the 1961 Annual Meeting (ICNAF, 1961a). For certain species, concern that fishing mortality on the given stock was approaching a value greater than that which would maximize the long-term yield or yield per recruit was first expressed by the Subcommittee on Assessments in 1963 and 1968 (ICNAF, 1963b and 1968b), prior to severe overfishing (Hennemuth, 1969; Brown and Hennemuth, 1971). As a consequence, at the Annual Meeting in June 1972 the Commission extended quotas to many of the heavily fished species-stocks[4] (ICNAF, 1972c). Large reductions in the current catches were necessary in order to begin to rebuild those stocks which were severely reduced. STACRES also recognized that the rapid expansion of fishing activity precluded timely and complete assessments of the effects of fishing, particularly when a multitude of species-stocks was being harvested.

More importantly, STACRES began considering in the late 1960's the question of whether the goals of

---

[1] National Marine Fisheries Service, Northeast Fisheries Center, Woods Hole, Massachusetts, USA, 02543.
[2] All references to catches in tons refer to metric tons.
[3] Throughout the text the abbreviation SA is used when referring to Subareas and Statistical Areas, and Div. is used for Divisions.
[4] A species-stock refers to an ICNAF regulatory management unit; *i.e.* some regulations apply to a single population that is a self-contained component of the species considered to have uniform growth and mortality rates, other apply to convenient geographical groups of such stocks, while still others to even a combination of species.

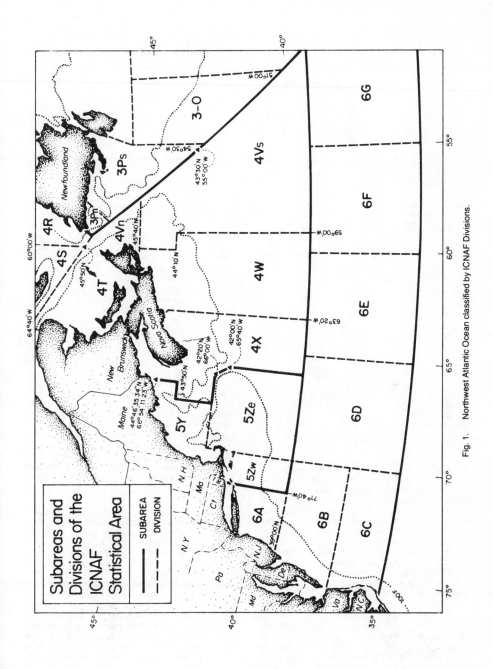

Fig. 1.   Northwest Atlantic Ocean classified by ICNAF Divisions.

TABLE 1. Number of vessels (USA, Others) fishing in SA 5 + 6 during 1959, 1965, and 1971 by tonnage class (A = less than 901 GRT[a]; B = 901 GRT[a] and above).

Year	Area	USA		Others		Total		Total
		A	B	A	B	A	B	Total
1959	SA 5	301	—	26	—	327	—	327
1965	SA 5	323	—	244	110	567	110	677
1971	SA 5	299 ⎱ 461	—	447 ⎱ 490	213 ⎱ 222	746 ⎱ 951	213 ⎱ 222	959 ⎱ 1173
1971[c]	SA 6	162 ⎰	—	43 ⎰	9 ⎰	205 ⎰	9 ⎰	214 ⎰

[a] GRT — Gross registered tons.
[b] Data from ICNAF List of Fishing Vessels; ICNAF 1961b, 1967c, and 1973d.
[c] This includes only vessels which were listed for SA 6 and not for SA 5. SA 6 listings were not available for 1959 and 1965.

TABLE 2. Subarea 5 nominal catches as reported in ICNAF Statistical Bulletin Vol. 10 and 22 (ICNAF 1962a, Table 1; ICNAF 1974a, Table 3).

Species	1960	1972
Cod	14,430	31,357
Haddock	45,801	6,669
Redfish	11,375	19,095
Yellowtail flounder	13,581	29,620
Winter flounder	6,953	10,505
Witch flounder	1,255	5,454
Scup	3,779	1,229
Pollock	10,397	12,989
Silver hake	46,688	107,113
Red hake	3,410	60,062
White hake	2,483	3,084
Groundfish not stated	19,110[a]	1,239
Herring	69,046	220,964
Mackerel	1,011	200,518
Alewife	8,669	8,656
Atlantic Saury	—	3,429
Angler	8	4,332
Sculpins	—	4,862
Argentine	—	32,707
Sharks[b]	801	13,154
Skates	128	8,735
Other fish not stated	—	21,661
Squid	741	26,111

[a] 15,320 listed as industrial presumed to be mainly red and silver hake based on USA national statistical studies.
[b] Includes dogfishes.

management could be achieved based on individual stock assessments and regulations. The difficulty of achieving these goals stemmed primarily from the lack of resources allocated to collect the necessary data and to make the required assessments within the required time period. In addition, the severe mixed-species nature of the current fisheries in SA 5+6 led to the difficult but necessary consideration of the fishing mortality caused by the by-catch, *i.e.* the catch of species other than that which is the main target of the fishery (Brown *et al.*, 1973). The mixed-species catches resulted primarily from the extensive use of bottom tending otter trawls which are quite unselective.

In SA 5 + 6 numerous species make up significant portions of the fishable biomass (Fig. 2) and, hence, the otter trawl fishery catch. The species mixture is illustrated by the catches in the USA and USSR joint bottom trawl survey in 1971 in the area off Southern New England (Fig.

3, Strata 1-12), where the mean number of species caught per tow was 12 for the USA vessel and 11 for the USSR vessel (Grosslein, 1973). The inevitable incidental catches in species-directed fisheries may be great enough to harvest the total surplus production of some stocks, and this creates conflicts in objectives of conserving stocks which are at low abundance levels or maintaining an existing directed fishery without overfishing, *e.g.* haddock and yellowtail flounder (Brown *et al.*, 1973). The Assessment Subcommittee of ICNAF estimated that, in 1971, 33% of the total fishing mortality in SA 5 + 6 was generated as by-catch of the major species-directed fisheries (ICNAF, 1973b). Finally, the current generation of fishery-yield models does not include terms which describe the effects of species interactions on long-term biological productivity. The interrelationships among species are not well understood, and considerable research is needed on this subject. However, consideration of basic ecological concepts, such as prey-predator and competitive relationships, underscores the need to examine the yield of this total ecosystem as an integrated whole rather than as just the sum of the individual components (Steele, 1974). In this paper, the interspecific effects of the finfish component of the ecosystem are included implicitly in analysis of the total sustainable yield of the finfish biomass to the extent that they have been significant in affecting total yield as measured over the period 1962-72.

The description of the status of the finfish biomass in SA 5+6 is based on analyses of total finfish catch, fishing activity, research vessel surveys, and a preliminary evaluation of the status of the biomass reported in Grosslein, *et al.* (1972). The finfish biomass was defined as all species of finfish plus squids, except menhaden (which are captured close to shore primarily in a single-species fishery in the most southerly part of SA 6) and large pelagic species (swordfish, sharks other than dogfish, and tuna). The large pelagics contribute minimally to the total catch in a quantitative sense and hence would not affect the calculations significantly. This is not to say, however, that the interactions of other fish with this component are not important, but that the results presented are provisional with respect to them. Species

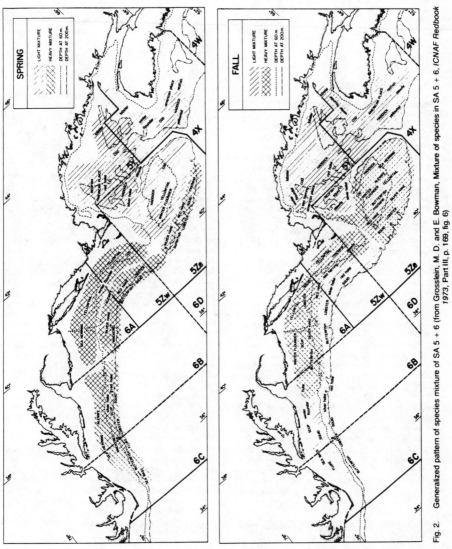

Fig. 2. Generalized pattern of species mixture of SA 5 + 6 (from Grosslein, M. D. and E. Bowman, Mixture of species in SA 5 + 6, *ICNAF Redbook* 1973, Part III, p. 169, fig. 6)

assessments based on analysis of commercial catch and effort data are combined to give one estimate of overall maximum sustainable yield (MSY). A Schaefer yield model (Schaefer, 1954) for total finfish and squid, using commercial catch and effort data, is also used to estimate a total MSY. The relationship of current effort to that providing MSY is discussed.

Standardization of Fishing Effort

Those indices of fishing effort, which purport to be directly proportional to instantaneous fishing mortality (F), exerted on fishery resources over a period of time, have traditionally been used to determine the status of fisheries. For this study, because of the diversity of gear employed and the availability of comprehensive statistics reported to ICNAF for the SA 5 + 6 fisheries, the reported effort data of the different country-gear-tonnage classes were standardized in an attempt to have equivalent fishing mortality generated per unit activity. Catch and effort data from 1961 to 1972 were obtained from Table 4 of ICNAF Statistical Bulletin Vol. 11-21 (ICNAF, 1963a-73a) and Table 5 of ICNAF Statistical Bulletin Vol. 22 (ICNAF, 1974a). USA catch data for SA 6, 1961-62, was taken from national statistics.

Numbers of days on grounds, days fished, hours fished and numbers of sets (trawl hauls or purse seine hauls) have all been reported with varying degrees of completeness to ICNAF. 'Hours fished' is probably the best of the effort units reported, in that it is likely to be more consistently proportional to F than 'days fished'. However, member countries have reported 'days fished' more often through the years than 'hours fished'. 'Days fished' was considered more closely related to fishing intensity than 'days on grounds' and it also appeared to be a more standard measure of fishing activity over all types of vessels and gears; for example, 'hours fished' definition may have differed greatly for purse seines depending on how searching time was recorded. Hence, 'days fished', as reported to ICNAF, was chosen as the basic unit of fishing effort for analysis.

In order to measure total fishing effort in standard units, catchability coefficients relative to an arbitrarily chosen standard class of vessel and gear were estimated for the various other classes, and used to convert the reported number of days fished for each respective category to the number in the standard equivalent. In all cases, the yearly total of catch and effort data for each class was the basic variable in the analysis.

Robson (1966) proposed a method for determining effort standardization coefficients using an analysis of variance model assuming no interaction. This model was selected for the present study and is defined as follows:

$$Y_{ijk} = m \times a_i \times b_j \times e_{ijk}$$

where Y_{ijk} = catch per effort of all fish for the ith country, j^{th} gear-tonnage class, and k^{th} year, *i.e.* Σ catch/Σ days, where the sum is of the appropriate entries of Table 4 or 5 of the ICNAF Statistical Bulletins *(op. cit.)* entries over each month of the year and each area (SA 5 + 6),

m = the mean catch per effort over all categories,

a_i = the ith country effect,

b_j = the jth gear-tonnage class category effect, and

e_{ijk} = the error for testing significance and precision, if the kth observation at the i-j level is such that *ln* (e_{ijk}) has a N $(0,\sigma^2)$ distribution.

The error term was measured by the year to year variations. A natural logarithmic transformation of the observations, Y_{ijk}, was performed so that the linear model analysis of variance procedures could be used. The cell coefficients $(a_i b_j)$ were estimated using an analysis of variance procedure outlined by Snedecor and Cochran (1967) for a row $\times$ column design with unequal cell frequencies and missing observations. In order to express these coefficients in terms of a standard cell (gear tonnage class-country category), the value $a_i b_j$ for the standard cell was subtracted from that of each of the remaining cells and the resultant sequence of numbers transformed by the exponential function. Since the $a_i b_j$ values are all estimated from the row and column totals, it was immaterial which cell was selected as the standard. For this paper, the USA side trawler 0-50 GRT was used for the standard cell.

Fishing gear for which data were used in the analysis of variance included stern, side and pair trawls, purse seines, drift gillnets, long lines, and hand lines. These gears accounted for approximately 80% of the total catch of the species considered. The remaining 20% of the catch was taken by a variety of gears (mostly inshore) for which no effort data were recorded. The standardized effort associated with this catch for each year was estimated in the last stages of analysis by dividing this catch by the overall annual catch per standardized effort.

Adjustment for learning

It may be logically asserted that the development of new fisheries in areas and on stocks not previously fished involves learning: how to conduct and distribute the fishing fleet over the grounds, particularly in relation to

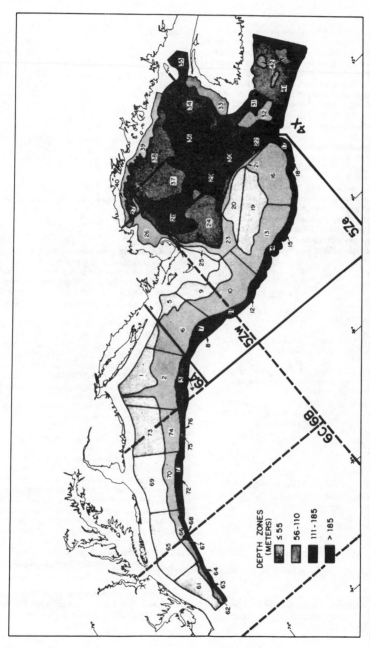

Fig. 3. Northwest Atlantic Ocean categorized by depth zone, sampling strata, and ICNAF Divisions. Southern New England area includes strata 1-12; Georges Bank area includes strata 13-25; Gulf of Maine area includes strata 26-30, 36-40; Mid-Atlantic area includes strata 61-76.

seasonal changes; how to deploy the different kinds of gear in relation to depth or bottom types, current, and weather patterns; and how best to utilize spawning or feeding concentrations (time and space) and migratory patterns. All these factors affected the efficiency of operations (for further discussion see ICNAF Report of Special Meeting of Experts on Effort Limitation (ICNAF, 1973c)) and consequently involved a learning period. The magnitude of this learning was assumed to be reflected in the catch and effort statistics for the various countries but not clearly separated from other causes of variation in catch. There undoubtedly were many other components of success involved with the development of each fishery. In this study, no attempt was made to define the learning factors in terms of explicit causes. Rather, the problem was approached by assuming that learning could be expressed as a monotonically increasing function of catch per effort through a continuous time period, which was not caused by changes in stock abundance. In order to estimate the magnitude of learning, a multiplicative learning function was hypothesized for a given fleet in a fishery. The model for learning was defined as:

$$L_i = \frac{O_i}{P_i} \qquad (1)$$

where L_i = learning gained by a fleet in the ith year in a fishery,

O_i = observed catch per effort by the fleet in the ith year in the fishery,

P_i = predicted catch per effort for the fleet in the ith year in the fishery assuming no learning,

P_1 = O_1

L_1 = 1

and $i = 1, 2, 3 \ldots$ The predicted catch per effort, P_i, was defined algebraically to be:

$$P_i = \left[\frac{(Z_i)}{(Z_i-1)} \right] P_{i-1} \qquad (2)$$

where Z_i is an independent estimate of the abundance of the species in the ith year in the fishery.

By recursion

$$P_i = \prod_{j=2}^{i} \left[\frac{(Z_j)}{(Z_j-1)} \right] P_1 \qquad (3)$$

$$= \left[\frac{(Z_i)}{(Z_1)} \right] P_1$$

$$= \left[\frac{(Z_i)}{(Z_1)} \right] O_1$$

and $P_1 = O_1$.

The observed catch per effort in the first year in the fishery, O_1, was taken to be the predicted catch per effort, P_1. The first year of presence in a fishery was taken as that year in which a fleet first caught 20% of its total catch in a particular fishery, *i.e.* 20% of the total catch of a fleet was of the species by which the fishery is identified.

It was assumed that if the catch of a single species exceeded 80% of the total catch by the fleet in an area for a particular year, a "directed fishing" effort had taken place, and all days fished for the fleet during the year were assigned to the species. If the catch of the species was between 20 and 80% of the total catch, the directed effort was estimated as proportional to the species catch in the nominal landings. If less than 20% of the total catch consisted of a specific species, a directed fishery was assumed not to exist in that year.

Since the catch per effort for each species used to study the presence of a learning factor was based on the total catch of that species directed fishery divided by the total effort for that record, it seemed pertinent to examine whether or not any relationship existed between the learning factor and the percent of that species in the catch. Examination of these values (Table 3) showed no consistent relationship; for example, approximately the same learning value between the first and second year existed for herring for Div. 5Z Poland (1.00 and 2.16), SA6 Poland (1.00 and 2.80) and Div. 5Z Romania (1.00 and 2.80) despite differences in the percentage herring made up of the catch which were respectively (90, 86; 98, 66; and 80, 57).

A further assumption made in applying a learning function was that learning ceased when the ratio (L) decreased from year i to year $i + 1$, i.e. when $L_{i+1} < L_i$.

An independent measure of the abundance of a species was provided by the mean catch (pounds per tow) of the annual USA bottom trawl surveys.

Certain "sets" of data were incomplete and could not be used to estimate a learning factor, *e.g.* no fishing effort (in "days fished" units) was recorded by the USSR for 1962, although there was fishing before, during, and after 1962. For certain fisheries, *i.e.* squid and mackerel, there was no time series of commercial catch and effort data available at the time of analysis. Therefore, only complete sets of data could be used (Table 3). A learning function, derived from situations where statistics were available, could then be used to adjust reported units for other fisheries where the data were not available.

In most cases where L_i could be estimated for 4 or 5 successive years, L_i declined in the fourth year in the fishery (Table 3). It was concluded, therefore, that in general the learning process was completed by the end of the third year in the fishery.

An exponential curve was fitted to a fleet's data for the first three years in the fishery (see Fig. 4).

$$L_i = \frac{O_i}{P_i} = [\exp(a(i-1))]\,e_i, \quad \text{where} \qquad (4)$$

where

$$P_i = O_1 \times \frac{Z_i}{Z_1}$$

O_i = the observed commercial catch per effort in the ith year in the fishery after entrance, where $i = 1, 2, 3 \ldots$

Z_i = the stock abundance in the same year

e_i = the residual error, where $\ln(e_i)$ has a $N(O, \sigma^2)$ distribution, and

a = constant

TABLE 3. Statistics used in development of learning model, by fleet, species, and area.

Data set	Year	Observed total catch per effort	% species in total catch	Research vessel abundance index[a]	Predicted catch per effort	i	L_i
Herring	1966	30.99	90	10.41	30.99	1	1.00
Div: 5Z	1967	20.98	86	3.26	9.70	2	2.16
Poland OTST	1968	28.13	60	1.36	4.05	3	6.94
> 1800 GRT	1969	22.96	42	1.14	3.39	4	6.77
	1970	27.21	45	.66	1.96	5	13.88
	1971	35.63	56	2.07	6.15	6	5.80
Cod	1964	6.00	82	7.62	6.00	1	1.00
Div: 5Z	1965	11.80	86	5.52	4.35	2	2.71
Spain	1966	19.25	88	4.84	3.81	3	5.05
P. trawl	1967	16.22	91	12.46	9.81	4	1.65
	1968	15.96	81	5.74	4.52	5	3.53
	1969	13.92	91	5.24	4.12	6	3.38
	1970	15.48	89	6.70	5.27	7	2.94
	1971	15.22	83	4.53	3.56	8	4.27
S. Hake	1963	6.13	33	9.90	6.13	1	1.00
Div: 5Z	1965	8.90	77	10.76	6.66	2	1.34
USSR OTSI	1966	10.56	53	5.84	3.62	3	2.92
151-500 GRT							
S. Hake	1964	8.65	73	8.16	8.65	1	1.00
Div: 5Z	1965	19.72	36	10.76	11.40	2	1.73
USSR OTSI	1966	16.03	28	5.84	6.19	3	2.59
501-900 GRT	1967	12.17	51	6.37	6.75	4	1.80
Herring	1968	12.20	98	17.40	17.40	1	1.00
SA 6	1969	10.23	66	6.40	4.49	2	2.28
Poland OTST	1970	12.02	65	1.20	.84	3	14.31
501-900 GRT	1971	8.71	54	3.70	2.59	4	3.36
Herring	1967	19.19	80	3.26	19.19	1	1.00
Div: 5Z	1968	22.42	57	1.36	8.01	2	2.80
Romania OTST	1969	12.03	54	1.14	6.71	3	1.79
> 1800 GRT	1970	13.95	25	.66	3.88	4	3.59
	1971	17.41	19	2.07	12.37	5	1.40

[a] Pounds/tow index as recorded by USA Research Vessel *Albatross IV* bottom trawl surveys: all autumn surveys except for SA 6 where spring surveys were used. Method used to calculate indexes is described by Grosslein (1971).

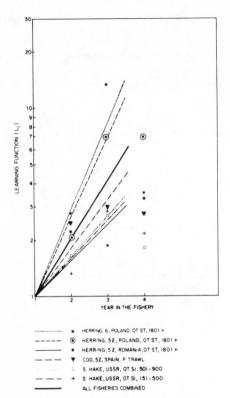

LEARNING FUNCTION (L_i)

YEAR IN THE FISHERY

....	•	HERRING, 6, POLAND, OT ST, 1801 +
— — — —	⊙	HERRING, 5 Z, POLAND, OT ST, 1801 +
	•	HERRING, 5 Z, ROMANIA, OT ST, 1801 +
— — —	▼	COD, 5 Z, SPAIN, P TRAWL
	⋮	S HAKE, USSR, OT SI, 501 - 900
— · — —	+	S HAKE, USSR, OT SI, 151 - 500
————		ALL FISHERIES COMBINED

Fig. 4.   Relationship of learning function (L_i) to year in the fishery (see text for explanation).

This curve was selected since the ideas underlying the model seemed to coincide with the underlying notion of learning: that the learning gained by time t_i was dependent on the learning gained by time t_{i-1} as well as the time interval t_i-t_{i-1}. Since the values of L_i did not appear to be grouped by country or tonnage class, pooled data were used to fit the curve. A least squares linear fit of ln (L_i) on i yielded the curve

$$L_i = 0.48 \, [exp \, (0.735 \, i)], \text{ for } i = 1 \ldots 3$$

with a coefficient of determination of 0.82. From this equation, $L_1 = 1.00$, $L_2 = 2.09$, and $L_3 = 4.35$. This is approximately equivalent to halving and quartering the units of effort on that species during this learning period.

The effort data were adjusted so that a unit of effort in the first two years in a fishery was made equivalent to a unit of effort in later years. The adjustment involved was:

$$X_i = \frac{O_i}{L_i} \times L_3, \text{ for } i = 1 \ldots 3 \qquad (5)$$

where X_i = adjusted catch per effort for the ith year in a fishery by a fleet, and O_i, L_i, and L_3 are as defined previously. The values of 1, 2 and 4 were used for L_1, L_2 and L_3 respectively. Adjusting data according to (5) essentially brought all entering fleets to the equivalent of the level of knowledge of the third year in the fishery. The data adjusted by (5) included data used in the development of the model (Table 3), as well as sets of data excluded because they were incomplete or where the abundance indices were not available. Table 4 lists these sets of data.

Application of Fishing Effort Standardization

Analysis of variance results

The assumptions of normality of errors and equality of error variances were investigated in both the original and logged catch per effort data. Both assumptions were more valid with the logged data. Due to the presence of many empty cells, tests for linearity on the logged data were not performed.

Standardization of effort was calculated with and without adjustments for learning. Both vessel class and country effects showed significance at the 0.01 probability level (Tables 5 and 6).

Inspection of the data to determine which levels of the two factors contributed most to the interaction sum of squares revealed that departures from main effect trends could be attributed mainly to USSR drift gillnet data. Considering the relatively minor contribution of this catetory to both total catch (0.08%) and effort (0.3%), the consequence of ignoring the interaction term was considered to be minimal. Relative catchability coefficients were calculated for the USA 0-50 GRT side trawler standard (Tables 7 and 8) which were present in the fishery during the years under consideration.

Estimation of total fishing intensity

Total fishing effort units in standard days fished (standardized effort) directed at finfish were estimated for 1961-72 for each country and gear combination, by multiplying the reported number of days fished by the relative catchability coefficients, with and without learning.

TABLE 4. Country gear-tonnage class categories where effort was adjusted for learning. Parenthesis indicates year in which greater than 20°₀ of the total catch by gear-tonnage class-country was taken in the given species, but for which days fished data was unavailable. (OTSI = otter trawl side; OTST = otter trawl stern; PS = purse seine; PT = pair trawl; DGN = drift gillnet.)

					Area							
			Div. 5Y				Div. 5Z				SA 6	
Species	Country	Gear	GRT	Years adjusted	Country	Gear	GRT	Years adjusted	Country	Gear	GRT	Years adjusted
Herring	FRG	OTSI	901-1800	1969, 70	FRG	OTST	901-1800	1967, 68	Poland	OTSI	501-900	1968, 69
		OTST	1800 +	1969, 70		OTST	1801 +	1967, 68		OTST	1800 +	1968, 69
	USA	PS	51 +	1965, 66	GDR[a]	OTST	1801 +	(1965), 66	USSR	OTSI	151-500	1967, 68
					Poland	OTSI	501-900	1967, 68		OTSI	501-900	1969, 70
						OTST	901-1800	1967, 68				
						OTST	1801 +	1966, 67				
					Romania	OTST	1801 +	1967, 68				
					USSR	OTSI	151-500	(1962), 63				
						OTST	1801 +	1961, (62)				
						PS	51 +	1968, 69				
						DGN	All	1961, (62)				
	GDR[a]	OTST	501-900	1968, 69					GDR[a]	OTSI	501-900	1968, 69
		OTST	1801 +	1968, 69								
Cod	Spain	PT	151-500	1969, 70	Spain	PT	151-500	1964, 65				
Haddock					USSR	OTSI	501-900	1965, 66				
Silver hake					USSR	OTSI	151-900	1963, 65				
						OTSI	501-900	1964, 65				
						OTST	1801 +	(1962), 63				
Mackerel					Poland	OTSI	501-900	1969, 70	Poland	OTSI	501-900	1969, 70
						OTST	1801 +	1968, 69		OTST	1801 +	1970, 71
					Romania	OTST	1801 +	1969, 70	USSR	OTSI	151-500	1968, 69
					USSR	OTSI	151-500	1969, 70		OTSI	501-900	1969, 70
						OTSI	501-900	1969, 70		OTST	1801 +	1970, 71

[a]German Democratic Republic.

TABLE 5. Analysis of variance of ln (catch per effort) data for SA 5 + 6, adjusted for learning.

Source of variation	Sums of squares	Degrees of freedom	Mean square	F
Total	547.38	299	—	—
Country (unadjusted)	174.18	9	—	—
Gear-tonnage class (unadjusted)	477.53	18	—	—
Country (adjusted)	15.58	9	1.73	4.08[a]
Gear-tonnage class (adjusted)	257.96	18	14.33	33.80[a]
Interaction	45.39	26	1.75	—
Error	69.84	246	0.28	—
Interaction plus error	115.23	272	0.424	—

[a]Significant at 0.01 level.

TABLE 6. Analysis of variance of ln (catch per effort) data for SA 5 + 6 not adjusted for learning.

Source of variation	Sums of squares	Degrees of freedom	Mean square	F
Total	473.42	299	—	—
Country (unadjusted)	124.08	9	—	—
Gear-tonnage class (unadjusted)	421.65	18	—	—
Country (adjusted)	11.55	9	1.28	3.90[a]
Gear-tonnage class (adjusted)	260.21	18	14.46	44.09[a]
Interaction	37.35	26	1.44	—
Error	51.78	246	0.21	—
Interaction plus error	89.13	272	0.328	—

[a]Significant at 0.01 level.

TABLE 7. Estimates of fishing power factors for given country and gear-tonnage class combinations for SA 5 + 6 data without adjustments for learning, 1961-72. (OTSI = otter trawl side; OTST = otter trawl stern; PS = purse seine; PT = pair trawl; LT = line trawls; HL = hand lines; DGN = drift gillnet.)

Gear	Tonnage class (GRT)	USA	Canada	USSR	Spain	Poland	FRG	GDR	Romania	Japan	Bulgaria
	0-50	1.00	0.67	—	—	—	—	—	—	—	—
	51-150	1.30	0.87	—	—	—	—	—	—	—	—
OTSI	151-500	1.77	1.19	1.44	—	—	1.72	—	—	—	—
	501-900	—	1.63	1.98	—	1.64	2.35	1.32	—	—	—
	901-1800	—	—	3.16	—	2.62	3.76	2.11	—	—	—
	0-50	3.33	2.24	—	—	—	—	—	—	—	—
	51-150	0.94	.63	—	—	—	—	—	—	—	—
OTST	151-500	1.75	1.18	—	—	—	—	—	—	—	—
	501-900	2.67	1.79	2.18	—	—	2.59	1.45	—	—	—
	901-1800	—	—	5.54	4.38	4.59	6.59	3.70	—	—	—
	> 1801+	—	—	7.65	—	6.34	9.09	5.10	3.53	3.19	6.47
PS	> 50	7.35	4.93	5.99	—	—	—	—	—	—	—
	< 50	0.95	—	—	—	—	—	—	—	—	—
PT	All	4.36	—	3.56	2.81	—	—	—	—	—	—
LT	All	0.46	0.31	—	—	—	—	—	—	—	—
HL	All	0.14	—	—	—	—	—	—	—	—	—
DGN	All	0.11	—	0.09	—	—	—	—	—	—	—
OTSI (pelagic)	501-900	—	—	—	—	—	—	0.82	—	—	—
	901-1800	—	—	—	—	—	—	0.29	—	—	—

TABLE 8. Estimates of fishing power factors for given country and gear-tonnage class combinations for SA 5 + 6 data with adjustments for learning, 1961-72. (OTSI = otter trawl side; OTST = otter trawl stern; PS = purse seine; PT = pair trawl; LT = line trawls; HL = handlines; DGN = drift gillnet.)

Gear	Tonnage class (GRT)	USA	Canada	USSR	Spain	Poland	FRG	GDR	Romania	Japan	Bulgaria
	0-50	1.00	0.69	—	—	—	—	—	—	—	—
	51-150	1.29	0.89	—	—	—	—	—	—	—	—
OTSI	151-500	1.73	1.20	1.77	—	—	2.69	—	—	—	—
	501-900	—	1.58	2.34	—	2.28	3.55	1.40	—	—	—
	901-1800	—	—	3.73	—	3.63	5.65	2.23	—	—	—
	0-50	3.34	2.31	—	—	—	—	—	—	—	—
	51-150	0.94	0.65	—	—	—	—	—	—	—	—
OTST	151-500	1.74	1.21	—	—	—	—	—	—	—	—
	501-900	2.41	1.66	2.47	—	—	3.74	1.47	—	—	—
	901-1800	—	—	6.56	5.63	6.38	9.94	3.92	—	—	—
	> 1801+	—	—	9.09	—	8.84	13.77	5.43	5.04	3.22	6.54
PS	> 50 GRT	7.96	5.51	8.17	—	—	—	—	—	—	—
	< 50 GRT	0.96	—	—	—	—	—	—	—	—	—
PT	All	3.94	—	4.04	3.47	—	—	—	—	—	—
LT	All	0.46	0.32	—	—	—	—	—	—	—	—
HL	All	0.14	—	—	—	—	—	—	—	—	—
DGN	All	0.12	—	0.13	—	—	—	—	—	—	—
OTSI (pelagic)	601-900	—	—	—	—	—	—	0.83	—	—	—
	901-1800	—	—	—	—	—	—	0.30	—	—	—

Finfish catch per standardized effort was then estimated for each year by dividing the total annual catch of the categories associated with this effort by the standardized effort. Finally, the total annual finfish catch over all categories, including those catches from gear-country combinations which were excluded from the analysis of variance (Table 9), was divided by the yearly catch per standardized effort to obtain the total standardized effort in each year for SA 5 + 6.

Standardized effort for SA 6 prior to 1968 for countries other than the USA was estimated by dividing the catch in SA 6 by the corresponding SA 5 catch per standardized effort for that year. This policy was justified by the fact that these countries fished primarily on stocks which migrate between SA 6 and SA 5. Comparison of SA 5 + 6 catch per effort by country and vessel categories for 1968-72 supports this contention (Table 10). The effort for the USA fishery in SA 6 for 1961-67 was estimated by

TABLE 9.   Subarea 5 + 6 catch data (tons) for which days fished was not reported.

| Country | \multicolumn{12}{c}{Year} |
	1961	1962	1963	1964	1965	1966	1967	1968	1969	1970	1971	1972
Canada	27	137	93	—	1,091	2,997	9,564	36,341	10,309	7,496	30,999	14,718
France	—	—	—	—	—	—	—	—	—	—	—	296
Iceland	—	—	—	—	—	—	—	292	12,786	—	—	—
Italy	—	—	—	—	—	—	—	—	—	—	—	4,000
Japan	—	—	—	—	—	—	331	7,212	16,251	28,795	27,673	—
Norway	—	—	—	—	—	—	—	—	1,224	—	—	—
Romania	—	—	—	—	1,982	3,433	—	—	—	—	—	—
Spain	—	—	—	—	—	—	—	—	—	—	4,197	7,546
USSR	—	201,224	7,960	22,393	34,464	—	—	—	—	—	—	75,704
USA	70,819	43,163	166,789	152,855	155,585	127,899	115,084	106,563	96,180	77,733	74,594	69,062
Others	—	—	2	10,213	5,722	—	202	263	6,516	8,998	108,180	2,381
Total (1)	70,819	244,524	174,844	185,461	198,844	134,329	125,181	150,671	143,266	123,022	245,643	173,707
Total catch[a] (2)	344,286	472,263	650,825	786,346	954,808	988,568	759,881	942,762	1,029,392	840,267	1,124,872	1,144,597
Percent[b] (1) (2)	21	52	27	24	21	14	16	16	14	15	22	15

[a] Total catch = catch of finfish (excluding sharks, billfishes, tunas, swordfish, menhaden), and squid.
[b] All years combined = 20%.

TABLE 10.   Catch per effort (days fished) in tons for SA 5 + 6.

| Year | Country | Vessel class | GRT | \multicolumn{2}{c}{Catch per effort} |
				SA 5	SA 6
1971	Poland	OTSI	501-900	7.3	8.5
	Poland	OTST	901-1800	32.6	35.4
	Poland	OTST	1801 +	35.6	42.3
	Romania	OTST	1801 +	17.0	25.4
	USSR	OTSI	151-500	6.2	5.8
	USSR	OTSI	501-900	8.2	8.5
	USSR	OTST	1801 +	33.6	38.1
1970	Poland	OTSI	501-900	7.9	12.0
	Poland	OTST	1801 +	27.2	40.9
	USSR	OTSI	151-500	5.9	7.2
	USSR	OTSI	501-900	8.7	9.2
	USSR	OTST	1801 +	34.2	47.6
1969	Poland	OTSI	501-900	7.9	10.2
	Poland	OTST	1801 +	23.0	26.2
	USSR	OTSI	151-500	5.6	6.9
	USSR	OTSI	501-900	27.5	9.1
	USSR	OTST	1801 +	41.3	31.5
1968	Poland	OTSI	501-900	11.4	12.5
	Poland	OTST	1801 +	25.6	21.6
	USSR	OTSI	151-500	6.8	8.1
	USSR	OTST	1801 +	42.9	50.1

TABLE 11. SA 5 + 6 total finfish plus squid catches, unadjusted effort, standardized effort with (without) learning, catch per effort and catch per standardized effort with (without learning) for the years 1961-72.

Year	Unadjusted effort	Total stand-ardized effort	Standard effort with learning adjustment	Total catch	Catch per effort	Catch per standard effort	Catch per standard effort (learning)
1961	42,348	61,590	53,879	344,286	8.13	5.59	6.39
1962	60,780	110,342	108,816	472,263	7.77	4.28	4.34
1963	66,683	123,262	108,834	650,825	9.76	5.28	5.98
1964	71,812	150,353	165,896	786,346	10.95	5.23	4.74
1965	70,884	161,558	169,895	954,808	13.47	5.91	5.62
1966	68,698	175,278	191,583	988,568	14.39	5.64	5.16
1967	68,892	137,411	143,104	759,881	11.03	5.53	5.31
1968	84,100	171,723	180,260	942,762	11.21	5.49	5.23
1969	86,576	210,941	221,137	1,029,391	11.89	4.88	4.65
1970	75,905	171,834	182,667	840,267	11.07	4.89	4.60
1971	81,749	230,035	267,190	1,124,872	13.76	4.89	4.21
1972	91,203	263,126	315,316	1,144,597	12.55	4.35	3.63

dividing the yearly catches by the 1968-70 average USA catch per standardized effort for SA 6. The stocks fished primarily by the USA in this area, *e.g.* fluke, scup, black sea bass, etc., were different from those in the major fisheries in Div. 5Z. If these stocks had been decreasing over this period, an overestimate of effort would result. However, this bias would have a minor effect on overall results, because the USA yearly catches in SA 6 were always small (between 75,000 and 124,000 tons) relative to the total. Fishing by the distant water fleets in SA 6 with accompanying by-catch of species sought by the USA was minimal in the period 1963-65 (nominal catches less than 50,000 tons). The combined results of the above computations are presented in Table 11 and Fig. 5.

Relationships Between Fishing Intensity and Yield

The relationships between fishing intensity and yield were examined in three ways: (1) relative changes in finfish biomass measured by research vessel surveys were related to relative changes in total fishing intensity estimated in this paper, (2) data from individual species assessments (based on commercial catch and effort data and research vessel survey data) were combined to estimate the total potential yield, and (3) annual total catch and total standardized effort as estimated herein were used in an equilibrium yield to describe the equilibrium relationship between catch and standardized effort.

Changes in biomass as estimated from research vessel surveys

Estimates of relative changes in biomass of ground-fish and flounder species for Gulf of Maine (Strata 26-30, 36-40), Georges Bank (Strata 13-25), and Southern New England (Strata 1-12) areas (Fig. 3) of SA 5 + 6 were calculated by comparing mean catch per tow in pounds (1

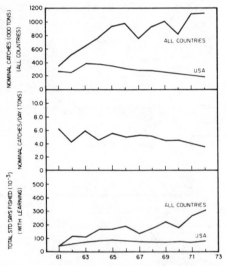

Fig. 5. Total commercial landings of finfish (excluding swordfish, tunas, billfishes, menhaden) plus squid, landings per day, and total standardized days fished (with learning) for SA 5 + 6 plotted against time.

pound = 0.454 kg) for USA autumn bottom trawl surveys in 1963-1965 with the mean for 1970-72 (see Clark and Brown MS 1975, and Grosslein MS 1971, for detailed statistics). With few exceptions, there were substantial declines in abundance (Table 12).

An estimate of the relative change in biomass for all of SA 5 + 6 was made by pooling the survey results for all areas. This set of sampling strata covered almost all of SA 5 and Div. 6A. However, since the bulk of the major stocks

TABLE 12.   Mean catch per haul (pounds) on USA *Albatross IV* autumn surveys for 1963-65 and 1970-72, and percent change from 1963-65 to 1970-72.[a]

Species	Southern New England[b]			Georges Bank[h]			Gulf of Maine[h]		
	1963-65	1970-72	% Change	1963-65	1970-72	% Change	1963-65	1970-72	% Change
Haddock	8.1	0.4	−95	134.1	14.2	−89	48.6	9.3	−81
Cod	3.8	2.1	−45	18.6	20.6	+11	23.9	20.9	−13
Redfish	—	—	—	4.4	7.6	+73	73.4	61.8	−16
Silver hake	13.6	8.2	−40	6.4	4.4	−31	30.7	8.6	−72
Red hake	13.3	10.2	−23	8.4	2.9	−65	4.9	2.3	−53
White hake	1.7	0.9	−47	1.9	4.9	+158	15.3	35.6	+133
Pollock	0.03	—	—	4.5	2.7	−40	14.7	12.8	−13
Yellowtail flounder	23.9	35.4	+48	16.3	7.9	−52	0.8	0.4	−50
Winter flounder	6.3	4.7	−25	4.3	5.3	+23	0.9	0.5	−44
All other flounders	8.3	4.5	−46	9.6	6.4	−33	17.9	12.1	−32
Sculpin	2.2	2.6	+18	6.2	4.4	−29	0.3	0.3	—
Ocean pout	1.0	0.3	−70	2.6	0.4	−85	—	0.2	—
Angler	12.0	10.3	−14	8.1	2.2	−73	5.2	6.3	+21
Other groundfish	3.5	2.1	−40	1.2	0.8	−33	4.1	5.7	+39
Skates	26.2	13.7	−48	50.0	29.4	−48	26.2	26.5	+1
Dogfish	263.5	83.4	−68	6.8	12.1	−78	59.3	34.9	−41
Squid	3.3	12.3	+273	1.5	3.1	+107	0.3	.7	+133

[a] Squid records were not adequately kept in the early years of the survey. Essentially none recorded in 1963, therefore the average for 1964-65 was used as a minimum estimate.
[h] See Fig. 3 for area description.

TABLE 13.   Comparison of mean catch per tow (pounds) on *Albatross IV* autumn surveys in Div. 5Z and Div. 6A for the two periods 1963-65 and 1970-72, the percentage change relative to the earlier period, and cumulative landings from 1962-72, for groundfish, skates, and herring.

Species	Mean catch per tow		% change	Cumulative landings for 1963 72 (000's tons)
	1963-65[a]	1970-72[a]		
Haddock	63.0	9.1	− 87	569
Cod	15.9	14.9	− 6	400
Redfish	28.9	25.3	− 12	121
Silver Hake	17.6	7.1	− 60	1,597
Red Hake	8.6	5.0	− 43	508
White Hake	6.8	14.8	+118	20
Pollock	7.0	5.6	− 20	98
Yellowtail Flounder	12.9	13.7	+ 6	387
Winter Flounder	3.7	3.3	− 11	135
Other Flounder	12.3	7.9	− 36	123
Sculpin	2.8	2.3	− 18	47
Ocean Pout	1.1	0.3	− 73	82
Angler	8.2	6.2	− 24	17
All other groundfish	3.0	3.0	—	242[b]
Skates	33.7	23.4	− 31	49
Dogfish	106.7	42.8	− 60	62
Total groundfish, dogfish, skates, and flounders	332.2	183.7	− 45	4.457
Squid	1.6	5.1	+215	111
Herring	—		−93	2,478
Mackerel	—		−37	1,160
	Weighted mean[c] percentage change		− 55	8,206

[a] Calculated by pooling the means shown in Table 12 for Gulf of Maine, Georges Bank, and Southern New England into a single stratified mean representing Div. 5Y, 5Z, and 6A.
[b] For the "all other groundfish" category, the component of "groundfish NS" in each ICNAF Statistical Bulletin was averaged for the years 1967-72 and prorated for the period. That average of 8,300 tons contrasts with the average of 63,000 tons for 1963-67 before more detailed species breakdowns were available.
[c] Weights equivalent to cumulative landings in 1963-72.

are found east of Hudson Canyon (which is close to the boundary line between Div. 6A and Div. 6B (Fig. 1)) in the autumn, the data were considered adequate to represent changes in the whole of SA 6. The pooled mean catch per tow of all but four of the species or species groups declined from 6 to 90% (Table 13). The four exceptions are the catches of white hake (118% increase), yellowtail flounder (6% increase), other groundfish (no change), and squids (215% increase). The small increase in catch per tow of yellowtail is due to a large catch in the 1972 survey. This may be anomalous since it was not consistent with commercial yellowtail catches nor with previous and subsequent survey abundance indices of the year-classes involved (Parrack, MS 1974). Silver hake and dogfish declined 60%; red hake, skates, and miscellaneous flounders all declined between 30% and 40%; cod and winter flounder dropped about 10%; ocean pout showed a decline of 73%, and anglers and miscellaneous groundfish declined approximately by one third. The overall decline of all of these species pooled was 45%. The squid abundance indices show no evidence of a trend during this period; this is not surprising since directed fisheries for squid did not begin until 1970.

An estimate of the decline for herring was made using herring abundance indices from USA spring surveys which first began in 1968 (Fig. 6). The spring surveys begin in March when herring are concentrated south of Cape Cod. The abundance indices[5] shown in Fig. 6

represent sampling strata 1-12 and 61-76 combined (Nantucket to Cape Hatteras). The slope (estimated by least squares linear regression) of the line was −1.95 (ln scale) which gave a decline of about 93% in the period 1963-1972. This estimate corresponded closely to that based on other data (ICNAF, 1972b).

An estimate of the decline of mackerel was based on the USA spring surveys of 1967-74 as analyzed by Anderson (MS 1974a). A least squares linear regression through stratified means of ln (pounds + 1) (Fig. 7) eliminating the outlier value for 1969, gave a slope of −0.078 indicating a decline of 37% since 1967. There was no observed decline in the mackerel population until after 1967 (Anderson, MS 1974a).

The decline in total biomass of finfish in SA 5 and Div. 6A was calculated by weighting the percentage decline of groundfish, herring, squid, and mackerel (Table 13) in proportion to the total landings of those species groups in the 10-year period 1963-72. The resulting weighted change indicated about a 55% drop in total biomass of these species during the last decade (Table 13). The nominal catches were considered the best available proportional measure of the contribution of such species in the biomass. The estimate of the overall decline thus derived could be less than the true decline because nominal catches of some miscellaneous groundfish

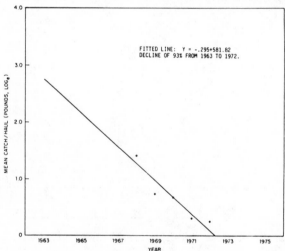

FITTED LINE: Y = −.295+581.82
DECLINE OF 93% FROM 1963 TO 1972.

Fig. 6.   Plot of herring abundance indices from USA *Albatross IV* spring bottom trawl surveys in strata 1-12 and 61-76 (Nantucket to Cape Hatteras).

[5] Abundance indices for herring were calculated using a mean of the natural logarithm (number of fish + 1).

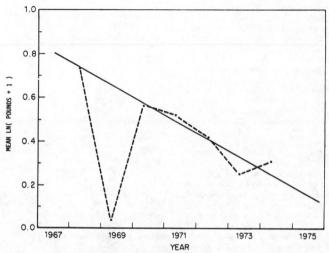

Fig. 7.   Least squares regression fit (mackerel, SA 5 + 6) of mean 1n (pounds + 1) from USA bottom trawl spring
surveys through time. Data for 1969 was excluded from the calculations (Anderson, 1974a).

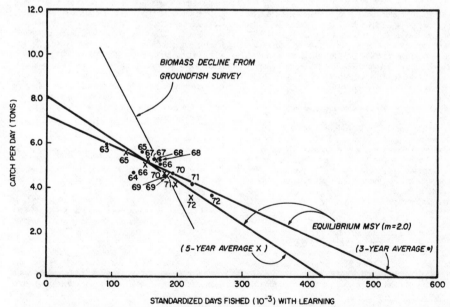

Fig. 8.   Catch per standardized effort plotted against standardized days fished (with learning) for data from SA 5 + 6, 1961-72 (linear regression
estimated using all years). Also, estimate of biomass decline of groundfish, skates, and herring from bottom trawl surveys (see Table 13).

species (particularly ocean pout, angler and skates) were not adequately reported in earlier years, either being included in categories such as other groundfish or other fish in Table 1 of the Statistical Bulletin Vol. 13 and 22 (ICNAF, 1965a and 1974a), or being unreported entirely (perhaps as discard), and these species showed major declines. The percentage declines were measured from an initial point of time (1963-1965) prior to which many of the stocks concerned had already been harvested in moderate to severe (say 25 to 100% of MSY) degrees. Thus, the overall decline from unfished abundance levels was greater than the 1963-72 decline.

It has been postulated, based on the Schaefer yield model, that MSY is obtained at a stock size about one-half the virgin stock size. The estimated decline of 55% since 1963, from research vessel surveys (Table 13), thus implied a significant degree of overfishing (Fig. 8). The average standardized effort estimates for 1963-65 and 1970-72 were used to position end points of the line with respect to the abscissa, and the line was fitted through the mean of commercial catch per standardized effort and standardized effort for the decade, to position it with respect to the ordinate. This implied a 65% decline in catch per effort between 1963 and 1972 concurrent with the increase in fishing intensity during the period.

An even greater rate of decline in biomass since 1967 was indicated by USSR autumn research surveys in the Southern New England area and by both USA and USSR autumn surveys since 1967 for the Mid-Atlantic area to the south (strata 61-76) (ICNAF 1973c) and provided further evidence of overfishing. USSR and USA autumn survey indices for all finfish for SA 6 declined about 80% and 70% respectively in this later period of years.

Individual stock assessments and total yield

Results from individual species assessment studies and review of historic catches were used to estimate a composite MSY for the combined finfish stocks in SA 5 + 6 (Table 14). The ICNAF Assessments Subcommittee Reports for 1962-74 (ICNAF, 1962b-74b) provided the general source for the estimates, other than yellowtail catches in the last ten years (1963-72). The first silver hake assessment was presented by Anderson for 1972[6] followed in 1974 by further reports (Anderson, MS 1974b and Rikhter, MS 1974a and b). The MSY for silver hake was taken to be equal to the recommended total allowable catches (TAC's) for the Div. 5Z and SA 6 stocks in 1973 and 1974 (ICNAF, 1973b, 1974b) plus the estimate of MSY for the Gulf of Maine (Div. 5Y) stock given in the 1972 Assessment Report (ICNAF, 1972b). The MSY for red hake was taken as the TAC recommended for 1973

TABLE 14.   Individual stock assessments for finfish stocks in SA 5 + 6.

Species	Estimate of MSY (000's tons)
Herring	335
Mackerel	310
Silver Hake	200
Squid	80
Red Hake	70
Haddock	50
Cod	45
Yellowtail Flounder	37
Redfish	30
Pollock	20[a]
Other Flounder	25
Other Finfish	150
Sum of species assessments	1,352

[a]MSY estimated to be 50,000 tons including Div. 4VWX (ICNAF, 1972b), 20,000 tons based on catch ratios assigned to SA5.

(ICNAF, 1973b). For pollock, cod (see also Brown and Heyerdahl, MS 1972) and redfish, estimates of MSY corresponded to the recommended TAC values for 1973 (ICNAF, 1973b). Estimates of MSY for haddock and yellowtail flounder were taken from the 1973 ICNAF Redbook (ICNAF, 1973b). The ICNAF Assessments Subcommittee provided a preliminary assessment of squid *(Loligo)* in 1972 and estimated an MSY of 50,000-80,000 tons. In order to include the yield of *Illex* squid, a value of 80,000 tons for the two genera combined was assumed (ICNAF 1973b). Individual assessments for herring in Div. 5Z and SA 6 (Schumacher and Anthony, MS 1972) and mackerel in Div. 5Z + SA 6 (Anderson, MS 1973) indicated MSYs of 285,000 and 310,000 tons respectively. The herring stock in Div. 5Y was estimated to have an MSY value of 50,000 tons (ICNAF, 1973b). (See also Anthony and Brown, MS 1972). Combining all MSY estimates for the entire species complement gave a total of 1,352,000 tons as a projected MSY value for the total finfish biomass.

The MSY estimates probably are too high for many of the species in SA 5 + 6 which have been subjected to heavy fishing mortality only recently. A high proportion of the available data represented an expanding fishery which was harvesting accumulated biomass rather than only yearly productivity. In addition, these single species assessment models have not explicitly accounted for species interactions. (See Pope and Harris, MS 1975 for discussion of implications of competition between species on yield, based on data of Stander and Le Roux (1968)).

These principles were perhaps of greatest significance in terms of the total biomass of herring and mackerel, for which the assessed MSY values were estimated during a time period when there were two

---

[6] Anderson, E. A. 1972. Mimeograph Report on file at National Marine Fisheries Service, Northeast Fisheries Center, Woods Hole, Massachusetts, USA 02543.

extremely good year-classes in the fisheries, and when a rapid monotonic increase in fishing effort occurred. Lett, *et al.* (MS 1975) have presented a model for the Gulf of St. Lawrence mackerel stock (at least some of the mackerel in SA 5 + 6 are spawned in the Gulf of St. Lawrence (Moores *et al.*, 1975) indicating that mackerel year-class strength may be density dependent on the combined abundance of herring and mackerel. Food studies indicate a possible high degree of competition between herring and mackerel (Maurer, MS 1975). Furthermore, herring and mackerel, at least in recent history, have not maintained a high biomass concurrently, but rather have fluctuated inversely, with the mackerel showing an increase in abundance while the herring have declined. The strong herring year-classes were 1960 and 1961, while those for mackerel were 1967 and 1968. Consequently, a more accurate description of the potential yield for the two species might be estimated by looking at their average combined catches. Table 15 presents the quantities of herring and mackerel landed by all countries over the period of the analysis. The average annual nominal catch for the two species combined (1961-72) is 336,000 tons. Substituting this combined figure for the individual assessment estimates resulted in reducing the projected MSY value for the total biomass to 1,043,000 tons.

TABLE 15. Total annual nominal catches (000's tons) from SA 5 + 6 for herring and mackerel, 1961-72.

Year	Herring	Mackerel	Total
1961	94	1	95
1962	224	1	225
1963	167	2	169
1964	159	2	161
1965	74	5	79
1966	172	9	181
1967	257	23	280
1968	436	60	496
1969	361	113	474
1970	303	210	513
1971	314	349	663
1972	237	387	624
Average	233	96	330

Surplus yield modeling

An estimate of MSY was calculated for the above selected finfish community as a whole in SA 5 and 6, using the generalized stock production model approach discussed by Schaefer (1954).

Fitted curves derived from this type of analysis were considered to represent the equilibrium, or long-term average, expected yields. However, as has been discussed earlier in this paper, a rather consistent and rapid increase in effort occurred in SA 5 + 6, particularly during the first part of the 1960's. When such large and consistent increases or decreases in fishing effort exist,

the fitted curves will tend not to describe the true situation unless the population can react instantaneously in adjusting its productivity to the new density structure. If the population cannot do this, the effects of fishing effort in any given year will be dependent upon the cumulative effect of previous years' effort. Gulland (1961) suggested that, in order to account for this effect, an average of effort over previous years should be taken as the effort applicable to the final year, where the averaging occurs over the mean number of years that a year-class contributes significantly to the catch. The number of years to be averaged is, therefore, a function of the total mortality rate, age at maturity, and changes in growth rate.

For the SA 5 + 6 fish stocks in an equilibrium state which provided maximum yields, an average year-class contributes significantly to the catch over about a 3-year period. However, for the period covered in this study, some significant events should be considered. For herring, two very good year-classes were spawned in 1960 and 1961, and these fish carried a major share of the fishery for 5-6 years (Schumacher and Anthony, MS 1972; Anthony and Brown, MS 1972). Haddock have existed virtually without any significant recruitment since the 1962 and 1963 year-classes, and thus these year-classes contributed significantly over 7-8 years (Hennemuth, MS 1969; ICNAF, 1972b-74b). The mackerel fishery has been harvesting principally the same two year-classes, 1966 and 1967, since the fishery increased in 1968 through 1972 (ICNAF, 1974b). Silver hake, with a more stable distribution, showed a 3-4 year pattern of contribution (Anderson, MS 1972), as did yellowtail flounder (Brown and Hennemuth, MS 1971). Response to changes in biomass through fishing are reflected by recruitment and thus the age at maturity is important. Most of the major species mature between 3 and 4 years of age and contribute to the spawning stock for several years. Growth rate and age at maturity can also change in response to changes in stock abundance. No changes have been reported in these parameters that would alter the use of 3- to 5-year averaging periods. Consequently, running averages of total effort were made over 3-, 4-, and 5-year lag or delay-time periods to cover the possible range of this effect.

Solutions of the Schaefer model were obtained by computing least squares linear regressions of catch per standardized effort in year i on an averaged standardized effort as defined above (both with and without learning), terminating with year i. A series of linear regressions were calculated corresponding to data sets beginning with 1968-72 and successively adding earlier years' data back to 1961 (Tables 16 and 17 and Fig. 8). Each regression was then expressed as a yield *versus* effort parabola to obtain the equilibrium catches and corresponding effort in terms of the USA side trawler 0-50 GRT standard days fished (Tables 16 and 17 and Fig. 9). Coefficients of

TABLE 16. Estimate of optimum standardized effort, MSY, catch per standardized effort, and coefficient of determination for SA 5 + 6 catch and standardized effort data applied to the Schaefer model. Gulland's averaging method to determine effort in year i was used on the basic data. [a]

		3 Years	4 Years	5 Years
Optimum	1963-72	271,857		
standardized	1964-72	271,681	291,031	
effort	1965-72	224,375	216,987	213,740
	1966-72	225,709	217,342	202,690
	1967-72	227,835	212,405	194,369
	1968-72	235,535	220,108	193,089
	1969-72	257,552	241,430	209,264
MSY	1963-72	981,474		
	1964-72	980,942	996,064	
	1965-72	931,365	898,352	865,270
	1966-72	931,772	898,458	859,465
	1967-72	901,001	898,705	860,987
	1968-72	931,451	896,762	861,988
	1969-72	940,004	899,972	852,617
Catch per	1963-72	3.61		
standardized	1964-72	3.61	3.42	
effort	1965-72	4.15	4.14	4.05
	1966-72	4.13	4.13	4.24
	1967-72	4.09	4.23	4.43
	1968-72	3.95	4.07	4.46
	1969-72	3.65	3.73	4.07
Coefficient of	1963-72	0.77		
determination	1964-72	0.67	0.57	
	1965-72	0.96	0.95	0.94
	1966-72	0.94	0.93	0.94
	1967-72	0.94	0.93	0.96
	1968-72	0.93	0.93	0.94
	1969-72	0.97	0.99	0.99

[a] Data adjusted for learning.

TABLE 17. Estimate of optimum standardized effort, MSY, catch per standardized effort, and coefficient of determination for SA 5 + 6 catch and standardized effort data applied to the Schaefer model. Gulland's averaging method to determine effort in year i was used on the basic data. No learning.

		3 Years	4 Years	5 Years
Optimum	1963-72	385,275		
standardized	1964-72	270,246	302,939	
effort	1965-72	227,747	222,918	218,769
	1966-72	236,249	229,864	210,140
	1967-72	248,347	238,370	213,249
	1968-72	263,172	253,553	216,526
	1969-72	338,600	322,727	260,661
MSY	1963-72	1,274,828		
	1964-72	998,504	1,081,719	
	1965-72	964,367	939,613	910,024
	1966-72	974,312	947,857	899,647
	1967-72	986,631	957,489	902,401
	1968-72	1,007,138	976,682	905,414
	1969-72	1,117,381	1,088,310	959,131
Catch per	1963-72	3.31		
standardized	1964-72	3.69	3.57	
effort	1965-72	4.23	4.21	4.16
	1966-72	4.12	4.12	4.28
	1967-72	3.97	4.02	4.23
	1968-72	3.82	3.85	4.18
	1969-72	3.33	3.37	3.67
Coefficient of	1963-72	0.42		
determination	1964-72	0.65	0.56	
	1965-72	0.87	0.89	0.94
	1966-72	0.82	0.85	0.93
	1967-72	0.82	0.82	0.90
	1968-72	0.76	0.76	0.85
	1969-72	0.61	0.74	0.87

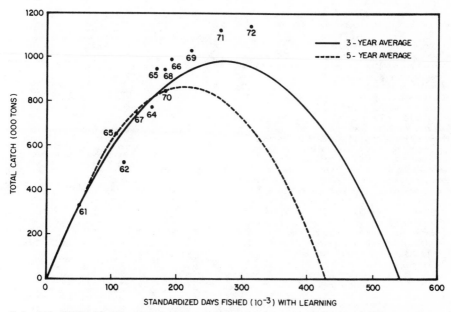

Fig. 9.   Total catch (finfish plus squid) vs. standardized effort (with learning) for SA 5 + 6, 1961-72, using a 3-year average over standardized effort (days fished), and a 5-year average over standardized effort. Original data points (catch vs. standardized days fished) are plotted.

determination for all data sets, adjusted for learning, ranged from 0.57 to 0.99 with 15 of the 17 values being above 0.9; for data sets not adjusted for learning, the coefficients ranged from 0.42 to 0.97 with 3 values above 0.9 and 11 above 0.8. The range of parameter estimates derived from the yield *versus* standardized effort parabolas was less for data sets adjusted for learning than for those sets that had not been adjusted. However, this would be expected since learning accounted for a major source of variation or bias in estimating population size. For both data sets, *i.e.* with and without a learning adjustment, the best fit to the Schaefer model occurred when data for the years 1965-72 and later were used. The years prior to 1965 were those for which data were less complete, and for which the consequential changes associated with learning had their greatest effect. In addition, in those years effort was directed towards fewer species than in later years.

Discussion

Results of these analyses demonstrated a rapid and substantial increase in fishing intensity (a factor of 6, with data adjusted for learning), and a concurrent marked decline in abundance (about 55%) for the offshore finfish community in SA 5 + 6 during the period 1961-72. Yield *versus* standardized effort parabolas, estimated using the Schaefer approach, indicated that fishing mortality since 1968 had exceeded that level which would result in sustaining a maximum yield for the fishery under equilibrium conditions. The average MSY for the data sets for 1965-72, using 3-year, 4-year, and 5-year averaging methods for fishing effort, was 898,329 tons for data adjusted for learning and 938,000 tons for data without adjustment for learning (Tables 16 and 17).

The projected MSY value from the Schaefer model, approximately 900,000 tons, was somewhat lower than the composite MSY estimated earlier from single assessment summations of 1,300,000 tons, but, as discussed in that section, it may riot be reasonable to assume that these individual assessments can be summed for the total biomass yield. It was similar to the 1,000,000 tons estimated from assessment summations after discounting for a hypothesized mackerel-herring interaction.

The estimated MSY values were for long-term equilibrium yields. Because the fishery had been subject to overfishing (as indicated in this case by the Schaefer model), the sustainable yield would be considerably less than the estimated MSY value.

The effort giving MSY was 218,367 standard days fished when adjusted for learning and 223,145 standard days fished without the learning adjustment. These are in the same order of magnitude as the respective efforts estimated for 1969, which were 221,137 and 210,914 standard days fished respectively (Table 11). The averages of catch and standardized effort for the years following 1968 (except for 1970) exceed the projected allowable values for maximum sustained yield of the fishery and hence indicated a condition of overfishing. For example, the percentage reductions in standardized effort from the 1972 observed levels required to reach the average MSY level resulting from the above fits to the Schaefer model ranged from 30.7% to 27.7% respectively for data with and without an adjustment for learning.

Using the survey cruise estimate of population decline of 8% per year for 1969-71 and assuming that the 1969 effort was equal to that giving the MSY, then the 1972 fishing effort was 22% in excess of that needed to take the total catch equivalent to the MSY.

References

ANDERSON, E. D. MS 1973. Assessment of Atlantic mackerel in ICNAF Subarea 5 and Statistical Area 6. *Annu. Meet. int. Comm. Northw. Atlant. Fish. 1973,* Res. Doc. No. 14, Serial No. 2916 (mimeographed).
        MS 1974a. Relative abundance of Atlantic mackerel in ICNAF Subrea 5 and Statistical Area 6. *Ibid, 1974,* Res. Doc. No. 10, Serial No. 3156 (mimeographed).
        MS 1974b. Assessment of the red hake in ICNAF Subarea 5 and Statistical Area 6. *Ibid.,* Res. Doc. No. 19, Serial No. 3165 (mimeographed).
ANTHONY, V. C., and B. E. BROWN. MS 1972. Herring assessment for the Gulf of Maine (ICNAF Division 5Y) stock. *Annu. Meet. int. Comm. Northw. Atlant. Fish. 1972,* Res. Doc. No. 13, Serial No. 2696 (mimeographed).
BROWN, B. E., J. A. BRENNAN, E. G. HEYERDAHL, M. D. GROSSLEIN, and R. C. HENNEMUTH. 1973. Effect of by-catch on the management of mixed species fisheries in Subarea 5 and Statistical Area 6, *Int. Comm. Northw. Atlant. Fish. Redbook 1973,* Part III: 217-232.
BROWN, B. E., and R. C. HENNEMUTH. MS 1971. Assessment of the yellowtail flounder fishery in Subarea 5. *Annu. Meet. int. Comm. Northw. Atlant. Fish. 1971,* Res. Doc. No. 14, Serial No. 2599 (mimeographed).
BROWN, B. E., and E. G. HEYERDAHL. MS 1972. An assessment of the Georges Bank cod stock (Div. 5Z). *Ibid., 1972,* Res. Doc. No. 117, Serial No. 2831 (mimeographed).
CLARK, S. H., and B. E. BROWN. MS 1975. Changes in biomass of finfish and squid in ICNAF SA 5 and 6, 1963-1974, as evidenced by *Albatross IV* autumn bottom trawl surveys. *Annu. Meet. int. Comm. Northw. Atlant. Fish. 1975,* Res. Doc. No. 65, Serial No. 3549 (mimeographed).
GROSSLEIN, M.D. MS 1971. Some observations on the accuracy of abundance indices derived from research vessels surveys. *Annu.*

Meet. int. Comm. Northw. Atlant. Fish. 1971, Res. Doc. No. 59, Serial No. 2598 (mimeographed).
        1973. Mixture of species in Subareas 5 and 6. *Int. Comm. Northw. Atlant. Fish. Redbook 1973,* Part III: 163-208.
GULLAND, J. A. 1961. Fishing and the stocks of fish at Iceland. *Fish. Invest., Lond. (2),* 23(4): 32 p.
HENNEMUTH, R. C. MS 1969. Status of the Georges Bank haddock fishery. *Annu. Meet. int. Comm. Northw. Atlant. Fish. 1969,* Res. Doc. No. 90, Serial No. 2256 (mimeographed).
ICNAF. 1961a. Report of Standing Committee on Research and Statistics. App. I. Working Group on Assessments. *Int. Comm. Northw. Atlant. Fish., Redbook 1961,* p. 3-17.
        1961b. List *Fish. Vessels int. Comm. Northw. Atlant. Fish., 1959.*
        1962a-1974a. *Stat. Bull. int. Comm. Northw. Atlant. Fish.,* Vols. 10-22.
        1962b-1974b. Report of Standing Committee on Research and Statistics. Subcommittee on Assessments. *Int. Comm. Northw. Atlant. Fish., Redbook,* Part I.
        1967c. List *Fish. Vessels int. Comm. Northw. Atlant. Fish.,* 1965.
        1972c Report of Meeting of Panel 5, Proceedings No. 7, App. I-XV, *Meet. Proc. int. Comm. Northw. Atlant. Fish. 1972,* 22nd Annu. Meet. and Sp. Meet. Herring, 1972, Part I: 33-73.
        1973c. Report of the Standing Committee on Regulatory Measures, Proc. No. 5, App. I-III, *Meet. Proc. int. Comm. Northw. Atlant. Fish. 1973,* Sp. Comm. Meet. Jan. 1973 and 23rd Annu. Meet. June 1973, Part II: 73-119.
        1973d. List *Fish. Vessels int. Comm. Northw. Atlant. Fish., 1971.*
LETT, P. F., A. C. KOHLER, and D. N. FITZGERALD. MS 1975. The influence of temperature on the interaction of the recruitment mechanisms of Atlantic herring and mackerel in the Gulf of St. Lawrence. *Annu. Meet. int. Comm. Northw. Atlant. Fish. 1975,* Res. Doc. No. 33, Serial No. 3512 (mimeographed).
MAURER, R. MS 1975. A preliminary description of some important feeding relationships. *Annu. Meet. int. Comm. Northw. Atlant. Fish. 1975,* Res. Doc. No. 130, Serial No. 3681 (mimeographed).
MOORES, J. A., G. H. WINTERS, and L. S. PARSONS. 1975. Migrations and biological characteristics of Atlantic mackerel *(Scomber scombrus)* occurring in Newfoundland waters. *J. Fish. Res. Bd. Canada,* **8:** 1347-1357.
PARRACK, M. L. MS 1974. Status review of ICNAF Subarea 5 and Statistical Area 6 yellowtail flounder stocks. Annu. Meet. int. Comm. Northw. Atlant. Fish. 1974, Res. Doc. No. 99, Serial No. 3335 (mimeographed).
POPE, J. G., and O. C. HARRIS, MS 1975. The South African pilchard and anchovy stock complex — an example of the effects of biological interactions between species on management strategy. *Annu. Meet. int. Comm. Northw. Atlant. Fish. 1975,* Res. Doc. No. 133, Serial No. 3685 (mimeographed).
RIKHTER, V. A. MS 1974a. A forecasting method and an approximate estimate of total allowable catch of red hake from Southern New England in 1974-1975. *Annu. Meet. int. Comm. Northw. Atlant. Fish. 1974,* Res. Doc. No. 64, Serial No. 3284 (mimeographed).
        MS 1974b. The estimation of total allowable catch of red hake from the Georges Bank for 1975. *Ibid.,* Res. Doc. No. 65, Serial No. 3285 (mimeographed).
ROBSON, D. S. 1966. Estimation of the relative fishing power of individual ships. *Res. Bull. int. Comm. Northw. Atlant. Fish.,* No. 3, p. 5-14.
SCHAEFER, M. B. 1954. Some aspects of the dynamics of populations important to the management of commercial marine fisheries. *Bull. Inter-Amer. trop. Tuna Comm.,* No. 1(2): 27-56.
SCHUMACHER, A., and V. C. ANTHONY, MS 1972. Georges Bank (ICNAF Division 5Z and Subarea 6) herring assessment. *Annu. Meet. int. Comm. Northw. Atlant. Fish. 1972,* Res. Doc. No. 24, Serial No. 2715 (mimeographed).
SNEDECOR, G. W., and W. G. COCHRAN. 1967. Statistical methods (6th edition). Iowa State University Press, Ames. Iowa, p. 484-493.
STANDER, G. H., and P. J. LE ROUX. 1968. Notes on fluctuations of the commercial catch of the South African pilchard *(Sardinops ocellata)* 1950-1965. *Invest. Rep. Div. Fish. S. Afr.,* No. 65.
STEELE, J. H. 1974. The structure of marine ecosystems. Harvard Univ. Press, Cambridge, Mass., 128 p.

The concept of the marginal yield from exploited fish stocks

J.A.Gulland

Journal du Conseil International pour l'Exploration de la Mer, **32,** 256-261, 1968.

and part of

Scientific advice on catch levels

J.A.Gulland and L.K.Boerema

Fishery Bulletin, **71,** 325-335, 1973. Pages 330-332 are reproduced here.

THE CONCEPT OF THE MARGINAL YIELD
FROM EXPLOITED FISH STOCKS

By

J. A. GULLAND*

Fisheries Depɑrtment, FAO, Rome.

When increasing fishing effort is applied to an exploited stock, the increase in total yield (the marginal yield) is less than might be estimated from the product of the increase in effort and the catch-per-unit-effort. The marginal efficiency, defined as the percentage that the actual increase is of the expected increase, decreases from near 100% for very lightly fished stocks to near zero or even negative for heavily fished stocks. The precise form of the marginal efficiency as a function of the catch is examined for two of the commonly used population models. The implication of the concepts of marginal yield and marginal efficiency for fisheries management, and the planning of fisheries development are discussed.

INTRODUCTION

An important part of the task of FAO is the provision of advice to national governments and international development agencies, such as the United Nations Development Program (UNDP), the International Bank for Reconstruction and Development (IBRD) on various plans for fishery development. Such plans commonly propose the construction of a number of modern vessels to supplement an existing fleet of smaller boats. Among the important questions to be studied in connection with the development plans, several of which relate to the costs of boat building, provision of harbour and marketing facilities etc., two are related to the state of the natural resources; these are: first, what will be the probable annual catch of each of the proposed vessels, and second, what is the likely effect of their introduction on the total catches, including those by the existing boats.

The two extreme situations are comparatively simple, and are also generally recognized. In a very heavily fished stock increased fishing will give no appreciable increase in total catch, and will perhaps result in a decrease; catches by additional new vessels will be wholly at the expense of those already fishing. On the other hand, for a very lightly fished stock the additional fishing will have a negligible effect on the stock, and on the catches by existing vessels. The situation in many fisheries is, however, an intermediate one, such that fishing is having a significant effect on the stocks, but increased fishing will result in an increase in total catch – perhaps by two or three-fold – though the catch-per-unit-effort will decrease.

---

* The views expressed in this paper are those of the author and not neccessarily those of the organization.

| J. Cons. perm. int. Explor. Mer | 32 | No. 2 | 256–61 | Copenhague, Novembre 1968 |

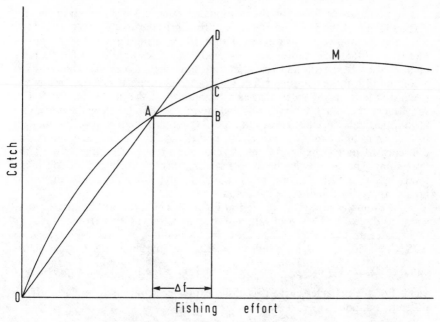

Figure 1. A general relation between total catch and fishing effort. The marginal increase in yield, BC, is less than the increase BD expected from the catch per-unit-effort.

THE MARGINAL YIELD

An example of such a situation is illustrated in Figure 1. The curve OACM describes a general relation between fishing effort and average long-term catch. An increase in effort Δf from the present situation A is represented by the line AB. The expected catch from this additional fishing effort, as calculated from the present catch-per-unit-effort, which is equal to the slope of the line OAD, will be BD. The actual increase in catch would be, however, BC which is less than BD. This increase in total catch, BC may be defined as the *marginal total catch*; the ratio BC/BD as the *marginal efficiency of the increase in effort*. The marginal efficiency tends to unity near the origin, is zero at the point of maximum catch (if there is a maximum in the particular yield curve), and will be negative for increases in effort beyond that which gives the maximum catch (when increased fishing will *decrease* the total yield). Figure 1 is drawn to show a finite increase in effort; the definition of marginal efficiency may readily be extended to relate to an infinitesimally small increase in effort, so that AC becomes the tangent to the yield curve at A. In concrete terms this means that if the marginal efficiency is 60%, and at the present level of fishing a certain type of boat catches on the average 100 tons of fish a year, then the addition of one extra boat will only increase the total catch by 60 tons. If 100 tons is only a small fraction of the total catch of, say, ten thousand tons, then the catch of the new boat will be only marginally less than 100 tons – perhaps 99 tons – but the catch by all the other boats together will decrease by just under 40 tons.

The quantitative value of the marginal efficiency has been examined for certain forms of the yield-effort curve, particularly those of SCHAEFER (1954) – a parabola – and of BEVERTON and HOLT (1957), using the latter's (BEVERTON and HOLT, 1966) yield tables. The marginal efficiencies were taken from a table computed by BEVERTON and HOLT, but not published. To express the results in a common form the marginal efficiency has been expressed as a function of the catch, the latter being calculated as a percentage of the maximum catch. For the Beverton and Holt curves this maximum has been taken as the maximum at that particular size at first capture, not taking into account possible changes in the size at first capture. Thus all the curves have two points in common, an efficiency of unity when the catch is zero, and of zero when the catch is the maximum. The curves also lie in two sectors, positive corresponding to effort levels less than those giving the maximum catch, and negative, for efforts greater than those giving the maximum.

Some examples results are plotted in Figure 2. The three broken curves correspond to three possible sets of values of the parameters in BEVERTON and HOLT's tables: a central value, $M/K = 1.25$, $C = .60$, and two extreme values, $M/K = 0.25$, $C = .30$, and $M/K = 5.0$, $C = .80$. The marginal efficiency, which is equal to $dY/df \cdot Y/f = \dfrac{dY}{Y} \Big/ \dfrac{df}{f}$, are calculated from the theoretical expression of yield. For the Schaefer curve it is easily expressed from his formulation (with different symbols):

$$Y = af(b - f) = abf - af^2,$$

where $a, b = $ constants.

$Y = $ catch
$f = $ effort
and $dY/df = ab - 2af$
and $Y_{max} = \frac{1}{4}ab^2$

so that $Y/Y_{max} = 4f(b - f)/b^2.$

The marginal efficiency is equal to the slope of the yield curve divided by the catch per unit effort, *i.e.*

marginal efficiency $= a(b - 2f)/a(b - f) = (b - 2f)/(b - f)$

Putting $Y/Y_{max} = c$, marginal efficiency $= x$, these equations reduce to $c = \dfrac{4(1 - x)}{(2 - x)^2}$. This is independent of a and b. That means for the Schaefer model the relation between the marginal efficiency and the catch (as a percentage of the maximum) is always the same. This relation is shown as the full curve in Figure 2.

This figure shows that if, for example, the present catch is only 30% of the maximum catch the marginal efficiency is, for the Schaefer model about 90%, and for the Beverton and Holt model between 70 and 85%. Thus, even at this level of exploitation, which might be thought to be sufficiently low that one need not be concerned about the effect of fishing on the level of stock, any additional effort will add between 10% and 30% less to the total catch than would be expected from a simple calculation based on the catch-per-unit effort just before the further effort was added.

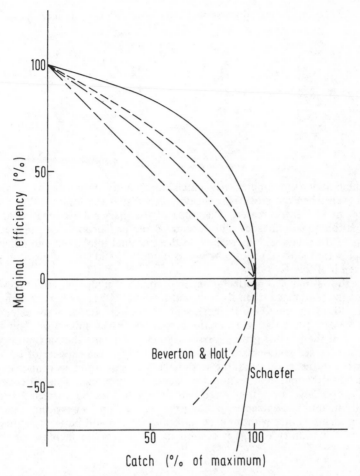

Figure 2. The marginal efficiency, as a function of total catch, for two models commonly used in fish population dynamics.

The difference between the efficiencies predicted on the basis of two models is quite large. This is due to differences in the shapes of the yield curves, and particularly to the greater curvature of the Schaefer curve near its maximum. So far as experience so far can help us distinguish between the shapes of the yield curves, it seems that the flatter curves of the Beverton and Holt model – especially near its maximum – accord better with reality.

INTERNATIONAL FISHERIES

The concept of the marginal yield can be applied also to the increase of the yield by one section of a fishery when the effort of that section alone is increased. An obvious example is an international fishery. In such a fishery a

TABLE 1. Marginal sectional efficiencies for different percentages
of the total yield

Marginal total efficiency (%)	Percentage of total yield		
	10%	50%	90%
80	98	90	82
60	96	80	64
40	94	70	46
20	92	60	28
0	90	50	10
−20	88	40	−8
−40	86	30	−26

particular nation may increase its catch significantly by increasing its fishing effort, even when the stock is so heavily fished that the total yield is not increased, or is even decreased. The value of the marginal sectional yield, and marginal sectional efficiency will depend on the proportion that the sectional yield is of the total yield, as well as on the marginal total yield and efficiency. Examples, for sectional yields making up 10%, 50% and 90% of the total yield, are given in Table 1.

This shows that if a particular section (country, company, etc.) takes only 10% of the total catch, then its marginal sectional efficiency is very high; that is, it could, by increased effort, increase its own catch almost to the extent predicted by direct calculation from the present catch-per-unit-effort. This holds true even if the level of total effort is beyond that giving the maximum yield, then its increased sectional catch would be wholly at the expense of the other sections of the fishery. Even if the sectional catch in question is as much as 50% of the total, the marginal sectional efficiency is quite high. This, of course, is why "overfishing" in a multi-sectional fishery cannot be corrected without virtually unanimous agreement among the sections on conservation (management) measures, since without such agreement it can still be in the immediate interest of a country (or other section of the fishery) to increase its fishing effort.

APPLICATION TO DEVELOPMENT PLANNING

The use of these concepts in studying actual problems of fishery development may be illustrated by considering a proposal, say, to introduce a number of modern trawlers into a fishery already exploited by other vessels. Such a proposal would, at least ideally, be based on studies of the resource which showed perhaps that the sustained catches could be approximately increased four-fold, so that increased fishing would appear desirable. On the basis of present catch rates, it might further be estimated that the probable catch of each trawler would be 100 tons per month, and that all costs (including interest on loans and depreciation) would be covered by catches of 85 tons per month. The building and operation of at least a limited number of trawlers would therefore appear a well-justified economic proposition.

In fact, as shown in Figure 1, the marginal efficiency predicted by the more conservative and probably more realistic Beverton and Holt model, at a catch of 25% of the maximum is between 75 and 85%. That is, the operations of a trawler will add to the total catch only 75–85 tons per month. If the whole catch is taken by the one country, then the desirability of the development of trawling for the national economy as a whole is in this situation questionable. If, however, the fishery is international, and the national share is only 50% of the total, then the marginal national catch is 85–90 tons per month, and the proposed development may again appear desirable from a national point of view, though not necessarily from a global view.

REFERENCES

BEVERTON, R. J. H.and HOLT, S. J., 1957. "On the dynamics of fish populations". Fishery Invest., Lond., Ser. 2, **19**: 533 pp.

BEVERTON, R. J. H. and HOLT, S. J., 1966. "Manual of methods for fish stock assessment. Part II. Tables of yield function". F.A.O. Fish. Biol. tech. Pap., (38) 10 + 67 pp. (revision 1).

SCHAEFER, M. B., 1954. "Some aspects of the dynamics of populations important to the management of commercial marine fisheries". Bull. inter-Am. trop. Tuna Commn, **1**: 26–56.

Scientific advise on catch levels

J.A. Gulland and L.K.Boerema

NATURAL FLUCTUATIONS

For fish, on the other hand, the lag effects on the average are less disruptive to the simple model than natural fluctuations, among which changes in year-class strength are the most striking. Where differences in year classes are very large, it is likely that when a strong year class enters the fishery the stock will increase whatever catch is taken (within practicable limits); when a succession of strong year classes is replaced by a run of poor ones, the stock may decrease even if fishing is cut back virtually to nothing. In this situation it is difficult to talk about a sustainable or maintainable yield.

However, it is precisely in the situation of a declining stock, when strong year classes are being replaced by weak ones, that concern about the management of the stock is likely to be greatest, and when scientists are often asked for advice (e.g., regarding herring in subarea 5 of ICNAF). Sometimes the advice is requested in general terms, allowing the scientists to describe the situation in detail, but leaving the decision as to the control measures (such as the level of catch quotas) to administrators. At other times the administrators cannot decide easily among themselves on the amount of catch that should be taken and ask the scientists for an explicit figure of the "correct" or "desirable" catch. This requires some objective basis for determining this, analogous to the sustainable or replacement yield for whales.

The simplest case is that in which the abundance of recruits (strength of the year class) is independent of the abundance of the parent stock. All that management can do is make the best use of whatever recruitment happened to occur, that is, to maintain fishing at whatever level is considered the optimum position on the yield-per-recruit curve. In the simplest situation the curve of yield per recruit as a function of fishing effort will have a clear maximum, for which the corresponding fishing mortality can be reasonably easily determined. This fishing mortality might then be considered as being one possible value of the optimum fishing mortality. However, in practice the yield curve will be quite flat in the neighborhood of the maximum. This may make determination of the position of the maximum difficult and will certainly mean that obtaining the absolute maximum (e.g., moving from the effort giving 99% of the catch to that giving 100%) will require a disproportionate increase in effort and hence in costs (Gulland, 1968a). Given particular values of the price of fish and costs of exerting unit fishing mortality, a mortality which will maximize the net economic yield can be calculated, and might be considered the optimum for the given economic situation. Different economic conditions will result in somewhat different optima, but for most conditions the optimum will lie within a fairly narrow range. It is, therefore, possible to define a target of the desirable fishing mortality in the middle of this range which will receive general acceptance, or at least, especially for a heavily fished stock, a minimum target at the upper end of this range. Whatever the precise objectives and economic conditions of any individual country participating in a fishery, the reduction of the amount of fishing to this target level should be desirable.

A more objective method of calculating a limiting value to desirable fishing mortality may be derived from considering the marginal yield, i.e., the increase in total yield achieved by adding one extra unit of effort (Gulland, 1968b). The marginal yield will be equal to the slope of the tangent to the curve of catch against fishing effort. It will always be less than the catch per unit effort, which is the slope of the line joining the point in the curve to the origin. The economic optimum, i.e., the greatest net economic return, occurs when the value of the marginal yield is equal to the marginal costs of a unit of effort. At the point giving the maximum physical yield the marginal yield will be zero. Clearly from any practical viewpoint, it would be undesirable to increase the amount of fishing beyond the level at which the value of the marginal yield is small compared with the costs of the extra unit of effort required to produce that yield. The question then arises as to what might be considered as small. An arbitrary figure, which has in fact been used in connection with the herring on Georges Bank (ICNAF, 1972) is a marginal yield equal to one-tenth of the original catch per unit effort in the very lightly exploited fishery. This is illustrated in Figure 3. The two straight lines through the origin show catches per unit effort in the nearly unexploited fishery—the tangent to the catch curve at the origin—and a catch per unit effort of 10% of this value. The limiting point beyond which any increase in fishing would certainly not be worthwhile—assuming a marginal yield of 10% of the initial catch per unit effort is not worthwhile—is where the tangent to curve is parallel to this 10% line. The selection of 10% is arbitrary but once the 10% figure is accepted the corresponding catch can be calculated objectively. Thus it can be used

to provide a Commission or other management body objective guidance based on scientific grounds.

Unless there are marked density-dependent changes in mortality or growth this target will be achieved by exerting some fixed rate of fishing mortality. It may be noted that the optimum fishing mortality, exerted at a constant rate during the whole of its life, above the age at first capture, is the same for any strength of year class, but slightly larger catches would in principle be obtained by fishing less hard when the fish are young, and harder when they are old, i.e., concentrating catches more at the age when the total biomass of the year class is at its maximum. The fishing mortality in a particular year which leads to the greatest catch over a period will be the mean of the best mortality for each year class present, weighted according to their strengths. If the year classes are equal, this will be equal to the optimum constant rate for a single year class during its entire life. If there is a big variation in year-class strength there would be some theoretical advantage, other things being equal, in fishing slightly harder when the strongest year class present is middle-aged (since they will not grow much more but will suffer losses by natural mortality), and less hard when the strong year

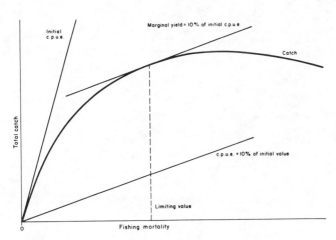

Figure 3.—Determination of an economic optimum position at which the marginal yield is 10% of the initial catch per unit effort.

classes have just recruited. However, the theoretical increases in yield are not likely to be great, and it is simpler to keep, as the objective, the same constant fishing mortality independent of year-class change. From an economic standpoint the optimum level of fishing mortality will increase slightly when strong year classes are present, because then it becomes more worthwhile to squeeze out an extra 1 or 2% of the maximum yield. For those few fisheries for which the investigators are fortunate enough to have immediately available a measure of total effort which provides a measure of fishing mortality consistent from year to year, the optimum level of fishing can be defined at once in terms of total fishing effort, without the need for year to year adjustments.

Usually difficulties of standardization in a multinational or multigear fishery, or changes in the effectiveness of a nominal unit of effort, will mean that the amount of fishing in each year will have to be controlled in terms of total catch. The scientists can, in principle, given adequate information, calculate what the magnitude of this catch should be, taking into account the strength of the year classes already present in the fishery, and those that will be recruited during the year in question. This catch might be defined as the catch for optimum harvesting rate.

Often a precise optimum level of fishing mortality cannot be defined or cannot be agreed upon. It is still possible to estimate the catches in each year which would be required to attain any prescribed value of fishing mortality. These mortalities may be those which occurred at some previous time when it was believed that the fishery was in better condition (in some general, unspecified sense) than the present, or some convenient figure which the scientists believe approximates to the optimum condition. In this way the scientists, without preempting the administrators' duty to decide on the objective of management, can provide some figures derived in a reasonably objective way, on which it may be possible for agreement to be reached. An example of such calculations are those made by the Assessments Sub-Committee of ICNAF for the cod stock at West Greenland. This stock undergoes moderately strong year-class fluctuations, and estimates have been made of the catches required to attain fishing mortalities of 0.8 and 0.6, as set out in Table 2 (from ICNAF, 1970).

It may be noted that for virtually all patterns of mortality, there is a drop in predicted catch from 1969 to 1970 and again from 1970 to 1971, due to the entry of weak year classes into the fishery.

Stock and recruitment

D.H.Cushing

Fisheries Biology:a Study in Population Dynamics, Second Edition, University of Wisconsin Press, pp. 142 – 171, 1981.

7

Stock and Recruitment

When fishermen speak of overfishing, they believe that the run of smaller fish is due to a reduction of stock to a level at which young are no longer produced in sufficient numbers to maintain the stock. Very often, at least among demersal fish, the decline in average size is due to growth overfishing alone (i.e., the fish are caught before they reach full growth), for the number of recruits to low stock can be as much as at high stock. Recruitment per unit stock then increases with declining stock. Reasoning thus, the early fisheries biologists denied the belief of fishermen in recruitment overfishing (i.e., reduction of the magnitude of recruitment because of fishing). The fishermen's question remains, Can recruitment fail because the stock has become thinned through fishing? There are two reasons why the fishermen's question has not been properly answered. The first is that the true relation between parent stock and subsequent recruitment is always obscured by the high variation in year class strength; one is left with the impression that observations should be collected by centuries. The second reason is that the failure of recruitment might result from a single event never to be repeated in quite the same frame of circumstances.

Changes in Catches in Some Fish Stocks

Some pelagic fisheries have collapsed sharply and often without explanation. Some of the dramatic changes in catch are shown in Figure 65. Over a period of half a century, catches of the Japanese sardine (*Sardinops melanosticta* [Temminck and Schlegel]) varied by a factor of fifteen, and changes in catches of the California sardine were of two orders of magnitude (Yamanaka, 1960). Catches of the Hokkaido herring (*Clupea harengus pallasi* Cuvier and Valenciennes) (Motoda and Hirano, 1963) and of the Norwegian herring (Devold, 1963) fluctuated to the same degree. Catches of the Peruvian anchoveta increased to about 12 million tons per year between 1959 and 1970.

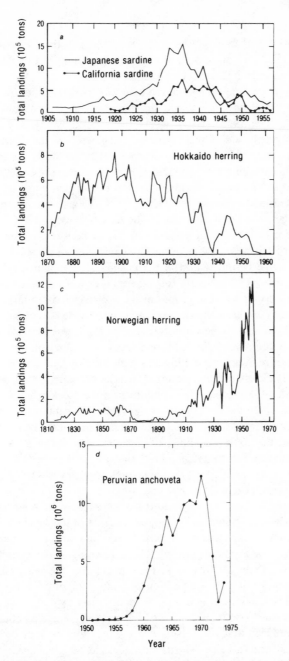

Figure 65. The catches over long periods of time for: (*a*) the Japanese sardine and the Californian sardine; adapted from Yamanaka, 1960; (*b*) the Hokkaido (or Sakhalin) herring; adapted from Motoda and Hirano, 1963; (*c*) the Norwegian or Atlanto-Scandian herring; adapted from Devold, 1963; and (*d*) the Peruvian anchoveta; adapted from Jordan, 1976.

The 1971 year class failed followed by poor recruitment in the "El Niño" years of 1972 and 1973 (see discussion of the El Niño current in Chapter 9); and from 1972–77 catches fell to about 2–3 million tons.

Changes of great magnitude have occurred over long periods. Those of the Norwegian herring stock have fluctuated for centuries, as recorded in the Icelandic sagas (Devold, 1963). Even during the 1950s, fishing mortality represented a minor proportion of total mortality, as shown from the extensive tagging results (Gulland, 1955b). During the late 1950s the stock declined, and Cushing (1968a) attributed the collapse to recruitment overfishing. The stock recovered somewhat, but in the middle 1960s Norwegian purse seiners discovered, exploited, and extinguished a shoal of Norwegian herring in the East Icelandic current, southeast of Iceland. The fishery ceased in 1967. On the other hand, there is not enough evidence to decide between fishing and natural causes for the Japanese sardine or the Hokkaido herring stocks. Figure 65 documents the range of fluctuation in the pelagic stocks, which is well known to the fishermen. Traditionally, the fishermen expect the herring-like fishes to appear and disappear. The causes of the violent variation are quite unknown save that they are basically changes in the magnitude of recruitment.

The failure of recruitment to the California sardine stock was very sudden. During the 1950s, a controversy in California revolved around the question whether the failure of recruitment in 1951 was generated by heavy fishing or whether a purely natural change had occurred (Clark and Marr, 1956). Murphy (1966) contrasted the reproductive conditions in the stock when it was exploited with those when the only cause of mortality was natural, and concluded that the stock suffered from recruitment overfishing. There was apparently a single event, the complete failure of recruitment with the 1949 year class, on which to make a judgment, and the change might have been irreversible, especially if a competitive replacement of sardine by anchovy (*Engraulis mordax* Girard) took place (Ahlstrom, 1966).

Stocks of sardines off South Africa and off Namibia also were replaced to some degree by anchovies (Newman, 1970), as was true of the Japanese sardine in the early 1940s (Zupanovitch, 1968). In recent years, the Japanese sardine has returned. The Plymouth herring (*Clupea harengus harengus* Linnaeus) stock collapsed in the 1930s with no recruitment after the year class of 1926; possibly the stock failed in competition with the stock of pilchards that succeeded it. The Plymouth herring stock was not heavily exploited, and its replacement was a natural one (Cushing, 1961). But when pelagic and opportunistic stocks are reduced to low levels by fishing, they may be replaced, as in the case of the California sardine. The pelagic stocks suffer great changes in catch from recruitment overfishing and may be replaced by pelagic competitors, particularly after heavy exploitation.

Beverton (1962) has calculated the trends in catches for certain North Sea stocks of demersal fish in a fairly long time series (Fig. 66). The North Sea cod catches remained steady for 51 years. Turbot (*Scophthalmus maximus* [Linnaeus 1758]) and plaice catches have increased slightly. Sole catches, however, have increased by ten times, and the haddock catches have fallen by an appreciable amount. During the 1960s a most remarkable change occurred. In 1962 a haddock year class was hatched which was twenty-five times larger than the preceding average, and year classes almost as large were hatched in 1967 and 1974. The recruitment to the whiting (*Merlangius merlangus* [Linnaeus]) stock increased to almost the same degree in the same years. Cod year classes were high in the years 1963, 1964, 1965, 1966, 1969, and 1970; Dickson, Pope, and Holden (1974) correlated recruitment to the North Sea cod stock with temperature and suggested that the gadoid outburst was related to cooler conditions, as the climate had become colder since the mid-1940s. Figure 84 (see below, Chapter 8) shows the increase in stock of all the North Sea gadoid species, the rate of which was common to all (Cushing,

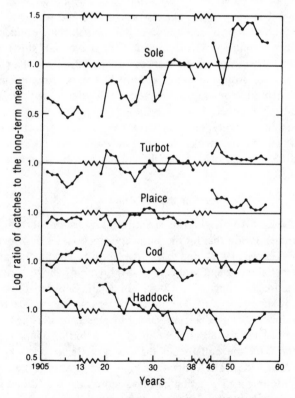

Figure 66. The logarithmic ratio of catches to the long-term mean (1906–57) of certain demersal species in the North Sea; the logarithmic ratio is used only to reduce the scale. Adapted from Beverton, 1962.

1980). Figure 66 shows that the catches of both cod and haddock were higher in the first two decades of the century, before the climate ameliorated in the 1920s and 1930s. As part of the general recovery of the gadoid stock, haddock returned to the southern North Sea and the English Channel in the 1960s, and in the Irish Sea the gadoid species recovered much as they did in the North Sea (Brander, 1977).

At a symposium in 1974 in Aarhus, Denmark, organized by the International Council for the Exploration of the Sea to discuss the gadoid outburst, other explanations were put forward. The North Sea herring stock declined between 1955 and 1968 from south to north (Burd, 1978); because it was a large stock yielding nearly one million tons in catch before collapse, the gadoids may well have taken the food released by the herring. In other words, the gadoid outburst represented a massive switch within the ecosystem under the pressure of heavy fishing, and was not an independent event associated with climatic change. No choice can yet be made between the two hypotheses, but the problem lies in the nature of the stock and recruitment relationship — how it can accommodate very large changes in numbers and whether links with other species are important. Some stocks of pelagic fish change suddenly and dramatically, but certain stocks of demersal fish rise or fall slowly for about half a century. There are no differences in larval habit or spawning behavior between the two groups of fish. The only difference is really one in growth: the pelagic fish grow quickly and demersal fish grow slowly. Pelagic fish tend to be small, and demersal ones tend to be large. The differences in growth are expressed in fecundity, large demersal fish being more fecund than small pelagic ones. The fecund stocks change slowly with time, as shown in Figure 66, but the less fecund ones suffer the dramatic changes as fisheries appear and disappear.

The Nature of the Problem

Recruitment to the East Anglian herring stock varies by factors of from 3 to 5 (Cushing and Bridger 1966); that for the arctic cod stock varies by about an order of magnitude (Garrod, 1967); and that of the North Sea haddock may be considerably greater (Sahrhage and Wagner, 1978). Figure 67 shows series of recruitments in time for a number of fish stocks in detail. Between 1887 and 1953, the year class strengths of the Karluk River sockeye salmon declined steadily. The recruitments to the herring stocks increased or declined with time, whereas those of gadoid stocks in four areas of the North Atlantic varied about a mean (except for the collapse of the Georges Bank haddock, not shown in Figure 67, in 1965). The variation is high, and most of it is a direct response to the environment. The arctic cod lays more than 4 million eggs, from which only two fish survive long enough to reproduce as adults. The implication is that there is a very fine adjustment of mortality, high stock numbers yielding about the same level of recruitment as low stock. In the

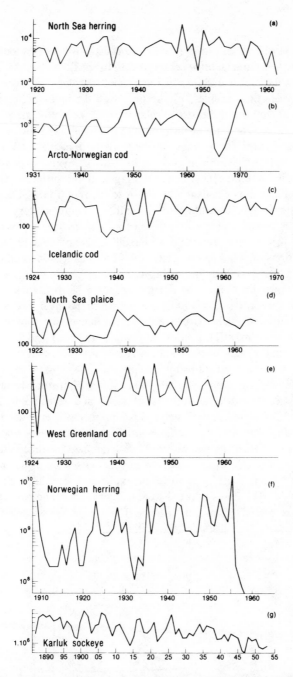

Figure 67. The variability of recruitment in numbers in long-time series for the following stocks: (*a*) North Sea herring; (*b*) Arcto-Norwegian cod; (*c*) Icelandic cod; (*d*) North Sea plaice; (*e*) West Greenland cod; (*f*) Norwegian herring; (*g*) Karluk River sockeye salmon. Recruitment is scaled in logarithms which compare variability between stocks. From Garrod, personal communication.

381

southern North Sea plaice, Beverton (1962) could establish no relationship between stock and recruitment for a series of 26 observations. The lack of correlation reflects the high variability of recruitment but does not establish independence between the variables. Figure 68 shows the index of survival (recruitment as a proportion of the stock) related to the weight of the plaice stock (Beverton, 1962). The ratio of maximum to minimum is about 20 to 1 for the survival-rate index for a range of about 10 to 1 in the total adult population weight. If survival is higher at low stock levels, a density-dependent mortality operates between hatching and recruitment. Consequently the stock has a mechanism allowing it to recover from setback or even from disaster; such a mechanism is called "compensatory." A mechanism of quite opposite character generates a "depensatory" mortality that increases at low stock and presents considerable danger of stock collapse. Indeed, in some forms of depensation (Clark, 1976) the changes in stock may be irreversible; they have not been shown in any fish stock, but the collapse of the California sardine, like that of the Atlanto-Scandian herring, took place very quickly. Another form of compensation is shown in the inverse relationship between the stock of pilchards in the western Channel and subsequent recruitment (Cushing, 1961). This stock is virtually unfished and is more abundant than the plaice stock of the southern North Sea. This fact suggests perhaps another form of compensatory expression — where density-dependent mortality among young fish is high enough at high stock levels to reduce the total recruitment. The problem of stock and recruitment is a difficult one because of the nature of this compensatory mechanism. But it is of very great importance, because it is perhaps at the root of the great fluctuations in the fisheries. Hence, its nature should be examined in some detail.

Hjort (1926) has suggested that fluctuations in year classes are the result of a "critical period" in larval life — a critical period being one of high mortality, due to starvation when the yolk is exhausted. There is not much evidence that this particular period is in fact critical. The larvae are sampled by

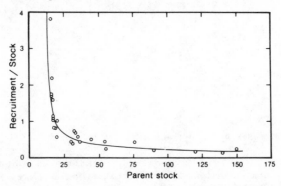

Figure 68. Index of recruit survival (or ratio of recruitment to parent stock) at different stock levels in weight of plaice in the southern North Sea. Adapted from Beverton, 1962.

plankton nets, and the larger ones may escape. Rarely has larval mortality been separated from the two loss rates due to escape from the nets and to diffusion of larvae from the spawning center. Marr (1956) has published a larval survival curve for the Atlantic mackerel derived by Ahlstrom and Nair using Sette's information. There is no difference in Sette's original data between the day and night catches of larvae, so the larvae did not escape the net in the daytime; therefore it is likely that this curve represents both mortality and diffusion. Egg mortality is less than the early larval mortality. Later, at the time of metamorphosis, there is a step-like increase in mortality, which may be due to an increase in the escape of larvae from the nets. In a laboratory study, Farris (1960) found that the high mortality of larval sardines starts well before the yolk is exhausted, and Strasburg (1959) and Marr (1956) have recorded catches of dead larvae of sardines and other species in plankton nets which shows that larvae may die of starvation before they are eaten. Figure 69 shows that the mortality of eggs and larvae of plaice in the southern North Sea

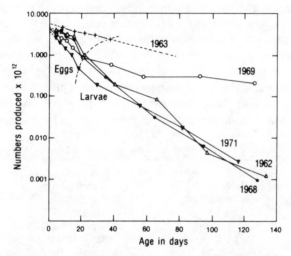

Figure 69. Mortality of eggs and larvae of plaice in the southern North Sea. Adapted from Bannister, Harding, and Lockwood, 1974.

do not differ. There is no evidence, in the studies cited, of Hjort's critical period at the time of yolk exhaustion, although there is some evidence that mortality increases significantly at the time of metamorphosis.

The Biology of Recruitment to the Parent Stock

Egg Stages. In the large populations of those species that support commercial fisheries, eggs are all about the same size, that is, about 1 mm across; exceptions are salmon, halibut, and lumpsucker (*Cyclopterus lumpus* Linnaeus), the eggs of which are 4–5 mm in diameter, up to two orders of magnitude larger by volume. They are nearly always pelagic, except those of salmon, herring, and sandeel (*Ammodytes marinus* Raitt) (of the commercial species) which stick to the gravel on the seabed. Fecundity is a function of weight; if the eggs are of the same size, bigger fish are more fecund than smaller ones. However, fecundity may not be a simple function of weight. Martyshev (1964) has shown that as carp grow older, toward middle age, their eggs and larvae are larger, but those of the oldest fish are not quite as big:

Age	4+	5+	6+	7+	8+	9+	10+	11+	12+	17+
Mean weight of eggs, in mg	1.38	1.61	1.93	2.12	2.19	2.19	2.27	2.10	2.10	1.80
Mean weight of fry, in mg	0.94	1.04	0.98	1.18	1.26	1.66	1.20	1.20	1.10	1.10

Nikolskii (1969) quotes much evidence of the same kind for different species. Older females tend to spawn earlier in the spawning season, which may last as long as three months, and so larger eggs with more yolk are hatched first. If the larvae are hatched into a food-rich production cycle (see Chapter 9), there is an advantage in releasing the larger eggs first, with their greater chance of survival. In Table 10, part (A) compares the fecundities of analogous species living in the Atlantic and in the Pacific, and part (B) those of pike and perch in the Aral and Caspian seas (Nikolskii, 1953). From parts (A) and (B), it appears that larval and juvenile mortality tends to be greater in the Pacific than

Table 10.   Fecundities of various species in thousands of eggs (Nikolskii, 1953)

Species or genus	Mean number of eggs per female	
(A)	Atlantic	Pacific
Mallotus	*M. villosus* Müller	*M. villosus*
	6.2–13.4	15.3–39.9
Limanda	*L. limanda* (Linnaeus)	*L. aspera* (Pallas)
	80.0–140.0	626.0–1133.0
Eleginus	*E. navaga* Pallas	*E. gracilis* (Tilesius)
	6.2–63.0	25.0–210.0
Hippoglossoides	*H. platessoides* (Fabricius)	*H. elassodon* Jordan & Gilbert
	240.0–370.0	211.0–241.0
Scomber	*S. scombrus* Linnaeus	*S. japonicus* Houttuyn
	350.0–450.0	400.0–850.0
Gadus	*G. morhua* Linnaeus	*G. macrocephalus* Tilesius
	570.0–930.0	170.0–600.0
Engraulis	*E. engrasicholus* (Linnaeus)	*E. japonicus* Schl.
	30.0	35.0
(B)	Aral Sea	Caspian Sea
Pike, 30–40 cm	8.3	17.6
Perch, 16–20 cm	18.3	32.9

in the Atlantic (but not for *Hippoglossoides* or cod) and greater in the Caspian Sea than in the Aral Sea. The important point is that the difference is not restricted to one species. Again, the stocks are probably not larger in the Pacific than they are in the Atlantic. Perhaps the greater numbers of larval fish support a larger number of predators. Blaxter and Hempel (1963) have shown persistent differences in fecundity and egg size between many herring stocks in the Northeast Atlantic, which implies differences in egg or larval mortality between regions of larval drifts. Bagenal (1973) reviews information on variation in fecundity of plaice, long rough dab, witch flounder (*Glyptocephalus cynoglossus* [Linnaeus]), haddock, and pike of the same weight from year to year. It is, however, uncertain whether such variations in fecundity from year to year have any adaptive significance.

When the eggs are laid in the sea, they take up water and increase their volume considerably (up to five times in the plaice, Fulton, 1897). The gonad volume is one-fifth or one-sixth of the total volume of the fish, and the eggs have to be laid in batches in order to take up water in stages. In temperate waters, the long spawning season is the result of two processes, the batching of eggs and the sequence of spawners in time from old to young females; in subtropical waters the batching of eggs spreads the spawning season to perhaps six months, and if spawning of subtropical species is linked to food, batches may be released into rich patches of food (Cushing, 1978b).

The death rate of pelagic eggs in temperate seas is high, 5 percent per day in mackerel (Sette, 1943) and 5–6 percent per day in North Sea plaice (Harding and Talbot, 1973); that of herrings on the seabed during three weeks of development is 12 percent (Runstrøm, 1941) or 10 percent (Outram, 1958), or only about 0.05 percent per day. Harding, Nichols, and Tungate (1978) examined the mortality of plaice eggs in the Southern Bight of the North Sea for eleven years; it ranged from 1.76–12.64 percent per day; in each year, it was a log-linear function of time, which suggests that the gross differences between years are due to differences in predation. The difference of a factor of 7.5 in mortality is associated with a difference of 3.5 times in the initial numbers of eggs, but the death rate is independent of numbers. The number of predators may vary from year to year, but it is hard to imagine how they can aggregate on densities of eggs as low as $1-5/m^3$ in a period limited to three weeks. The lowest mortality rates (1.76 percent per day in 1963 and 3.81 percent per day in 1947) were found in the cold winters that occur about once in each decade in the Southern Bight of the North Sea, when they took twice as long to develop; hence they were vulnerable to potential predators for longer, but they died less, perhaps because the predators were also cold, with a reduced metabolic demand.

Larval Stages. During the first days of life in the sea, the larva subsists on its yolk. The larva of the Norwegian herring weighs two or three times as much as that of a Buchan herring (which spawns in the northern North Sea) and it survives for about twice as long before the yolk is exhausted, about thirty days. When the yolk has gone, larvae of either stock must have food within fourteen days; any subsequent food is of no use (Blaxter and Hempel, 1963). During its first days in the sea, the anchovy larva learns to feed. It swims intermittently and forms an S-shape laterally to discharge an attack. Its feeding success in attack in the first three days is only 10 percent, and it takes thirty days to reach 100 percent. Hence the larva can only grow successfully toward the middle of its larval life (Hunter, 1972); the Ricker-Foerster (1948) thesis that growth and mortality are linked could only work when the larvae had learned to feed and grow successfully. An anchovy larva of 1 cm in length searches about 1 liter/hour and one of 1.5 cm about 4 liters/hour (Hunter 1972), and these quantities are close to those for herring larvae (Blaxter and Hempel, 1963; Rosenthal and Hempel, 1970) and for plaice larvae (Blaxter, 1968). Lasker (1975) has shown that the first feeding anchovy gathers on thin layers of the dinoflagellate *Gymnodinium splendens* Lebour, close to the coast of Southern California. Hunter and Thomas (1974) simulated the mechanism of aggregation by anchovy larvae in the laboratory with artificial patches of the same dinoflagellate. Such mechanisms are needed because fish larvae cannot survive on an average density of food in the sea (Jones and Hall, 1973). O'Connell and Raymond (1970) showed that anchovy larvae die if they take only one nauplius/ml/day, but survive on 4 nauplii/ml/day. Wyatt (1972) analyzed the dependence of *Oikopleura* pellets/gut in plaice larvae upon the number of encounters/m. He showed that plaice depended on a low density of *Oikopleura* whereas sandeels subsisted on a high one, and so competition is limited to densities at which the distributions overlap. This short biology of fish larvae summarizes part of the extensive work now being done to illuminate the problems of stock and recruitment.

The mortality of fish larvae is sometimes difficult to determine because the baby fish can dodge the sampling nets at some point in their growth. With adequate sampling methods, it has been observed that plaice larvae die at 5 percent per day (Bannister, Harding, and Lockwood, 1974), haddock at 10 percent per day (R. Jones, 1973b), mackerel at 12 percent per day (Sette, 1943), cod at 10 percent per day (Cushing and Horwood, 1977), and herring at 4 percent per day (Graham, Chenoweth, and Davis, 1972). Larval mortality continues from that of the eggs at the same rate (Fig. 69). If that is true, there are differences in predation from year to year common to eggs and larvae. From the few observations on the plaice larvae in the Southern Bight of the North Sea, there is no evidence of stock-dependent (or density-dependent) mortality, which is a conclusion expected from a mortality common to eggs and larvae.

A good measure of larval mortality has been made by Pearcy (1962) in a study of the winter flounder in the estuary of the Mystic River in Connecticut.

Because the sizes of larvae were shown to be the same by day and by night, it is unlikely that the bigger larvae were escaping from the nets during the daytime. There is a daily loss of about 30 percent of the estuary volume, mostly near the surface. Since Pearcy found that about 85 percent of the larvae remain close to the bottom, where the eggs are laid, the daily loss by translocation to the sea is about 3 percent. The curve of larval survival with age is markedly concave, survival increasing sharply with age. From physical measurements the translocation rate (percentage removed by seaward movement) was estimated as

$$\phi = 1 - (1 - \rho'm)\psi, \tag{122}$$

where ϕ is the translocation rate,
$\quad m$ is the proportion of larvae in the upper 2 meters (where the main seaward translocation takes place),
$\quad \rho'$ is the exchange ratio per tidal cycle, and
$\quad \psi$ is the number of tidal cycles per day.
Since the percentage of both total loss and daily translocation loss are known, the remaining mortality can be attributed to natural mortality:

		Percentage	
Age		*Loss by*	*Loss by*
(in days)	*Total loss*	*translocation*	*natural mortality*
9–25	0.248	0.041	0.207
26–53	0.112	0.015	0.097

Thus, the natural mortality rate is five times the translocation rate. As numbers became reduced, the natural mortality was halved in the second period as compared with the first. It is a general principle that natural mortality which starts at such high rates must decline with age, as Cushing (1975a) has shown for the plaice on the coarser scale of years; the surprising point is that the decline in mortality rate starts in larval life, which is confirmed by recent observations on plaice larval mortality (Harding, Nichols, and Tungate, 1978).

The material from the observations on plaice larvae in the Southern Bight can be ranked in order of magnitude by years (Bannister, Harding, and Lockwood, 1974).

Eggs, stage I	Eggs, stage V	Larvae, stage 1	Larvae, stage 4
1. 1963	1. 1963	1. 1963	
2. 1962	2. 1962	4. 1969	4. 1969
3. 1968	3. 1968	3. 1968	5. 1971
4. 1969	4. 1969	2. 1962	2. 1962
5. 1971	5. 1971	5. 1971	3. 1968

The differences in numbers at hatching are small ($\times$ 1.8), but those at the last stage of larval life are considerable ($\times$ 2000); if recruitment were a density-dependent function of stock, such changes in ranking would be expected. Moreover, differences in numbers imply dome-shaped relationships between the stages sampled. If there were no compensation, the ranks should remain unchanged, as they do between stage I and stage V eggs, where differences in numbers are less than at later larval stages. The same point may be seen from the material collected on larval and adolescent cod in the Danish Belt seas (Fig. 70). Yolk-sac larvae were collected with a Petersen's young-fish trawl, and fish in age groups 0, I, and II were caught with an eel tog, which is a small trawl, not very efficient but capable of indicating gross changes in abundance. The complete data were published by Poulsen (1930a, 1930b). The material on the survival of five year classes of cod in the Baltic can be arranged in rank order of magnitude as follows:

Stock	Larval drift	0 group	II group
1924	1923	1923	1925
1927	1925	1925	1923
1923	1926	1926	1926
1925	1927	1927	1927
1926	1924	1924	1924

The same general point is made in this tabulation, and an additional point as

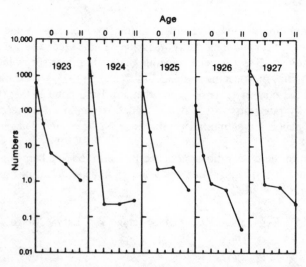

Figure 70. The larval and adolescent mortality of cod in age groups 0, I, and II in the Belt seas of central Denmark. Larvae were collected with a Petersen young-fish trawl, and fish from the 0, I, and II groups were caught with an eel tog, a form of small trawl. Data from Poulsen, 1930a, 1930b.

well — that some control can occur after metamorphosis, as will be shown below.

During the drift of eggs and larvae there is no evidence yet of a critical period such as an increase in mortality during the yolk-sac or first-feeding stages. Nor is there yet any evidence that the death rate of either eggs or larvae depends upon their numbers, that is, no evidence of density-dependent mortality generated by the aggregation of predators. However, by the Ricker-Foerster (1948) thesis, density dependence would be expressed in the time for which mortality endures according to the available food, briefly in rich food and for a long time in scarce food.

0-group Fishes. When fish larvae metamorphose, their morphology and behavior change radically. 0-group plaice turn on their sides and settle on the beaches; herring acquire fins and are caught in the waves on the shoreline; and codling swim deeper in the open sea when they start to swim like fish. Pearcy (1962) worked on the winter flounder in the Mystic River in Connecticut;

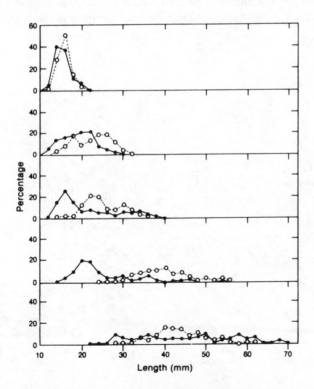

Figure 71. Changes in percentage length distribution of 0-group plaice on their nursery grounds in Firemore Bay, Loch Ewe, Scotland. Closed circles represent material from 1965; the open circles are observations from 1967. Adapted from Steele and Edwards, 1969.

Riley and Corlett (1965) examined 0-group plaice in Port Erin Bay in the Isle of Man. Macer (1967) studied the 0-group plaice population in Redwharf Bay in Anglesey, off the northwest coast of Wales. Steele and his co-workers (Steele and Edwards, 1969) analyzed the population processes in Firemore Bay in Loch Ewe in northwest Scotland. Lockwood (in press) worked on the 0-group plaice in Filey Bay in northeast England. The average mortality in these observations on flatfish in the month after metamorphosis was about 40 percent per month (1.7 percent per day), which reduced to about 10 percent per month (0.34 percent per day) six months later.

In Firemore Bay, the plaice fed on polychaete tentacles and *Tellina* siphons. Figure 71 shows how length distributions changed sharply in a very short time period. In three months, the growth of the larger individuals was considerable, but many of the smaller fish vanished, perhaps eaten by plaice (or other flatfish) of the brood one year older. The increment of growth in the short period may be an apparent one due to heavy predation. Steele and Edwards (1969) showed that the energy intake of the population remains the same, so the growth rate of individuals depends upon the mortality rate of the population. Figure 72 shows the dependence of weight increment (per unit of

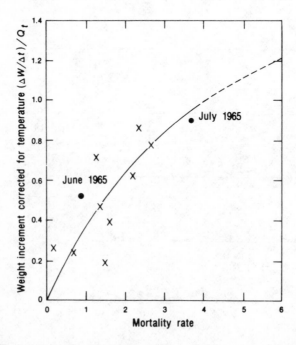

Figure 72. The dependence of weight increment of 0-group plaice, corrected for temperature — i.e., $(\Delta W/\Delta t/Q_t)$ — upon mortality in Firemore Bay, Loch Ewe, Scotland. Adapted from Steele and Edwards, 1969.

energy metabolized) on mortality rate: the greatest rate of energy transfer occurred when the death rate was highest. Not only are growth and mortality closely linked, but the proportion of growth to mortality increases with decreasing mortality rate, as time passes.

Thus growth and mortality are linked both in larvae and in 0-group fish, and there is probably a density-dependent component in both. If density dependence is a constant proportion it dies away with age, and we would expect the predominant effect to occur during larval life — but not exclusively so, as suggested in the ranking tables given above. However, the "postage stamp" plaice are attacked by cannibals a year older, among other predators.

The Theoretical Formulation of the Problem

The Malthusian principle was put by Darwin: The amount of food for each species must on average be constant, whereas the increase of all organisms tends to be geometrical, and in a vast majority of cases at an enormous ratio. The large increase is lost in the struggle for existence, and insofar as they are known, the numbers of wild populations are roughly stable. Howard and Fiske (1911) wrote that the implied density-dependent control of numbers was indicated either in fecundity or in mortality. There was a controversial literature on the nature of density dependence (Solomon, 1949; Andrewartha and Birch, 1954), but in fisheries science the principle was implicit in the use of the logistic curve, for recruitment was explicitly limited by the "carrying capacity" of the environment. Haldane (1953) and Moran (1962) expressed the necessity of density dependence as follows:

$$\Delta N = B - D' + I'' - E'', \qquad (123)$$

where B is the number of births/year,

D' is the number of deaths/year,

I'' is the number of immigrants/year, and

E'' is the number of emigrants/year.

Then

$$\Delta N/N = b - d' + i - e \qquad (124)$$

in relative rates of change; if $\Delta N/N$ is to approach zero in a number of generations, some or all such rates must be functions of N.

For the reasons given in Chapter 4, density-dependent fecundity is unlikely to play a major part in control, although it may have a secondary role. If, then, mortality is the major agent in stabilization, most little fishes must be eaten by predators; no other form of mortality could satisfy rates as high as 5–10 percent per day. Predation that is the proximate cause of death may include a minority of deaths due to parasitism and disease and perhaps a majority due to varying degrees of food lack. Ricker (1954) distinguished three forms of predation: (a) the predators take a fixed number, which generates the depensatory mortality referred to above; (b) the predators take a fixed fraction of prey numbers by random encounter, a mechanism which generates the usual instantaneous coefficient of mortality; (c) the predators take all in excess of a minimum number. Some bird predators that gather rapidly from great distances may be classed in the third category, but most fish predators fall into the second, although some that attack the salmon redds or fish shoals may generate the depensatory mortality characteristic of the first category.

Because many processes in the sea must be density dependent, it is useful to distinguish them from stock-dependent ones (Harris, 1975). Between the ages of hatching and recruitment, mortality may occur at any time, and it may depend on the density at that time. A stock-dependent mechanism controls through the initial numbers and hence is effective from generation to generation, and so a density-dependent mechanism in the earlier stages of the life history is also a stock-dependent one. The major formulations may be developed as follows:

$$dN/dt = -MN,$$

where M is the instantaneous coefficient of natural mortality.

$$R = N_0 \exp\left[-M(t_r - t_0)\right], \tag{125}$$

where N_0 is the number of animals at t_0, the time of hatching, and
R is the number of recruits at t_r, the age at which recruits accede to the spawning stock.

Let

$$N_0 = f^*P_e,$$

where P_e is the number of eggs spawned, and
f^* is the number of eggs/adult. Then

$$R = f^*P_e \exp\left[-M(t_r - t_0)\right]. \tag{126}$$

This equation forms a straight line in log numbers, and we are interested in the degree of departure of reduced recruitment from it at high stock.
Let

$$dN/dt = -(M_1 + M_2N)N, \tag{127}$$

where M_2 is the instantaneous coefficient of density-dependent mortality, and M_1 is that of density-independent mortality.
Integrating,

$$R = 1/[A' + (B'/P_e)], \tag{128}$$

where $A' = (M_2/M_1)\{\exp[M_1(t_r - t_0)] - 1\}$, and
$B' = 1/f^*\{\exp[M_1(t_r - t_0)]\}$.

This is the first equation of Beverton and Holt (1957), where the density-dependent mortality between t_0 and t_r becomes stock dependent at a critical stage in the life history. If mortality were constant with age, this period could occur at any age, but the death rate of larvae decreases with age (Pearcy, 1962), and to be stock dependent, the critical period must occur at an early stage in the life history. Recruitment is an asymptotic function of stock in which the initial slope is $(1/B')$, the density-independent mortality, and the asymptote is $(1/A')$, which expresses the greatest density-dependent mortality. If the density-dependent mortality is high, the asymptote is reached at a relatively low level of stock, which displays an apparent independence of recruitment from parent stock. A few observations might justify such a conclusion when the true density-dependent mortality was in fact much less.

$$M = M_0 + k_1 N_0; \tag{129}$$

then,

$$R = N_0 \exp [- (M_0 + k_1 N_0) (t_r - t_0)]. \tag{130}$$

Recall that

$$N_0 = f^* P_e,$$

$$\alpha = f^* \exp [-M_0 (t_r - t_0)],$$

and

$$\beta = k_1 f^* (t_r + t_0),$$

where α is density-independent survival, and
β is the coefficient of density-dependent mortality.

Then

$$R = \alpha P_e \exp (- \beta P_e). \tag{131}$$

This is the Ricker equation, where the density-dependent mortality is explicitly a function of the initial numbers and hence is stock dependent. Cannibalism was the first form of stock-dependent mortality proposed by Ricker; later he expressed it as an aggregation of predators — for example, brown trout may gather on patches of salmon fry as they leave a lake. The curve of recruitment on parent stock is dome-shaped, or potentially so; the initial slope is (α) and the maximum recruitment is at $(1/\beta)$.

The equation can be derived in another way. Let

$$N_c = N_0 \exp [- \mu_1 (t_c - t_0)], \tag{132}$$

where t_c is the time to reach the critical size at which a larva leaves a predatory field, and

μ_1 is the coefficient of density-independent mortality.

$$R = N_0 \exp \{ - [(\mu_1 - \mu_2) t_c + \mu_2 t_r - \mu_1 t_0] \}, \qquad (133)$$

where μ_2 is the coefficient of density-dependent mortality.

Let $(\mu_1 - \mu_2) t_c = \beta P_e$; let $\alpha = \exp(- \mu_2 t_r - \mu_1 t_0)$;

$$\therefore R = \alpha P_e \exp(- \beta P_e). \qquad (134)$$

This is Beverton and Holt's (1957) second equation. It expresses the Ricker-Foerster (1948) thesis, which states that a larva which feeds well and grows well passes through a predatory field quickly, and then t_c is small; with more larvae, there is less food and t_c is longer. Such a mechanism is density dependent at any stage in the life cycle, and is stock dependent in the early stages.

Beverton and Holt put the Ricker-Foerster thesis as follows: $t_c \alpha (1/W_\infty)$ α (1/food eaten) α density αN_0. Harris (1975) examined the relationship between t_c and $(1/W_\infty)$ by differentiating dt_c/dL_∞ for small values of L, compared with L_∞:

$$dt_c/dL_\infty = -L/KL_\infty^2. \qquad (135)$$

Integrating, $t_c = L/KL_\infty$ or $t_c \alpha (1/L_\infty) \alpha (1/W_\infty^{1/3})$

$\therefore t_c \alpha N_0^{1/3}$, i.e., the initial distance apart of larvae in the sea;

$$\therefore R = \alpha P_e \exp(- \beta P_e^{1/3}). \qquad (136)$$

This brief theoretical treatment is taken from Harris (1975). The equations used are continuous, but difference equations might be equally useful in that the relationship can be expressed as the ratio of recruitment to parent stock for each year of stock (Clark, 1976); this is the index of return used by W. F. Thompson for the Fraser River stock of sockeye salmon. May (1976) demonstrated that if the dome becomes critically steep, oscillations occur leading to "dynamic chaos," from which we conclude that the net increase rate of fishes is low.

There are three proposed mechanisms: the aggregation of predators, cannibalism, and food limitation (either in the Ricker-Foerster sense or as simple starvation; as will be shown below, the distinction between starvation and food limitation under predation, in the Ricker-Foerster sense, tends to disappear). For practical purposes there is a single equation which expresses any of the three mechanisms in a stock-dependent manner. There has been some argument whether a dome exists in the stock and recruitment relationship, but present evidence, as shown below, suggests that it exists in gadoids. Beverton and Holt (1957) thought that a dome-shaped curve was unstable, but if the right-hand limb of the dome is less than 45 percent to the abscissa, a perturbation generates a stable limit cycle. The choice of curve really depends on the nature of the mechanism, whether mortality can be shown to be a function of initial numbers or whether it endures for longer at high density.

The Ricker equation was developed for the more or less single-age stocks of the Pacific salmon fisheries. Ricker maximized Equation (131):

$$R/R_m = (P/P_m) \exp [1 - (P/P_m)], \tag{137}$$

where R_m is maximal recruitment, and
P_m is the stock that yields R_m.
P_r is the replacement stock at which $R_r = P_r$, i.e., where the curve cuts the bisector. Then

$$R/R_r = W' \exp [a (1 - W')], \tag{138}$$

where $W' = P/P_m$, and
$a = P/P_r$.
When $P_r > P_m$, density-dependent mechanisms predominate; when $P_r < P_m$, they are less important than the density-independent ones. Figure 73 (Ricker, 1958a) shows such curves for various conditions of recruitment per unit stock and for various rates of exploitation. The bisector is an essential part of the system because it indicates the point at which recruitment and stock are equal; as exploitation increases, the bisector shifts to the left. At a relatively low stock level there is high recruitment and at a high stock level there is low recruitment, which condition expresses the compensatory principle of density dependence.

Ricker (1950) explained the dominant cycle in the Fraser River sockeye salmon stock after the Hell's Gate landslide when he noticed that the I group fish of the dominant cycle fed on the fry of the succeeding one. Neave (1953) showed that the survival of chum salmon fry increased with increasing numbers and so the mortality was depensatory. The mechanism has been investigated by Larkin and Hourston (1964), Ward and Larkin (1964), and Larkin and MacDonald (1968), and is expressed generally in a relationship as follows (Larkin, Raleigh, and Wilimovsky, 1964):

$$(R/R_r) = W' \exp [a (W' - 1)]. \tag{139}$$

A model was constructed by Larkin and MacDonald on the basis of five stages in the life history of the sockeye, of which the most important was the depensatory mortality. For the Adams River stock (part of the Fraser network of rivers), the model was run for many simulated years, and the subdominant cycle emerged just after the dominant one. Thus, in hypothesis, observation, and simulation, depensatory mortality exists, and the question arises whether

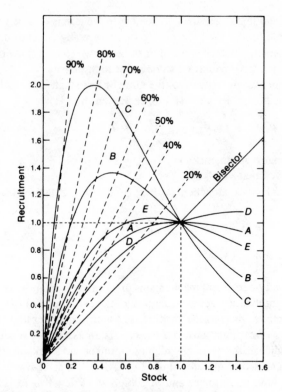

Figure 73. The stock and recruitment curves of Ricker for various conditions. The curves *A–E* represent different "stock and recruitment relationships." They pass through the bisector (where recruitment and stock are equal, and hence recruitment can replace stock) at the same point, an equilibrium point. The dashed lines represent different rates of exploitation. Adapted from Ricker, 1958a.

it occurs in marine fish stocks. Burd and Parnell (1973) have established a low concave dependence of numbers of herring larvae on their parent stock, which may indicate a form of depensatory mortality in the sea. The original explanation of the phenomenon was that herring eggs were destroyed by trawling, and destroyed most effectively at low density (Cushing and Bridger, 1966), which is of course a form of depensatory mortality.

Recently, Ricker (1973) has shown how the Ricker equation may be developed to estimate the proportion of density-dependent to density-independent mortality:

$$Z_c = \beta P_e,$$

where Z_c is the instantaneous coefficient of density-dependent mortality.

The survival rate at replacement, $S_r = (R_r/P_r)f^* = 1/f^*$,

$$\therefore Z_r = - \ln (1/f^*) = \ln f^*.$$

$$P_r = (\ln \alpha)/\beta, \therefore Z_c \text{ at replacement} = \ln \alpha; \qquad (140)$$

$$\therefore Z_i = \ln f^* - \ln \alpha,$$

where Z_i is the instantaneous coefficient of density-independent mortality
at all levels of stock, and

Z_r is total mortality between hatching and recruitment.

For many practical reasons Ricker's initial equation is useful because its
twin derivation embraces all three of the proposed biological mechanisms.
The three major dynamic processes are also expressed — the magnitude of the
stock, stock-dependent mechanisms, and those due to environmental changes
only. Cushing and Harris (1973) fitted this curve by least squares, i.e., by
minimizing $\Sigma [R - \alpha P_e \exp (- \beta P_e)]^2$, and 95 percent confidence limits
were set to the fitted curve; the method of plotting $\ln (R/P)$ on P is then avoided. This method was used because the replacement stock, P_r, is inaccessible
in a multi-age stock, while the trend of natural mortality with age remains unknown. Further, in a multi-age stock the annual recruitment is only a small
proportion of the stock and is equivalent to the annual loss by death. Then the
stock is at its replacement point at any stable population level. Indeed, the
cohorts may be considered to lead independent lives during their critical
growing periods.

Stock and Recruitment Relationships and the Models That Sustain Them

Observations. Cushing (1971b) analyzed stock and recruitment relationships
by groups of species (salmon, clupeids, flatfish, and gadoids) with a simple
equation, $R = k^*_2 P_n{}^{b_1}$ (or $R/P_n = k^*_2 P_n{}^{(b_1 - 1)}$), where b_1 is an index of density dependence. The object was only to compare the index between groups,
for the equation should not be used outside the range of observations. This
index was inversely related to the cube root of the fecundity, which suggests
that density dependence is linked to the distance apart of the eggs or larvae in
the sea. More than 90 percent of the mortality between hatching and recruitment takes place during larval life, and it is reasonable to suppose that control
occurs then, or not very long afterward.

The Ricker curve, or its analogue, Equation (134), represents the simplest
theoretical expression which subsumes three proposed mechanisms: the
aggregation of predators, cannibalism, and food limitation. Stock is expressed
as number of eggs produced and recruitment as numbers at the age of recruitment to the adult stock; i.e., the process lies in the mortality from egg to
recruit.

Figure 74 shows Ricker stock and recruitment curves fitted by least squares (Cushing and Harris, 1973). There is one chance in twenty that the fitted line lies outside the dashed ones. Two stocks of sockeye salmon, Karluk River (Rounsefell, 1958) and Skeena (Shephard and Withler, 1958), are represented; they are single-aged stocks, in which growth is probably density dependent. Recruitment varies about the curve by a factor of 2.5 in the Karluk stock (Fig. 74a) and by 1.5 in the Skeena (Fig. 74b), and the variability appears to decrease with declining stock. The Buchan herring is a multi-aged stock in which the first summer's growth is probably density dependent (Fig. 74c). Recruitment varies by a factor of 2.0 about the curve, and there are some high year classes at low stock, but no very poor ones. Figure 74d shows a different form of stock and recruitment curve for the chub mackerel, Scomber japonicus Houttuyn (Tanaka, 1974); eggs spawned by offspring are plotted on eggs spawned by parents. Both stock and recruitment are in the same units, expressing the replacement from generation to generation as in the single-aged salmon stocks. The variability about the curve is low.

Figures 74e and f show stock and recruitment curves for flatfish — Pacific halibut off Alaska and North Sea plaice (Cushing and Harris, 1973). Both are multi-aged stocks in which the variability of recruitment is rather low. Density-dependent growth has perhaps been detected in the stock of Pacific halibut at the age of three, but not among older fish (Fig. 45, Chapter 4). In Figure 74, g and h are given stock and recruitment relationships for gadoids, the Arcto-Norwegian cod (Cushing and Harris, 1973), and the North Sea haddock (Sahrhage and Wagner, 1978). Cushing and Horwood (1977) demonstrated the existence of density-dependent growth in the Arctic cod at the age of four, but not among older fish. The variation of recruitment about the curve is high (a factor of more than 25), and it is greater at low stock.

As shown above, the Ricker curve forms a dome, and the observations group themselves about it, on the left hand limb, about the middle, or across it. In the two salmon stocks the data lie on the left-hand limb of the dome; in the herring and mackerel stocks they are grouped in the same way, with perhaps a little more density dependence. The observations on the flatfish stocks lie in mid-dome, and in the gadoid stocks they lie across it. With increasing fecundity, there is greater density dependence; the gadoid dome is the most pronounced example of the trend. The estimates of average density dependence are as follows:

	$-\beta \bar{P}_e$
Pink salmon	0.61
Sockeye salmon	0.77
Herring	0.88
Flatfish	1.02
Gadoids	1.75

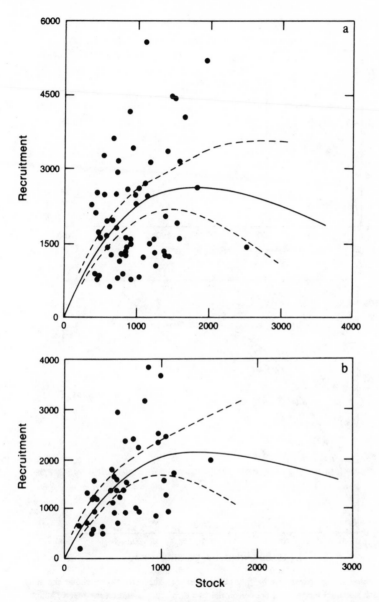

Figure 74. A gallery of stock and recruitment curves: (*a*) Karluk River sockeye salmon. Adapted from Cushing and Harris, 1973. (*b*) Skeena sockeye salmon. Adapted from Cushing and Harris, 1973. (Figure 74 continues on pages 166–68.)

399

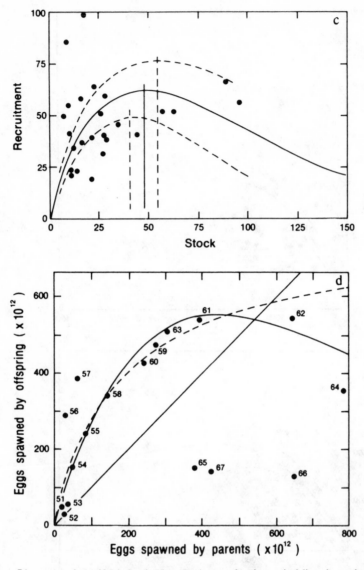

Figure 74, continued. (c) North Sea herring (Buchan stock); the vertical line shows the stock levels at which maximal recruitment occurs on the two error curves and on the mean. Adapted from Cushing and Harris, 1973. (d) Japanese mackerel; numbers beside observations are years they were made. Adapted from Tanaka, 1974.

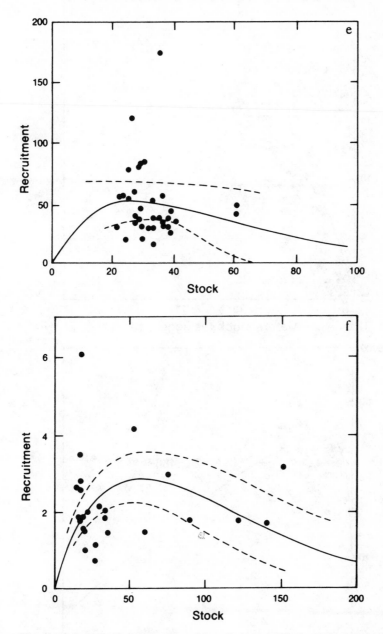

Figure 74, continued. (*e*) Pacific halibut. Adapted from Cushing and Harris, 1973. (*f*) North Sea plaice. Adapted from Cushing and Harris, 1973.

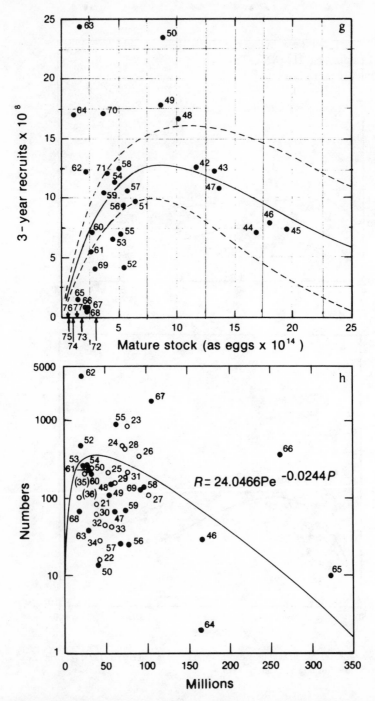

Figure 74, continued. (*g*) Arcto-Norwegian cod. Adapted from Cushing and Harris, 1973. (*h*) North Sea haddock. Adapted from Sahrhage and Wagner, 1978. Numbers beside observations are years they were made.

These estimates include a number of stocks not included in Figure 74, but listed in Cushing and Harris (1973).

In Chapter 4, density-dependent fecundity was described in the perch of Windermere. Holden (1973) thinks that density-dependent fecundity might occur in Elasmobranchii and that, because they lay few eggs, such stocks should only be exploited lightly. Conversely, the fecund cod stocks are exploited heavily because their density dependence provides them with considerable resilience. Figure 73 shows how a dome-shaped curve restores a stock more quickly than a near-linear one; a cod stock should therefore recover more quickly than a herring stock. Over the centuries, the Atlanto-Scandian herring fishery has appeared, grown, and vanished in the Norwegian Sea with a periodicity of about 110 years (Cushing and Dickson, 1976), which means that recruitment must vary by several orders of magnitude through the centuries. In contrast, the Arcto-Norwegian cod fishery in the same region has never disappeared, even if it has fluctuated considerably. In face of a similar environment, recruitment varied much less than that of the Atlanto-Scandian herring.

Harris (1975) concluded that density-dependent mortality was a function of $N_0^{1/3}$, and above it was linked to the cube root of the fecundity. The distance apart of eggs or larvae in the sea may play a crucial part in the control of marine fish populations. The parent fish may spawn at a fixed distance apart, and the density of eggs or larvae in the sea is probably a function of the number of eggs per batch spawned. If this explanation is true, a decrease in numbers in the population is associated with a reduced area of spawning, and vice versa, as shown in the decline of the California sardine population (Ahlstrom, 1966) or the expansion of the southern North Sea plaice population (Harding et al., 1978).

Models. The generation of recruitment during larval life can be simulated. In the model of R. Jones (1973b) and R. Jones and Hall (1973), fish larvae die if they fail to capture a specified number of food organisms: haddock larvae grow at 12 percent per day, die at 10 percent per day, and need at least 0.2–0.7 *Calanus* per day. The least number for survival depends on mean numbers, the search volume, and the mortality rate. As numbers vary, the minimum food requirement leads directly to a density-dependent mortality. Differences in survival of up to three orders of magnitude (i.e., in year class strength) were generated by rather small differences in the quantity of available food. An essential part of the model is that the fish larvae should grow with their food. R. Jones wrote that there would be considerable advantage in the existence of a single mechanism able to influence both growth and mortality simultaneously and hence control the balance between them in the long term.

Cushing and Harris (1973) and Cushing (1975) constructed an analogous model in which predatory mortality depended on the ratio of the cruising speed of the larvae to that of the predator. Growth rate depended on food, cruising speed on size, so predatory mortality was relatively high in poor food and relatively low in rich food; in this way both density-dependent growth and mortality were generated. The model is a development of the Ricker-Foerster thesis and links mortality and growth as did Ware (1975). Jones's model differs in that growth rate is fixed and the death of larvae is attributed to food lack, although predators may well eat the starving animals. Common to both models is the fact that density-dependent mortality is linked to food limitation during larval life.

A different form of model was made by Cushing and Horwood (1977) to investigate the form of the stock and recruitment curve. Density-dependent and density-independent mortalities were estimated from the origin to the maximum of the dome of the Ricker curve for the Arcto-Norwegian cod stock (Fig. 74g). A growth rate of 10.35 percent per day was used for the first 70 days of life, 2.4 percent per day from 70 to 168 days, and 0.71 percent per day from 168 days to three years of age. The growth rate for the rest of the life cycle was estimated with the von Bertalanffy equation, because W_∞ can be estimated from the weight at three years of age. A competition parameter, dependent on numbers, modified growth during larval life to the limit of density dependence observed at four years of age. Mortality and growth could be linked or not. Recruitment was generated at different levels of stock; the analysis in four parts is summarized in Table 11.

Table 11. The effects of density-dependent growth upon the nature of a stock and recruitment model (Cushing and Horwood, 1977)

Growth	Mortality	Dome	Density-dependent growth
1. Density independent	Not linked to growth	No	No
2. Density dependent	Not linked to growth	No	Yes
3. Density dependent	Linked to growth	Yes	Yes
4. Density independent	Not linked to growth; stock-dependent component added	Yes	No

Only the third combination fitted the observed presence of a dome and of density-dependent growth. But the dome was flat, and any greater mortality would have demanded a higher degree of density-dependent growth than observed. A dome could only be made by adding a component of stock-dependent mortality. The dome was generated in effect by a stock-dependent component of mortality, which may well have been cannibalism by adults — which can be stock dependent at any stage in the life history of the recruiting year class; gadoids eat young gadoids (Daan, 1975).

The observations of recruitment on parent stock may be fitted by the Ricker curve or its Beverton and Holt analogue because it expresses the three possible biological mechanisms in a convenient way. If the argument presented above on cannibalism is true, the asymptotic curve of Beverton and Holt, Equation (128), would be useful for noncannibalistic stocks. Many other expressions are conceivable, but those cited here at the least encapsulate the biological mechanisms. The models suggest that density-dependent processes in food limitation or cannibalism are sufficient, but they do not establish them nor do they deny the possibility that they could be generated by the aggregation of predators on eggs or larvae.

Summary

Many stocks of pelagic fishes have collapsed, whereas those of most demersal stocks vary to different degrees over long periods. An examination of time series of recruiting year classes shows that those of demersal stocks tend to vary about a mean, but do not always do so, and that those of pelagic stocks tend to show rising or falling trends. The survival of recruits is much less at high stock than at low stock, which is a compensatory process; on some occasions the reverse is true, and then the process is called depensatory. No decisive evidence has emerged that Hjort's critical period exists, but larval mortality may increase at metamorphosis.

The biology of fecundity — of eggs, larvae, and 0-group fishes in the sea — has been described insofar as it is known. Estimates of growth and mortality have been given for some species. The formulation of the stock and recruitment relationship in the theoretical terms of Ricker and Beverton and Holt have geen given, together with developments by Larkin and by Harris. Models to account for the process of larval growth and mortality have been described insofar as they are realistic. A gallery of stock and recruitment curves for different groups of fishes shows the high variability of recruitment about the thinly established curves; however, the density-dependent mortality is greater among the more fecund fishes.

The simplest way to account for the apparently highly variable processes by which the magnitude of recruitment is established when the cohort has reached a level of numbers "acceptable" to the adult stock is to imagine that the two processes — the generation of recruitment and the stabilization of numbers — are linked. The nature of the single process is unknown, but one might imagine that growth and mortality were both density dependent until there was enough food for the cohort and the animals could grow at their best metabolic rates.